Dynamics of mechanical systems

Dynamics of mechanical systems

J. M. PRENTIS, M.A., Ph.D.

Longman

LONGMAN GROUP LIMITED
LONDON

*Associated companies, branches and representatives
throughout the world*

© Longman Group Ltd 1970
All rights reserved. No part of this publication may be
reproduced, stored in a retrieval system, or transmitted in
any form or by any means, electronic, mechanical,
photocopying, recording, or otherwise, without the prior
permission of the Copyright owner.
First published 1970
SBN 582 44730 5

Set in Monophoto Imprint and printed by offset in
Great Britain by William Clowes and Sons, Limited
London and Beccles

CONTENTS

Preface   xi
Chapter 1 SIMPLE MECHANISMS I   Structural analysis
1.1   Introduction   1
1.2   Classification of mechanisms   3
1.3   Plane mechanisms with class I pairs   5
1.4   Plane mechanisms with class II pairs   9
1.5   Space mechanisms   10
1.6   Structural analysis of plane mechanisms   12
1.7   Rigid frames   12
1.8   Plane mechanisms with one degree of freedom   15
1.9   Alternative relationship between number of joints and links   18

Chapter 2 SIMPLE MECHANISMS II   Displacement analysis
2.1   Introduction   22
2.2   Cams   23
2.3   Interference (reciprocating cams)   25
2.4   Disc cams   26
2.5   Interference (disc cams)   28
2.6   Rolling contact   39
2.7   Cam kinematics — general comments   42
2.8   Gear tooth geometry   42
2.9   Involute gearing   43
2.10   The involute rack   48
2.11   Interference (involute gears)   48
2.12   Internal gears   51
2.13   Helical gears   53
2.14   Non-involute gears   54
2.15   Non-circular gears   55
2.16   Gear drives between non-parallel shafts   57
2.17   Gear trains   59
2.18   Linkages   65
2.19   Kinematics of rigid body motion in respect of finite displacements   66

vi Contents

2.20 Design of linkage mechanisms	68
2.21 Synthesis of a plane four-bar linkage for correlated crank displacements	69
2.22 Freudenstein's method	73
2.23 Synthesis of a plane four-bar mechanism for specified coupler displacements	75
2.24 Use of a plane four-bar mechanism as a curve generator	77
2.25 Euler-Savary equation	79
2.26 Roberts' theorem	84
2.27 Symmetrical coupler curves	86
2.28 Classification of four-bar mechanisms	88

Chapter 3 FORCE RELATIONSHIPS IN MECHANISMS I Transmitted forces and friction effects

3.1 Introduction	94
3.2 Forces in mechanisms without friction	94
3.3 Friction	98
3.4 Use of virtual work with friction	103
3.5 Friction effects in particular mechanisms	107
Appendix 3.1 Calculation of upper and lower bounds on the efficiency of a mechanism with friction	116

Chapter 4 VELOCITIES AND ACCELERATIONS

4.1 Introduction	118
4.2 Vector representation of velocity and acceleration	118
4.3 Relative velocity (plane motion)	123
4.4 Velocity diagrams for plane mechanisms	129
4.5 Velocity relationships in space mechanisms	133
4.6 Velocity diagrams for epicyclic gear trains	139
4.7 Relationship between velocity and displacement diagrams	142
4.8 Relative accelerations (plane motion)	142
4.9 Acceleration diagrams for plane mechanisms	150
4.10 Relative accelerations (space mechanisms)	163
4.11 Algebraic analysis of special mechanisms	164

Appendices

4.1 Equivalent mechanisms	171
4.2 Velocity and acceleration relationships for the slider-crank mechanism	173

Chapter 5 FORCE RELATIONSHIPS IN MECHANISMS II — Inertia forces

5.1 Introduction	177
5.2 Rigid body mechanics	177
5.3 D'Alembert's principle	178
5.4 Rotary balancing	180
5.5 Gyroscopic effects — circular rotor	194
5.6 Angular momentum	199

5.7	Balancing of reciprocating masses	207
5.8	Inertia effects due to plane motion of a rigid body	215
5.9	Transmission of inertia forces	217
5.10	Dynamic response of mechanisms	220
5.11	Linear systems	226
5.12	Flywheel design	227
5.13	Inertia stresses	232

Chapter 6 LINEAR SYSTEMS I First order systems

6.1	Introduction	237
6.2	Linear systems defined	237
6.3	Transfer relationships	240
6.4	Proportional elements	241
6.5	Integrating elements	242
6.6	Simple exponential lag	244
6.7	Goodness of theory	246
6.8	Standard inputs	247
6.9	Step response of a first order system	249
6.10	Integration by analogue computer	252
6.11	Superposition	253
6.12	Relationships between impulse, step, and ramp responses	257
6.13	Harmonic response of a first order system (1)	260
6.14	Vector diagram solution of $(1 + T\mathrm{D})x_0 = X_i \cos \omega t$	261
6.15	Representation of harmonic response data	264
6.16	Use of the Argand diagram	267
6.17	Harmonic response of first order systems (2)	269

Chapter 7 LINEAR SYSTEMS II Second order systems

7.1	Introduction	273
7.2	Equation of motion for a simple position control system (servo-mechanism)	273
7.3	Equation of motion for a spring-mounted mass	275
7.4	Solution of the equation of motion — the complementary function	277
7.5	Step response	279
7.6	Other transient responses	282
7.7	Harmonic response	282
7.8	The particular integral of $(T^2 \mathrm{D}^2 + 2cT\mathrm{D} + 1)\theta_0 = \Theta_i \cos \omega t$	283
7.9	The direct harmonic response locus	284
7.10	Variation of amplitude and phase with frequency	287
7.11	Comparison with first order systems	289

Chapter 8 AUTOMATIC CONTROL

8.1	Introduction	292
8.2	Elements in series	294
8.3	Closing the loop	298

viii Contents

8.4	Remote position control system	300
8.5	Derivative control	304
8.6	Velocity feedback	308
8.7	Speed regulation of an engine	309
8.8	Integral control	315
8.9	(a) Stability — Routh's criterion	318
8.9	(b) Stability — proof of Routh's criterion	320
8.10	Stability — Nyquist's criterion (I)	323
8.11	Open-loop harmonic response loci	327
8.12	Relationship between open-loop and closed-loop harmonic response loci	334
8.13	Nyquist's criterion (II)	335
8.14	Design criteria	336
8.15	Constant M and ϕ contours	339
8.16	Closed-loop steady-state performance	342
8.17	Improvement of system performance	344
8.18	Non-unity feedback	349
8.19	Logarithmic plotting — Bode diagrams	356
Appendix 8.1	Relationships between harmonic and transient responses	361

Chapter 9 MECHANICAL VIBRATIONS

9.1	Introduction	367
9.2	Transient vibrations of a single spring-mounted mass	368
9.3	Harmonic response of a spring-mounted mass — constant excitation	369
9.4	Harmonic response of a spring-mounted mass — inertia excitation	374
9.5	Vibration isolation	375
9.6	Harmonic response of a spring-mounted mass — seismic excitation	377
9.7	Vibration measurement	379
9.8	Undamped vibrations	381
9.9	The phase plane	383
9.10	Vibration of a system with two degrees of freedom	394
9.11	Vibration absorber	401
9.12	Vibration of two rotors connected by a flexible shaft	402
9.13	Geared systems	409
9.14	Principal coordinates — coupling	411
9.15	Equal natural frequencies	414
9.16	Two degrees of freedom — concluding comments	414
9.17	Torsional oscillations of three rotors on a flexible shaft	415
9.18	Symmetrical systems	417
9.19	Holzer's method	418
9.20	Composite systems	423

9.21 Vibrations of beams	426
9.22 Energy method and Rayleigh's principle	429
9.23 Continuous systems	436
9.24 Energy method applied to continuous systems	443
9.25 Critical speeds of shafts	451
9.26 Gyroscopic vibrations	454
Exercises	458
Index	537

PREFACE

This book provides the basis for a three-year course in the study of mechanical systems. The present title is preferred to the more traditional one of theory of machines because of the rather restricted class of mechanisms which the latter title conjures up. The earlier chapters of the book are devoted to conventional machine elements, linkage mechanisms, cams, gears, etc., but in the study of automatic control systems reference is made to many devices which are outside the scope of traditional mechanics.

It is essential that in the kinematics, design of mechanisms should be related to dynamic effects and vibration theory, and it is important to appreciate the close relationship between vibration theory itself and the theory of automatic control systems, and this is the reason for bringing such broad issues together in a single book, but the teaching of these individual topics need not be as sequential as the chapters in the book imply. In organizing courses teachers might well wish to run the study of chapter 6 and onwards in parallel with the earlier chapters, provided the students have sufficient facility in the use of, as well as knowledge of, the mathematical techniques which are used. The level of mathematical knowledge needed is not high; it is assumed that the reader has a grounding in elementary calculus, and is familiar with the D operator method of solving linear differential equations.

At the time of going to press Britain is in the process of 'going metric', and as a matter of national policy the foot/pound system of units is to be superseded by the International System of Units, usually abbreviated to SI units. Accordingly this book uses SI units throughout. The three main basic units which are used are; the *metre* (m), which is the unit of length; the *kilogramme* (kg), the unit of mass; and the second (s), the unit of time. There are three other basic units, namely, the unit of electric current, the interval of temperature, and the unit of luminous intensity, but these are of less interest in the present context. All six basic units are defined by international agreement. The units of length and time are defined in terms of the wavelength and frequency of specified radiations, and the unit of mass by reference to an international prototype which is kept at Sèvres near Paris.

Three derived units of particular interest are: the unit of plane angle, which is the radian (rad); the unit of frequency, the hertz (Hz); and the unit of force, the newton (N), which is the force necessary to give a mass of 1 kilogramme an acceleration of 1 metre per second.

The weight of a body of mass m is mg where g is the local acceleration due to gravity, because this force when applied to the mass causes it to have an acceleration g. At the surface of the earth g has the value of 9·81 m/s² approximately, so that the weight of a mass of 10 kg is 98·1 N. Alternatively the weight of a body can be expressed in kilopond (kp), or its equivalent kilogramme force (kgf), 1 kp (or kgf) being the force of gravity on a mass of 1 kg at the surface of the earth. Formally, 1 kp is defined as being equal to 9·806 65 N. It should be noted, however, that in this context the quantity 9·806 65 is a conversion factor only and has no units, it would be quite wrong to write 1 kp = g N. The student is advised to avoid the use of units of force other than the newton, and in particular to avoid the use of the kilogramme force with its inevitable confusion with the kilogramme mass. Where data in relation to dynamic systems is given in terms of kp or kg the first step in any calculation should be the conversion of all forces to newtons.

Most of the problems set as exercises in this book have either been taken directly from examinations set in the Engineering Department of the University of Cambridge, or are based on questions set in these examinations. The author is grateful to the Syndics of the University Press for permission to use these questions. Direct reference to particular examinations has not been made because it is felt that with recent changes in syllabus and in the examination system such references might be misleading. Furthermore, many problems were originally set in foot/pound units, and where conversion to SI units has been made it is the present author, and not the original examiners, to whom any errors should be attributed.

Cambridge   J. M. Prentis
April 1969

1

SIMPLE MECHANISMS
I Structural analysis

1.1 Introduction

A mechanical system may be defined as a device or number of interconnected devices for which it is possible to define a specific relationship between a disturbance applied to the system and the response of the system to that disturbance. With the simplest devices the response is directly proportional to the disturbance. Countless examples can be offered, but a few simple ones will suffice. In the case of the rigid lever in fig. 1.1(a) a disturbance which depresses the left-hand end a distance x causes the other end to rise a distance $y = kx$ where k is the lever ratio. A force P applied to the end of a light spring [fig. 1.1(b)] causes that end to move a distance $e = P/k$ where k is the stiffness of the spring. In fig. 1.1(c) the slowly varying pressure p applied to a manometer causes the level of liquid in the limb to move an amount $z = p/\rho g$ where ρ is the density of the liquid in manometer and g is the acceleration due to gravity. Figure 1.1(d) gives an analogous electrical system, the voltage drop across a pure resistance R is $v = iR$ for a current i.

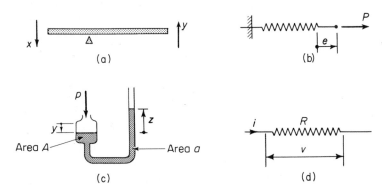

Figure 1.1

It will be noticed that in each of the above examples restrictive conditions are stated: the lever is rigid, otherwise it would bend under the action of any forces which may be applied to it and the basic equation would no longer apply; the spring is 'light', otherwise the force-displacement relationship would be complicated by inertia terms when P is varied. In the case of the manometer, we cannot conceive of a massless liquid since the weight of the liquid is fundamental to the operation of the device, so we specify that when p varies it shall do so slowly, the slower the variation the smaller are the inertia terms and the more accurately does our simple relationship apply. If the resistance in the final example is non-inductive the current i can vary at any frequency without upsetting the simple relationship stated; if the resistance were in fact inductive we should have to restrict the rate of change of i. Thus each of the above expressions must be recognized as being an idealization; in each case the expression will be sufficient for many practical purposes. In this book we shall study the circumstances under which our idealized expressions do apply and see how the relationships are modified by the conditions under which the particular device operates. Each of these devices is by itself a system and all are members of one particular family of systems. This family is characterized by the form of the equation which describes the behaviour of its members. In this case the equation is

$$\theta_o = K\theta_i \qquad (1.1)$$

θ_i is the disturbance or input (x, P, p, and i in the above examples) and θ_o is the response or output (y, e, z, and v).

The input and output do not necessarily have the same physical form. In the case of the lever both x and y are displacements so that K is simply a number. But with the spring, p is a force and e is a displacement, and $K = l/k$ has the units m/N.

The term mechanism is sometimes used as freely as the term system but in the study of mechanics its use is generally restricted to apply to devices for which the input and output are both displacements and for which the relationship between the input and output can be expressed irrespective of any forces which may be transmitted by the device. The latter restriction means that all the members of the mechanism are essentially rigid though they need not be in the purely practical sense. For example, in a pulley drive the belt must be flexible otherwise it would not wrap round the pulleys. It is, however, essentially a rigid member if it does not stretch at all due to the tension imposed on it by the pulley loads. With such an idealized belt the angular displacement of one pulley is always proportional to the angular displacement of the other provided that there is no slip between either pulley and the belt.

Returning to the systems in fig. 1.1, it can be seen that the lever is a mechanism in the sense indicated above while the spring is not. The manometer can be regarded as a mechanism since the two liquid levels are related in a way which does not depend on the pressure p; provided that the manometer walls are rigid and the liquid is incompressible $z = yA/a$. On the other hand, the purpose of a manometer is not to transmit motion but to convert a pressure into a displacement.

1.2 Classification of mechanisms

We will define a mechanism as an assemblage of essentially rigid elements or links in which the relative motion between the individual elements is constrained.

The immediate consequence of this definition is, that the relationship between the motions of the various parts of a mechanism may be studied and specified, without reference to any forces which may be transmitted. This does not imply that the relationship between the motion of a mechanism and the forces applied to it is of small importance, but simply that it can be treated as a separate matter.

Mechanisms can be classified in a number of different ways. One method is the purely descriptive, linkage mechanisms, gear devices, cams, and so on. Such a division is convenient and we shall subsequently study individual mechanisms under these headings, but for the moment we will take a broader view.

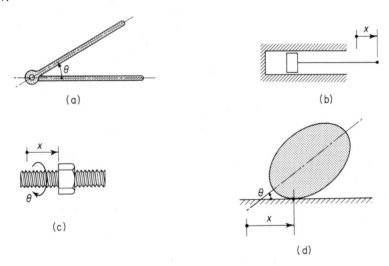

FIGURE 1.2

It is obvious that motion can only be transmitted through a series of elements if there is contact between successive elements. Furthermore, if we have a string of elements we require a certain minimum number of coordinates to define the configuration of the mechanism. Figure 1.2 gives a number of examples of *pairs* of contacting elements. In (a) two rods are hinged together, in (b) a piston slides in a cylinder. In both these cases a single variable defines the relative positions of the two elements of the pair, θ in the case of the hinged rods, and x in the case of the piston and cylinder; assuming that the piston cannot rotate. In the case of the nut and bolt, either x or θ serves to define the configuration, we do not need both because one determines the other. In (d) an elliptic disc is in contact with a plane surface. Here two variables are needed to define the configuration, x to fix the point of contact on the plane, and θ to fix the attitude of the disc. A pair is said to have as many degrees of freedom as the number of coordinates which are needed to define the relative position of the two members of the pair. Thus in fig. 1.2 (a), (b), and (c) are all examples of pairs with one degree of freedom; example (d) has two degrees of freedom, and the ball and socket joint in fig. 1.3 has three degrees of freedom. We shall refer to these different types as class I, class II, and class III pairs respectively. The maximum number of degrees of freedom which a pair can have is five. Such a pair consists of two members which are in contact only at a single point.

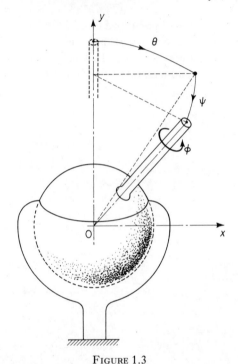

FIGURE 1.3

If the two members of a pair are of arbitrary shape, the number of degrees of freedom F is related to the number of points of contact p by the equation

$$F = 6 - p$$

This expression is, however, of limited usefulness because the shapes of the members of a pair are not usually arbitrary. In the case of the ball and socket joint the two parts touch at an infinite number of points, but as all the points lie on a spherical surface the pair still has three degrees of freedom. Similar observations apply to the examples of pairs depicted in fig. 1.2.

The number of degrees of freedom possessed by a mechanism depends on the types of joint, the numbers of the different types of joint, and the number of members. Although we can readily derive expressions for the number of

SIMPLE MECHANISMS I Structural analysis 5

degrees of freedom of a mechanism there are certain difficulties in interpreting these expressions correctly. These difficulties are most easily understood in relation to plane mechanisms, i.e., mechanisms in which all motion is either in or parallel to a fixed plane. Such mechanisms are far more common than ones in which the motion is three-dimensional so that there is, in any case, good reason to study them in their own right. In the first instance we shall consider plane mechanisms in which all the joints have but one degree of freedom.

1.3 Plane mechanisms with class I pairs

The two rods AB and CD in fig. 1.4(a) each require three coordinates to define their positions with respect to the axes Oxy. The system consisting of these two rods has therefore a total of six degrees of freedom, three for each rod. If the

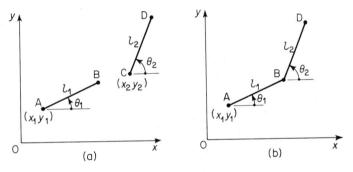

FIGURE 1.4

system consisted of n disconnected rods it would have $3n$ degrees of freedom. If the two rods are hinged together at B and C respectively then CD loses two of its degrees of freedom since C is no longer free to move independently in the Oxy plane but has to coincide with B although CD is still free to rotate about B. The system as a whole now has $6 - 2 = 4$ degrees of freedom, fig. 1.4(b). If j hinges are formed in a system consisting of n rods the system has $(3n - 2j)$ degrees of freedom. Note that although four coordinates are needed to specify the configuration in fig. 1.4(b), coordinates other than those chosen may be used. Thus polar coordinates can be used instead of cartesian coordinates to fix A, or we could choose x_2, y_2, θ_1, and θ_2 instead of x_1, y_1, θ_1, and θ_2. On the other hand there is not complete freedom of choice, e.g., x_1, x_2, θ_1, and θ_2 are not a possible set of coordinates. This is because x_1, x_2, and θ_1 are related by the equation

$$x_2 = x_1 + l_1 \cos \theta_1$$

Even if this equation is satisfied the positions of the rods are not fully determined because the position in the y direction is not defined. The number of degrees of freedom possessed by a system therefore equals the number of independent coordinates needed to define its configuration.

Normally a mechanism is not free to move bodily in space but is tied to the ground by fixing one member. This destroys three degrees of freedom as can be seen by imagining AB in fig. 1.4(b) to be fixed. Thus the number of degrees of freedom of a plane mechanism consisting entirely of class I pairs is

$$F = 3n - 2j - 3$$
$$= 3(n-1) - 2j \qquad (1.2)$$

A system is classed according to the value of F. The usual titles for different values of F are given in table 1.1

Table 1.1

$F = 1$	Simple mechanism
2	Differential
0	Simply stiff or rigid frame
-1	Structure with one redundant member
$-n$	Structure with n redundancies

Consider first the application of eqn. (1.2) to mechanisms which consist entirely of turning pairs. The simplest possible device consists of a single pair

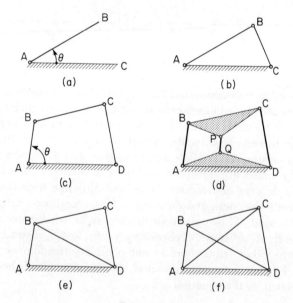

FIGURE 1.5

of elements, one of which is fixed. It is obvious that the system has one degree of freedom, and this checks with the above equation as $n = 2$ and $j = 1$ giving $F = 1$. Three bars hinged together as in fig. 1.5(b) form a rigid frame for which $F = 0$, and four bars form a mechanism with one degree of freedom, fig. 1.5(c).

The four-bar mechanism is of considerable importance and will be studied in detail subsequently.

When we come to add a fifth link we have a choice; either we can break one of the existing joints and insert the extra link to form a polygon which has two degrees of freedom, or we can add the fifth link between two arbitrary points P and Q to form a rigid framework. It is tempting to add the fifth link between B and D as in fig. 1.5(e) so that the system appears to have five links and four joints and making $F = 4$. If, however, proper allowance is made for the fact that the joints at B and D are double, thus giving six joints in all, eqn. (1.2) gives $F = 0$.

In fig. 1.5(f) $j = 8$, $n = 6$, and $F = -1$, and the rigid frame has been converted into a redundant structure. It will be seen that it is possible to do this only if the sixth link AC is exactly the right length. This is the first time that we have used the term length in this discussion. Hitherto we have tacitly assumed that some general restrictions must be placed on the dimensions of the

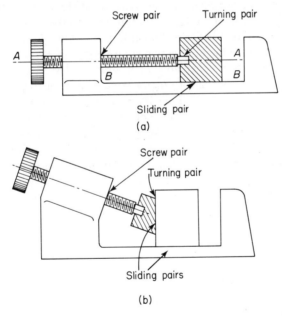

FIGURE 1.6

members, e.g., in fig. 1.5(b) it was taken for granted that $(AB + BC) > AC$, but apart from this the lengths are arbitrary. We can take it therefore that, if the application of eqn. (1.2) leads to the result $F = -1$, there must be a restriction on the dimensions of the elements of the device considered. By imposing geometrical restrictions we can make mechanisms which have one degree of freedom even though they are rigid frames according to eqn. (1.2). A familiar example is the screw clamp shown in fig. 1.6(a) which has one screw pair, one sliding pair and one turning pair. As there are three members eqn.

(1.2) gives $F = 3(3 - 1) - 2 \times 3 = 0$. Nevertheless, this is a very common device which definitely possesses one degree of freedom. Further thought shows this to be true only if the screw axis A–A is colinear with the axis of the turning pair and parallel to the sliding surface B–B. Otherwise, the device jams and is in fact a rigid frame. The restriction that A–A and B–B should be parallel can be relaxed only if another member is introduced as, for example, in fig. 1.6(b) where there are four lower pairs and four elements. The device in fig. 1.6(a) can be regarded as a degenerate version of that in fig. 1.6(b).

A somewhat different situation arises with the device shown in fig. 1.7(a) where two wedges are constrained by a third member which is fixed to the ground. Provided that proper contact is maintained between the members the configuration is fully defined by the variable x and the device has one degree of freedom. As $n = 3$ and $j = 3$, $F = 0$ and once again eqn. (1.2) apparently fails to give the right picture. But further consideration shows that the above description of the device is correct only if $\alpha + \beta = \gamma$. If this is not so, we do not have surface contact between the elements of all three pairs; there may, for example, be line contact at Q so that if the geometrical restriction that $\alpha + \beta = \gamma$ is relaxed the device no longer has three class I pairs but two class I and one class II pairs. The next section shows how eqn. (1.2) must be modified to make it applicable to mechanisms incorporating class II pairs.

In practice, then, a mechanism may have a greater degree of freedom than that indicated by eqn. (1.2). This happens when appropriate geometrical restrictions are made. An extra degree of freedom thus obtained may be dearly bought and precariously held. The clamp in fig. 1.6(a) must be more accurately made than the alternative version and may therefore be more costly even though it saves one member. It is also more likely to fail due to distortions of the bedplate whether from violence or heat. If the necessary angle condition is not met in respect of the wedge mechanism in fig. 1.7 there may be unforeseen

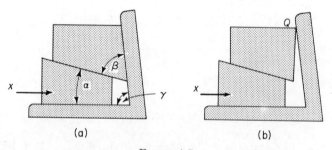

FIGURE 1.7

failure due to the high stresses generated at Q if a substantial load is applied to the device.

A further example of a system which has one degree of freedom, although according to eqn. (1.2) $F = 0$, is given in fig. 1.8 where it is obvious that the angle ϕ is not fixed. Here the sub-system BCDE has one degree of freedom because the member needed to convert it into a rigid frame has been built into the sub-system ABEF which is overconstrained; ABEF would still be a rigid frame even if member AE were removed. Once more we have the situation where the system can be constructed only if a geometrical constraint is imposed. In this case, a rigid link AE can be fitted into its present position only if it is of a certain length.

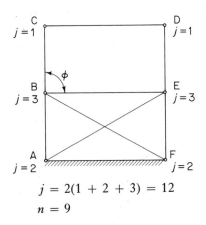

$$j = 2(1 + 2 + 3) = 12$$
$$n = 9$$

FIGURE 1.8

A pathological system of the type just considered can be detected without recourse to experiment by imagining it to be built up link by link and by considering at each stage whether any geometrical constraint is set upon the member about to be incorporated.

1.4 Plane mechanisms with class II pairs

All the general ideas developed above in respect of plane mechanisms which consist of class I pairs only apply equally to mechanisms which incorporate class II pairs. It is, however, necessary to write eqn. (1.2) in a more general form to allow for some of the pairs having more than one degree of freedom.

As before, we start with n disconnected elements having a total of $3n$ degrees of freedom. If some of the elements are connected together to form j_1 pairs each allowing one degree of freedom, $2j_1$ degrees of freedom are lost. If now a further j_2 pairs are formed, each of which allows two degrees of freedom, an additional j_2 degrees of freedom are lost. Finally, we fix one member to destroy three more degrees of freedom.

The system is then left with

$$F = 3n - 2j_1 - j_2 - 3$$
$$= 3(n - 1) - (2j_1 + j_2) \qquad (1.3)$$

degrees of freedom.

The cam and follower mechanism in fig. 1.9 has three elements: the cam, the follower, and the frame, two class I pairs A and B, and a class II pair C. We have therefore

$$F = 3(3 - 1) - (2 \times 2 + 1) = 1$$

and the system has one degree of freedom.

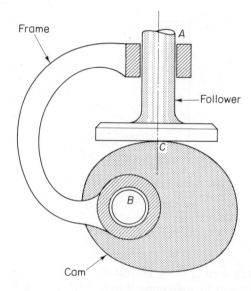

FIGURE 1.9

1.5 Space mechanisms

In space a body has six degrees of freedom, three of translation and three of rotation. As each pair is formed, one to five degrees of freedom are lost and for the whole mechanism we have

$$\begin{aligned} F &= 6(n - 1) - (5j_1 + 4j_2 + 3j_3 + 2j_4 + j_5) \\ &= 6(n - 1) - \sum (6 - r)j_r \end{aligned} \quad (1.4)$$

where j_r is the number of joints with r degrees of freedom.

Equation (1.4) is a general equation which applies to all mechanical systems composed of rigid elements. Equation (1.3) is a restricted version of eqn. (1.4) and eqn. (1.2) is an even more restricted form.

It must be understood that in general eqns. (1.4) and (1.3) do not give the same answer. Consider, for example, the four-bar mechanism in fig. 1.5(c) for which eqn. (1.2) or (1.3) gives $F = 1$. At each hinge $i = 1$ and there are four elements and four hinges so that from eqn. (1.4)

$$F = 6(4 - 1) - 5 \times 4 = -2$$

Since it is obviously possible to construct this mechanism so that it does have one degree of freedom we must conclude that certain geometrical restraints are imposed. In applying eqns. (1.2) and (1.3) it was tacitly assumed that all hinge axes were perpendicular to the plane of the paper. Equation (1.4) makes no such assumption.

If the connections at B and C are replaced by ball and socket joints each allowing an extra two degrees of freedom

$$F = 6(4 - 1) - (5 \times 2 + 3 \times 2) = 2$$

One of these degrees of freedom arises from the ability of BC to turn about its own axis and, depending upon the use of the mechanism, is probably of no consequence.

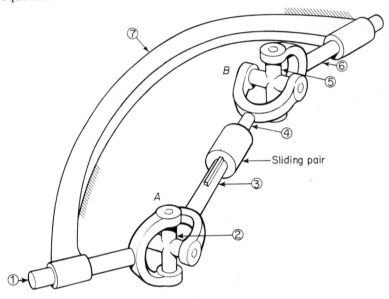

FIGURE 1.10

It is of interest to see what is the smallest number of elements which can be assembled into a single degree of freedom space-linkage which uses only class I pairs. Assuming that $j = n$, which has been the case for nearly all the mechanisms so far considered

$$F = 1 = 6(n - 1) - 5n$$

so that $n = 7$

Figure 1.10 shows a common means of transmitting motion from one shaft to another when their axes are neither parallel to nor intersect each other. There are seven members, with six turning pairs and one sliding pair. The sliding pair on the main connecting link can be dispensed with if the axes of the two turning pairs at A intersect on the axis of member (1), and if a similar restriction is made at B. In many practical mechanisms some geometrical constraints are imposed to reduce the number of links employed.

1.6 Structural analysis of plane mechanisms

We have so far been concerned almost entirely with the numerical aspects of the kinematics of mechanisms. The main problem considered has been that of determining the number of degrees of freedom possessed by a mechanical system which consists of a number of rigid links connected together with specified degrees of freedom between adjacent links. This can be considered as the first problem of structural analysis. The next question which arises is: given a specified number of links, and the number of degrees of freedom allowed to the assembled system, what are the possible ways in which the various elements can be connected?

Two types of system will be considered: plane rigid frames and plane mechanisms with one degree of freedom, using turning pairs only.

1.7 Rigid frames

In most of the systems so far considered the individual elements have each formed pairs with two others only. Where turning pairs only have been used this has meant that each member has had two hinges. Such elements will be called *binary* links and the number of such links in a system will be denoted by n_2. When a member is directly in contact with three others, i.e., it has three hinges, it will be called a *tertiary* link and the number of such links denoted by n_3. And so on. Note that we specifically exclude the trivial case of a member with only one hinge.

For a rigid frame $F = 0$ so that eqn. (1.2)

$$j = \tfrac{3}{2}(n - 1) \tag{1.5}$$

where n is the total number of links and

$$n = n_2 + n_3 + \cdots + n_i \tag{1.6}$$

Since we must have a whole number of hinges, eqn. (1.5) shows immediately that a rigid frame must have an odd number of links.

We next consider what is the maximum number of hinges which can be carried by any one member of the frame. Let one member, fig. 1.11(a), carry any number of hinges and connect to it two other links to form the rigid frame ABC. We cannot connect the next member directly between B and D for two reasons:

1. it would have to be of just the right length thereby imposing a geometrical constraint.
2. a redundant structure would be created as can be seen by inserting $n = 4$ and $j = 5$ (two at B) in eqn. (1.2) to give

$$F = 3(4 - 1) - 2 \times 5 = -1$$

Thus two members BE and ED must be added, fig. 1.11(b). The next two members could be attached to B and E but since we wish to determine the maximum number of connections which can be made to link ACD we connect EG and FG as shown.

By continuing in this manner, link ACD always has its largest possible share of the hinges and it can be seen that this share is $\frac{1}{2}(n+1)$.

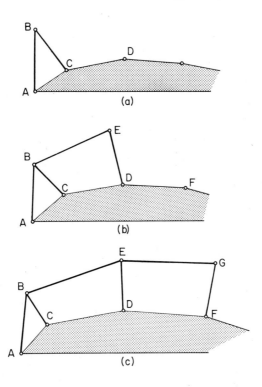

FIGURE 1.11

Remembering that a hinge joins two members together the total number of hinges is given by

$$j = \tfrac{1}{2}(2n_2 + 3n_3 + \cdots + in_i) \tag{1.7}$$

where $\quad i = \tfrac{1}{2}(n+1)$

Substituting from eqns. (1.6) and (1.7) in eqn. (1.5) we have

$$\tfrac{1}{2}(2n_2 + 3n_3 + \cdots + in_i) = \tfrac{3}{2}(n_2 + n_3 + \cdots + n_i - 1)$$

or $\quad n_2 = 3 + n_4 + 2n_5 + \cdots$

$$n_2 = 3 + \sum_{i=4}^{i=\frac{1}{2}(n+1)} (i-3)n_i \tag{1.8}$$

which shows that a plane rigid frame must have at least three binary links.

14 Dynamics of Mechanical Systems

The numerical properties of plane rigid frames ($F = 0$) can now be summarized:

1. There must be an odd number of links n.
2. At least three of these must be binary links.
3. The number of joints $j = \frac{3}{2}(n - 1)$.
4. The maximum number of these joints which can be carried by any one member is $\frac{1}{2}(n + 1)$.

These rules, together with eqns. (1.6) and (1.7), enable the possible forms of a plane rigid frame to be determined for any number of links.

Example Five-bar rigid frame

$$j = \tfrac{3}{2}(5 - 1) = 6$$
$$i = \tfrac{1}{2}(5 + 1) = 3$$

From eqn. (1.7)

$$2j = 12 = 2n_2 + 3n_3$$

From eqn. (1.6)

$$n = 5 = n_2 + n_3$$

Solving

$$n_2 = 3 \quad \text{and} \quad n_3 = 2$$

There are only two basically different ways of connecting these links together, as shown in fig. 1.12(a) and (b). The distinction between these alternatives disappears if hinges are merged together as in (c) and (d). If each element is to take the form of a rod with hinges only at the two ends, fig. 1.12(d) shows the only possible form for a five-bar frame.

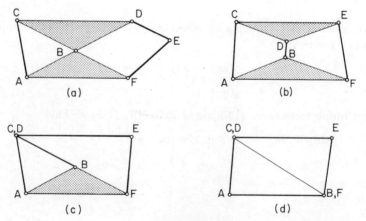

FIGURE 1.12

SIMPLE MECHANISMS I Structural analysis

Example Seven-bar frame

$$j = \tfrac{3}{2}(7 - 1) = 9$$
$$i = \tfrac{1}{2}(7 + 1) = 4$$

Equation (1.7) gives

$$2j = 18 = 2n_2 + 3n_3 + 4n_4$$

Equation (1.6) gives

$$n = 7 = n_2 + n_3 + n_4$$

If n_3 is eliminated from the two equations [*vide* eqn. (1.8)]

$$n_4 = n_2 - 3$$

This equation does not have a unique solution but a finite number of possible solutions.

If n_2 has its minimum possible of 3, $n_4 = 0$ and since $n = 7$, $n_3 = 4$. With $n_2 = 4$, $n_4 = 1$, and $n_3 = 2$. Finally, if $n_2 = 5$, $n_4 = 2$, and $n_3 = 0$. These results are most conveniently displayed in a table thus:

n_2	n_3	n_4
3	4	0
4	2	1
5	0	2

Some of the ways in which the links can be connected to satisfy these relationships are shown in fig. 1.13.

The number of possibilities increases with the number of links. With the nine-bar chain (see exercise 1.3) different combinations of n_3 and n_4 are possible for the same value of n_2.

1.8 Plane mechanisms with one degree of freedom

The reasoning in the case of plane mechanisms is very similar to that given above for plane frames and the main steps need be given only briefly. Starting again with the basic equation

$$F = 3(n - 1) - 2j$$

and putting $F = 1$ we have

$$j = \tfrac{1}{2}(3n - 4) \qquad (1.9)$$

the total number of links is again given by eqn. (1.6)

$$n = n_2 + n_3 + \cdots + n_i$$

and eqn. (1.9) shows that this time n must be even.

16 Dynamics of Mechanical Systems

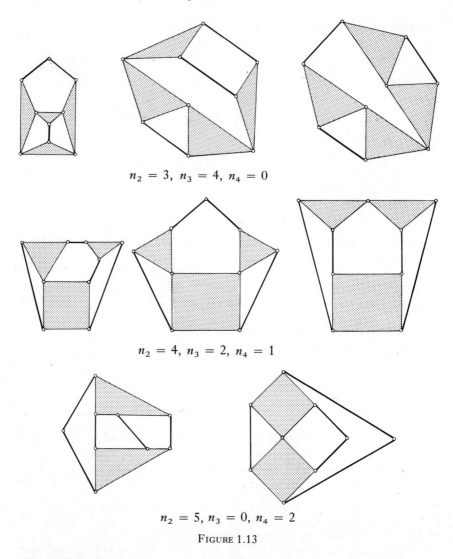

$n_2 = 3, \ n_3 = 4, \ n_4 = 0$

$n_2 = 4, \ n_3 = 2, \ n_4 = 1$

$n_2 = 5, \ n_3 = 0, \ n_4 = 2$

FIGURE 1.13

Figure 1.14 shows how a mechanism can be built up to have one link which carries the greatest possible number of hinges. This maximum is evidently $n/2$.

Equation (1.7),

$$j = \tfrac{1}{2}(2n_2 + 3n_3 + \cdots + in_i)$$

applies as before except that this time $i = n/2$.

Eliminating j and n_3 from eqns. (1.9) and (1.7) we find that

$$n_2 = 4 + \sum_{i=4}^{i=n/2} (i - 3)n_i \qquad (1.10)$$

and hence that the minimum possible number of binary links is 4.

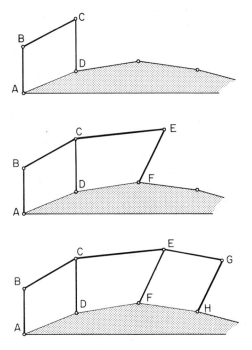

FIGURE 1.14

The numerical properties of plane mechanisms ($F = 1$) are:

1 The number of links n must be even.
2 At least four of these must be binary links.
3 The number of joints is $j = \frac{1}{2}(3n - 4)$.
4 The maximum number of joints which can be carried by one link is $n/2$.

In addition to these rules, we use eqns. (1.6) and (1.7).

Example Six-bar mechanism

$$j = \tfrac{1}{2}(3 \times 6 - 4) = 7$$
$$i = \tfrac{6}{2} = 3$$

From eqn. (1.7)

$$2j = 14 = 2n_2 + 3n_3$$
$$n = 6 = n_2 + n_3$$

Solving these two equations we have

$$n_3 = 2$$
$$n_2 = 4$$

The two possible basic forms of six-bar mechanism are shown in fig. 1.15.

18 Dynamics of Mechanical Systems

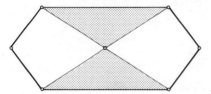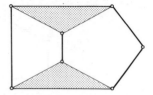

FIGURE 1.15

1.9 Alternative relationship between number of joints and links

Textbooks on the theory of structures[1] usually give a relationship between the number of links and the number of joints which differs from eqn. (1.2). The simplest form of this relationship is obtained when the structure is a pin-jointed plane frame, in which all the members are struts and ties and the joints are incapable of transmitting any bending moment. The term joint in this context holds a slightly different meaning from that which we have so far used in this chapter where joint, or pair, has referred to the contact of two members only. When three members have been in contact at one point, e.g., at C, D in fig. 1.12(d), we have been careful to refer to it as a double joint. In the structural sense the term joint does not imply any restriction on the number of members meeting. We will distinguish between the two situations by using the term *node* instead of joint in the analysis of structures.

Consider one member AB of a plane structure to be fixed in space. In order to fix a third point C not on AB we need two more members AC and AB, fig. 1.16(a). To fix another point D two more members are required CD and BD; to fix E two more are needed, and so on. In fig. 1.16(a) there are three members and three nodes; in fig. 1.16(b) two members and one node have been added. It can be deduced that the following relationship exists between the number of members n and the number of nodes g.

$$(n - 3) = 2(g - 3)$$

or $\qquad n + 3 - 2g = 0$ †

If an extra member such as AD is added n is increased by 1 but g is unaltered and the structure is said to be statically indeterminate. The addition of CF makes it twice times statically indeterminate and so on. In general, the degree of statical indeterminacy or redundancy is given by

$$R = n + 3 - 2g \qquad (1.11)$$

For a space frame with ball and socket joints at the nodes

$$R = n + 6 - 3g \qquad (1.12)$$

If eqn. (1.2)

$$F = 3(n - 1) - 2j$$

† This equation can also be derived by considering the conditions for equilibrium of each of the nodes, and of the structure as a whole, under a set of external loads. There are $(2g - 3)$ independent equations which must equal the n unknown bar forces.

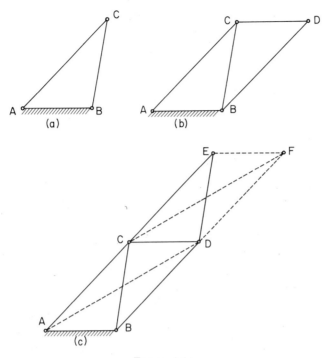

FIGURE 1.16

which is valid for all plane linkages using lower pairs only, and eqn. (1.11) are applied to the same structure it will be found that $F = -R$. In the example shown in fig. 1.17,

$$R = 10 + 3 - 2 \times 6 = 1$$

and
$$F = 3(10 - 1) - 2 \times 14 = -1$$

With this structure it is much more obvious that $g = 6$ than that $j = 14$ and clearly that eqn. (1.11) is much more convenient and safer to use than eqn. (1.2).

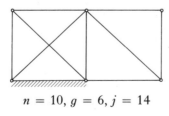

$n = 10, g = 6, j = 14$

FIGURE 1.17

In the case of the system in figs. 1.18, however, which is typical of the systems met in studying mechanisms we get into difficulty in applying eqn. (1.11) until we replace the tertiary and quaternary links by rigid frameworks of three and five links respectively. Now

$$F = 3(8 - 1) - 2 \times 10 = 1$$
$$R = 16 + 3 - 2 \times 10 = -1$$

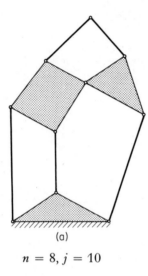
(a)
$n = 8, j = 10$

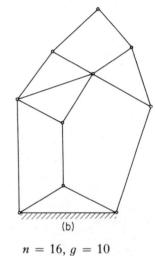
(b)
$n = 16, g = 10$

FIGURE 1.18

Although eqns. (1.2) and (1.11) are both generally valid each is the most convenient in its own field of application. Both are special cases of more general equations.[2]

$$R = 3(n - g + 1) - f \qquad (1.13)$$

for plane frames and

$$R = 6(n - g + 1) - f \qquad (1.14)$$

for space frames.

In both equations, the quantity f is determined by first considering each node in the system to be completely rigid and then summing up the total number of degrees of freedom which must be introduced in order to convert the completely rigid system into the actual system. In the particular case of pin-jointed plane frames, there will be $(m - 1)$ constraints removed at each node where m is the number of members meeting at that node. Since every member connects two nodes and since there are g nodes altogether

$$f = \sum (m - 1) = 2n - g$$

Equation (1.13) must be used when the connections at the nodes are other than by pin-joints. Equations (1.13) and (1.14), which are the counterparts of eqns. (1.3) and (1.4) respectively, may be applied only to systems in which all the elements span between two nodes with no intermediate connection to other members.

REFERENCES
1 CASE, J. and CHILVER, A. H. *Strength of Materials*, Arnold, 1959.
2 MORICE, P. B. *Linear Structural Analysis*, Thames & Hudson, 1959.

2

SIMPLE MECHANISMS
II Displacement analysis

2.1 Introduction
The main requirement of any mechanism is that it should transmit motion in a particular way. With an ideal mechanism the displacement of the output member is completely determined by that of the input member and the relationship between the two is independent of:

(i) the forces transmitted by the members of the mechanism,
(ii) the speed of operation.

In practical mechanisms, the input to output relationship is modified by both these considerations. The forces which are transmitted affect the motion because the members are not infinitely stiff. When the load P, which is transmitted by the simple push-rod in fig. 2.1, is zero the displacements of the two ends of

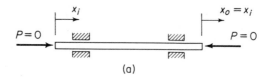

(a)

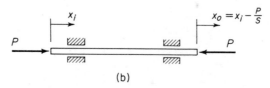

(b)

FIGURE 2.1

the rod are equal. If P is not zero the rod compresses by an amount P/s, where s is the stiffness of the rod, and the displacements of the two ends are no longer equal.

Even if the output force is zero, x_o will not equal x_i if x_i is altered rapidly because the mass of the rod gives rise to inertia forces and strains, and hence to vibrations of the rod which may persist even after the input displacement is complete. Although the transmitted force is zero a force must still be applied to the left-hand end to balance the inertia force of the rod. When additional forces are transmitted the forces applied to the two ends of the rod differ by the amount of this inertia force. Mechanisms may be classed as low-speed or high-speed systems, according as to whether or not the inertia effects are significant.

Leaving the effects of flexibility and inertia on one side, the relationship between the input and the output of a mechanism depends only on the geometry of the mechanism, and in this connection two problems are evident:

(i) that of determining the input to output relationship of a given mechanism,
(ii) that of designing a mechanism to achieve a specified input to output relationship.

For the sake of brevity, the input to output relationship will be referred to as the *transfer relationship* of a mechanism, and if we exclude direct measurement on the actual mechanism, there are two ways of determining it. The first is by mathematical analysis, the second by drawing the mechanism to scale in a sequence of positions and measuring the relevant displacements. There are relatively few mechanisms which have a simple mathematical expression for their transfer relationship. Generally, the equations are cumbersome, and numerical evaluation of them involves much computation. Graphical methods are usually much easier and, provided that the limitations on accuracy are acceptable, universally applicable.

The inverse problem of designing a mechanism to have a prescribed transfer relationship is more interesting and in general more difficult. The problems both of analysis and synthesis will now be considered in respect of three classes of mechanism: cams, gears, and linkage mechanisms.

2.2 Cams

Figure 2.2 shows a selection of different types of cam mechanism. The most common function of a cam is to impart a cyclic motion to an output member with the input shaft turning at constant speed. Disc cams impart to their followers either a translation in a direction perpendicular to the axis of the input shaft or a rotation about an axis parallel to that of the input shaft. With the cylindrical cam the translational motion of the follower is parallel to the axis of rotation of the input axis. The reciprocating cam can be regarded as an unwrapped cylindrical cam. It is clear that a flat-footed follower, fig. 2.2(b), cannot be used with either the cylindrical or the reciprocating cams and that it can be used with a disc cam only if the profile is convex outwards at all points.

The principles involved in deriving a cam profile which satisfies a given transfer relationship, such as that in fig. 2.3(a), can be most simply illustrated in respect of a reciprocating cam. We require that when the cam is displaced laterally a distance x_i the follower shall have risen a distance x_o. In the case of a knife-edged follower, as indicated in fig. 2.3(a), the cam profile is geometrically similar to the graph of the transfer relationship. A knife-edged follower is not normally used because it is too easily damaged and wears away rapidly

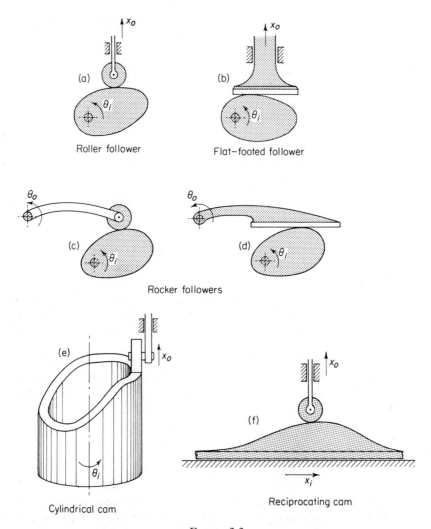

FIGURE 2.2

under all but the lightest of loads. With a roller follower there is much less wear and substantial loads can be transmitted. The method of determining the profile for a roller follower is shown in fig. 2.3(b): the roller is drawn in a series of positions relative to a base line drawn on the cam. The vertical displacement x_o of the centre of the roller equals the displacement required when the cam and follower are displaced laterally relative to each other by an amount x_i. The locus of the centre of the roller relative to the cam base thus reproduces the required lift curve, and the cam profile is the envelope of the roller circles.

The principle involved in holding the cam still and moving the follower relative to it is known as *kinematic inversion*. We have, so to speak, to stop the cam moving in order to study its shape and geometrical properties.

SIMPLE MECHANISMS II Displacement analysis

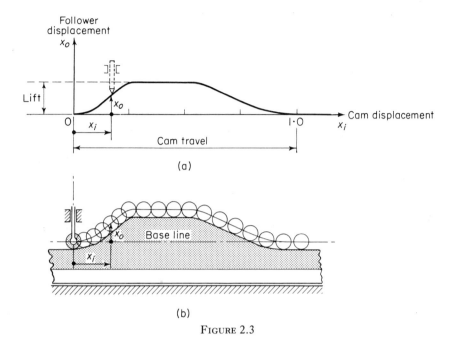

FIGURE 2.3

2.3 Interference reciprocating cams

The minimum diameter of a roller follower is usually determined by the size of bearing needed to carry the load which must be transmitted and also from a consideration of the contact stresses between the roller and the cam. Kinematic considerations place an upper limit on the diameter of the roller which can be used when the transfer relationship is predetermined. Suppose that $A_1C_1C_2A_2$ in fig. 2.4 is a desired lift curve, and suppose also that the roller

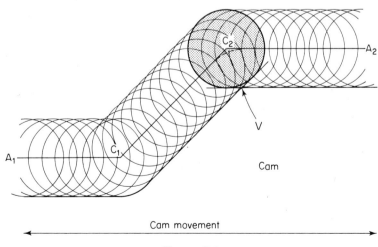

FIGURE 2.4

diameter has already been decided. Using the technique described in the previous paragraph to determine the cam profile, it can be seen that the profile required by section C_1C_2 of the lift curve is incompatible with the profile required by section C_2A_2 in the region V, where the shaded roller fails to make contact with the resultant profile. The two sections of the profile are said to *interfere* at V. The lift curve due to the resultant profile is shown dotted at C_2. The radius of curvature of the modified portion of the lift curve equals the radius of curvature of the roller as the roller pivots about the sharp corner of the cam at V. It can be seen that, if interference is to be avoided, the radius of curvature of the lift curve must not be less than the radius of the roller. This restriction applies, however, only to those sections of the transfer relationship where the cam profile is convex outwards. At a concave section such as at C_1, there is no such restriction on kinematic grounds although the adoption of the profile depicted would lead to undesirable dynamic effects (see exercise 9.11).

2.4 Disc cams

The profile of a disc cam is determined in a manner similar to that for a reciprocating cam and the general method is the same whatever the type of follower employed. Consider first a knife-edged follower, whose tip moves along a line which passes through the axis of rotation of the cam.

The lift curve drawn in polar coordinates immediately gives a possible cam profile for a knife-edged follower, fig. 2.5(a), the displacement of the tip of the follower being relative to a base circle of arbitrary radius. Space considerations usually require that the cam shall be made as small as possible. The size of shaft on which the cam is mounted obviously places a lower limit on the size of base circle which may be adopted. Friction is another factor which must be taken into account.

If the force against which the follower operates is P, and there is no friction between the follower and the cam, then there must be a sideways thrust Q on the tip of the follower, fig. 2.5(b), such that

$$Q \sin \phi = P \cos \phi$$

or

$$Q = P \cot \phi$$

where ϕ is the angle between the follower axis and the tangent to the cam profile. It is shown in chapter 3 that when friction is included of an amount $\mu = \tan \lambda$, the sideways force is given by

$$Q = P \cot(\phi \pm \lambda) \tag{2.1}$$

the negative sign being taken when the follower is rising, and the positive sign when it is falling. We need not go further into the problems of friction and force transmitted at this stage, the point has been made that the sideways thrust Q on the follower increases as ϕ is decreased, becoming very large when $\phi \approx \lambda$. This leads to rapid wear at the tip of the follower and may lead to jamming as a result of increased friction in the follower guides.

SIMPLE MECHANISMS II Displacement analysis

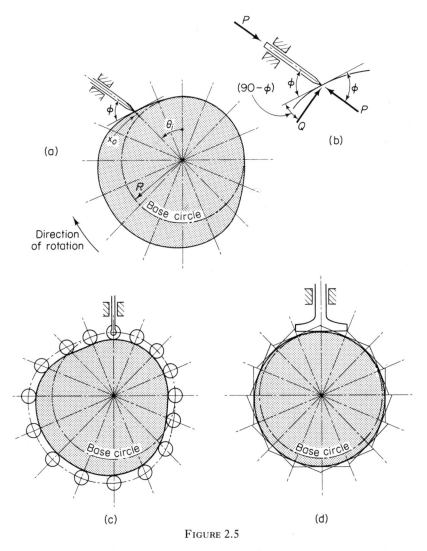

FIGURE 2.5

An expression for the angle ϕ between the tangent and the radius vector is derived by considering a small rotation $\delta\theta_i$ of the cam about its axis O which causes the point of contact to move from A to B, fig. 2.6. OA is of length r and OB is $(r + \delta r)$. OG is set off equal to OA so that $GB = \delta r$, $AG = r\,\delta\theta_i$, and $\angle AGB \approx 90$ degrees. From triangle AGB we have $\delta r/(r\,\delta\theta_i) = \cot\phi$, so that as $\delta\theta_i \to 0$, we have

$$\cot\phi = \frac{1}{r}\frac{dr}{d\theta_i} \qquad (2.2)$$

The distance from the centre of the cam to the profile is

$$r = R + x_o$$

where R is the radius of the base circle, and x_o is the displacement of the follower. Hence

$$\frac{dr}{d\theta_i} = \frac{dx_o}{d\theta_i}$$

so that eqn. (2.2) becomes

$$\tan \phi = (R + x_o) \bigg/ \frac{dx_o}{d\theta_i} \qquad (2.3)$$

If at any point ϕ is found to be too small it can be increased by increasing the value of R.

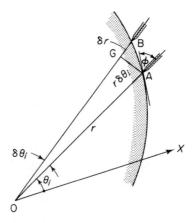

FIGURE 2.6

With a roller follower eqn. (2.3) still applies, but in this case a much smaller value of ϕ is acceptable because the effect of friction between the cam and its follower is very much reduced.

2.5 Interference (disc cams)

The method of setting out the profile of a disc cam for a roller follower is indicated in fig. 2.5(c). Just as with the reciprocating cam, interference occurs if the radius of curvature of the locus of the centre of the follower relative to the cam is anywhere less than the radius of the roller. The equation of this curve is

$$r = R + x_o$$

where R is the radius of the base circle, and x_o is the lift and a known function of the angle of cam rotation θ_i. The radius of curvature at any point of a curve whose equation, as in this case, is expressed in polar coordinates is given by a standard formula in differential calculus. It will, however, be convenient to give the derivation of this formula here.

Let OX in fig. 2.7 be a direction fixed with respect to the curve (i.e., drawn

SIMPLE MECHANISMS II Displacement analysis

on the cam which has O as centre). Any point A on the curve is defined by the coordinates r and θ. Let the tangent to the curve at A make an angle ψ with OX, and ϕ with OA. Let B be a short distance δs from A such that OB $= r + \delta r$

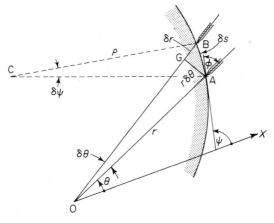

FIGURE 2.7

and $\angle AOB = \delta\theta$. The tangent at B makes an angle $(\psi + \delta\psi)$ with OX so that if AC and BC are normals to the curve at A and B respectively, C is the centre of curvature, and $\angle ACB = \delta\psi$. Thus we have

$$\delta s = \rho\, \delta\psi$$
$$= \rho(\delta\phi + \delta\theta), \quad \text{because } \psi = (\phi + \theta)$$

If a perpendicular AG is dropped from A to OB,

$$AG \approx r\, \delta\theta, \qquad BG \approx \delta r$$

and
$$\delta s = [\delta r^2 + (r\, \delta\theta)^2]^{\frac{1}{2}}$$

Equating these two expressions for δs and dividing through by $\delta\theta$ we have

$$\frac{\delta s}{\delta\theta} = \rho\left(\frac{\delta\phi}{\delta\theta} + 1\right) \approx \left[\left(\frac{\delta r}{\delta\theta}\right)^2 + r^2\right]^{\frac{1}{2}}$$

As $\delta s \to 0$ this equation becomes

$$\rho\left(\frac{d\phi}{d\theta} + 1\right) = \left[\left(\frac{dr}{d\theta}\right)^2 + r^2\right]^{\frac{1}{2}} \tag{2.4}$$

From $\triangle GAB$

$$\tan \phi = \operatorname*{Lt}_{\delta\theta \to 0} \frac{r\, \delta\theta}{\delta r} = r\frac{d\theta}{dr} = \frac{r}{dr/d\theta}$$

30 Dynamics of Mechanical Systems

Differentiating with respect to θ

$$\sec^2 \phi \frac{d\phi}{d\theta} = \left\{\left(\frac{dr}{d\theta}\right)^2 - r\frac{d^2r}{d\theta^2}\right\} \bigg/ \left(\frac{dr}{d\theta}\right)^2$$

Remembering that $\sec^2 \phi = 1 + \tan^2 \phi$ this becomes

$$\frac{d\phi}{d\theta} = \left\{\left(\frac{dr}{d\theta}\right)^2 - r\frac{d^2r}{d\theta^2}\right\} \bigg/ \left\{\left(\frac{dr}{d\theta}\right)^2 + r^2\right\}$$

Now substitute for $d\phi/d\theta$ in eqn. (2.4) and we have

$$\frac{1}{\rho} = \frac{r^2 + 2\left(\frac{dr}{d\theta}\right)^2 - r\frac{d^2r}{d\theta^2}}{\left[\left(\frac{dr}{d\theta}\right)^2 + r^2\right]^{\frac{3}{2}}} \qquad (2.5)$$

On replacing θ by θ_i in eqn. (2.5) and substituting $R + x_o$ for r we obtain ρ as a function of θ_i with the base circle radius R as a parameter. This done, the resulting expression can be differentiated with respect to θ_i to determine the minimum value for ρ. If this is equated to the least allowable value, we obtain an expression for the smallest possible value for R. The whole calculation is bound to be long and difficult even if relatively simple transfer relationships are used. Fortunately, eqn. (2.5) can often yield useful information without such an extended analysis. This is so when it can be seen by inspection that the radius of curvature is a minimum at the same point on the profile as where r is a maximum, that is where $dr/d\theta = 0$. In this case eqn. (2.5) becomes, on replacing θ by θ_i,

$$\frac{1}{\rho_{min}} = \frac{1}{r} - \frac{1}{r^2}\frac{d^2r}{d\theta_i^2} \qquad (2.6)$$

Example

A disc cam is to impart simple harmonic motion to a reciprocating roller follower. The amplitude of the motion of the follower is $\pm a/2$ about the mean position and three complete cycles are made for a single revolution of the cam. Determine the minimum size of cam which can be used if the radius of the roller follower is $1\cdot 2a$.

The base circle provides a datum relative to which the cam radii are defined, and although the base circle radius is associated with the minimum distance from the cam centre to the cam profile in fig. 2.5(a) and the minimum distance to the roller centre in fig. 2.5(b), any other circle drawn on the same cam could equally well be defined as the base circle. In the case of the problem just given, the most convenient base circle is that which coincides with the position of mean lift, for then we have

$$r = R + \frac{a}{2}\cos 3\theta_i$$

By sketching this profile for an arbitrary value of $R > a/2$ the reader will readily see that the curve has its minimum radius of curvature at the point where r is a maximum and equal to $(R + a/2)$, the corresponding values of θ_i being 0, 120, and 240 degrees. The conditions necessary for eqn. (2.6) are thus fulfilled and since

$$\frac{d^2 r}{d\theta_i^2} = -4 \cdot 5 a \cos 3\theta_i$$

we have, on substituting in eqn. (2.6) and putting $\cos 3\theta = 1$,

$$\frac{1}{\rho_{min}} = \frac{1}{R + \dfrac{a}{2}} + \frac{4 \cdot 5 a}{\left(R + \dfrac{a}{2}\right)^2}$$

ρ_{min} = radius of roller follower = $1 \cdot 2 a$. Substituting this and re-arranging the equation, we have

$$\left(R + \frac{a}{2}\right)^2 - 1 \cdot 2 a \left(R + \frac{a}{2}\right) - 5 \cdot 4 a^2 = 0 \tag{2.7}$$

so that

$$\left(R + \frac{a}{2}\right) = \tfrac{1}{2}\{1 \cdot 2 a \pm \sqrt{(1 \cdot 44 a^2 + 21 \cdot 6 a^2)}\} = (0 \cdot 6 a \pm 2 \cdot 4 a)$$

Ignoring for a moment the negative root the minimum possible value for R is $2 \cdot 5 a$. The cam profile for this value of R is constructed as indicated in fig. 2.8.

We have assumed that the roller runs on the outside of the cam profile but occasionally internal cams are used and the negative root, $R = -2 \cdot 3 a$, of eqn. (2.7), gives the solution to the above problem for an internal cam. The outline of the cam is given in fig. 2.9 to the same scale as fig. 2.8.

If contact stresses limit the minimum radius of curvature of the cam profile to a specified value e, and the roller radius is f, then ρ_{min} in eqn. (2.6) equals $(e + f)$. The calculation of the base circle radius then proceeds as before. Alternatively the profile in fig. 2.8 may be unacceptable on account of the concave portions since these limit the size of grinding wheel which may be used in the manufacture of the cam. Equation (2.5) can be used to calculate the size of cam necessary to restrict the radius of curvature of the concave sections of the profile to any specific value. It can be seen by inspection of fig. 2.8 that the radius of curvature has a limiting value when r is a minimum so that $dr/d\theta_i = 0$, and again

$$\frac{1}{\rho} = \frac{1}{r} - \frac{1}{r^2} \frac{d^2 r}{d\theta_i^2}$$

For the profile to be convex outwards ρ must be positive and in the limiting case, when the profile is just straight, ρ is infinite giving the relationship

$$r_{min} = \frac{d^2 r}{d\theta_i^2}$$

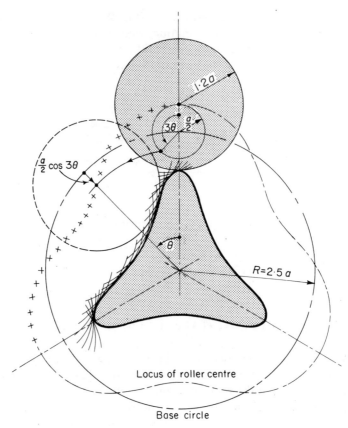

Minimum cam for the transfer relationship $x_0 = \frac{a}{2}\cos 3\theta_i$

FIGURE 2.8

In the case of the worked example

$$\frac{d^2 r}{d\theta_i^2} = -4\cdot 5 a \cos 3\theta = 4\cdot 5 a$$

for $\theta = 60°$, and $\quad r_{\min} = R - \dfrac{a}{2}$

The minimum possible base circle radius if the cam profile is not to be concave outwards is thus $R = 5\cdot 0 a$; the resultant form is shown in fig. 2.10. Note that figs. 2.8 and 2.10 are drawn to different scales, the actual size in fig. 2.10 being approximately $2\frac{1}{2}$ times of that in fig. 2.8.

In our study of disc cams with roller followers it has so far been assumed that the motion of the centre of the follower is along a straight line which passes through the centre of vibration of the cam. When the centre line of the follower is offset, or when a rocker follower is used, as in figs. 2.2(a) and (c), the analysis

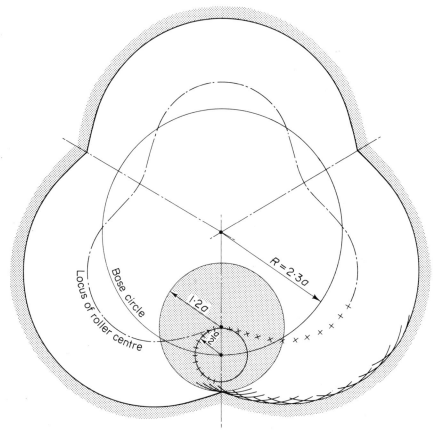

Minimum internal cam to generate $\theta_o = \frac{a}{2}\cos 3\theta_i$

FIGURE 2.9

is relatively complicated. As no new principles are involved we will not study these types in detail.

Disc cam with a flat-footed follower
Although the flat-footed follower might be regarded as a roller follower of infinite radius we cannot readily apply the foregoing analysis of interference to this case. The reason for this is clear: the variable r has so far been used to denote the distance from the centre of the cam to the centre of the roller follower, this being the same as the distance to the profile only when a knife-edged follower is used. With a flat-footed follower we go to the other extreme, r is infinite for all θ and so is ρ.

The method of setting out a cam profile on the drawing board was indicated in fig. 2.5(d). Consider now two positions of the follower with points of contact at A and B respectively, where A and B are an infinitesimal distance δs apart.

34 Dynamics of Mechanical Systems

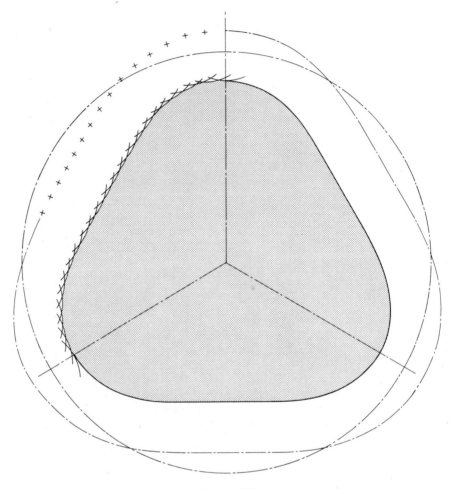

FIGURE 2.10

The normals at A and B, fig. 2.11, intersect at the centre of curvature C. O is the centre of the cam and OX is a datum drawn on the cam. Let OM, the perpendicular from O to the tangent AM at A be of length p. AM which makes an angle ψ with OX is of length t. BN the tangent at B, and ON its perpendicular are of lengths $(t + \delta t)$ and $(p + \delta p)$ respectively. The angle MON $= \delta\psi$, which is also the angle between AM and BN, and between CB and CA.

Let ON and AM intersect at L, then

$$\delta p = LN = BN\, \delta\psi = t\, \delta\psi$$

As $\delta\psi \to 0$ this equation becomes

$$t = \frac{dp}{d\psi} \tag{2.8}$$

SIMPLE MECHANISMS II Displacement analysis

Also we have
$$t + \delta s = (t + \delta t) + LM$$
$$= t + \delta t + p\,\delta\psi$$
and hence
$$\delta s = \delta t + p\,\delta\psi$$
But
$$\delta s = \rho\,\delta\psi$$
so that
$$\rho = \frac{dt}{d\psi} + p$$
and on substituting for t from eqn. (2.8)
$$\rho = p + \frac{d^2 p}{d\psi^2}$$

Now there is an important difference between figs. 2.6 and 2.11. In fig. 2.6 the angle of cam rotation θ_i from datum position is the angle AOX. This is not the case in fig. 2.11 where the angle of cam rotation is given by $(\psi - \pi/2)$ if OX

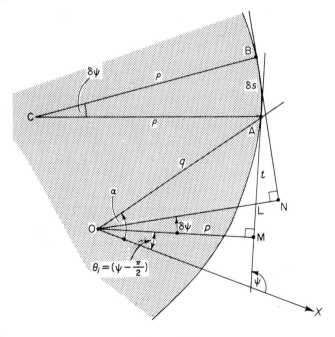

FIGURE 2.11

is the direction of follower motion when the cam is in its datum position. We can therefore write
$$\rho = p + \frac{d^2 p}{d\theta_i^2} \tag{2.9}$$

If R is the radius of the base circle

$$p = R + x_o \qquad (2.10)$$

and

$$\rho = R + x_o + \frac{d^2 x_o}{d\theta_i^2} \qquad (2.11)$$

Interference is avoided by making the base circle radius R sufficiently large for ρ to be positive for all values of θ_i. The lower limit on R is set at that value which just allows ρ to be zero at one or more points in the cycle of motion. Note that this condition automatically precludes the possibility of the profile being concave outwards.

Example
A cam with a flat-footed follower is required to generate $x_o = a \cos 2\theta_i$. Determine the minimum value for the base circle radius.

$$\begin{aligned} \rho &= R + x_o + \frac{d^2 x_o}{d\theta_i^2} \\ &= R + a \cos 2\theta_i - 4a \cos 2\theta_i \\ &= R - 3a \cos 2\theta_i \end{aligned}$$

For $\rho \geqslant 0$ for all θ_i we require $R \geqslant 3a$. Given a specific value of R the usual construction of the cam profile follows. In practice, it is difficult to draw in a satisfactory envelope when only the tangents which represent successive positions of the base of the follower are given, if the precise points of contact between the profile and the tangent are not known. The positions of the points of contact are, however, readily determined from eqn. (2.8), which may be modified to

$$t = \frac{dx_o}{d\theta_i} \qquad (2.12)$$

Once p and t are calculated from eqns. (2.10) and (2.12) for successive values of θ, the cam profile can be drawn without difficulty.

If OX defines the x-axis, the coordinates of A are

$$\left. \begin{array}{l} x = p \cos \theta_i - t \sin \theta_i \\ y = p \sin \theta_i + t \cos \theta_i \end{array} \right\} \qquad (2.13)$$

In polar coordinates

$$\left. \begin{array}{l} q = (t^2 + p^2)^{\frac{1}{2}} \\ \alpha = \theta_i + \tan^{-1}(t/p) \end{array} \right\} \qquad (2.14)$$

With either system of coordinates the equation to the curve is found by eliminating θ_i. In general, the results are very cumbersome. One exception to this is a cam which generates the simple harmonic motion

$$x_o = a \cos \theta_i$$

Substituting in eqns. (2.10) and (2.12) we have

$$p = R + a \cos \theta_i$$

and
$$t = -a \sin \theta_i$$

Inserting these expressions for t and p in eqn. (2.13) we find

$$\left. \begin{array}{l} x = R \cos \theta_i + a \\ y = R \sin \theta_i \end{array} \right\}$$

These are the equations in parametric form to a circle of radius R with its centre at $(a, 0)$.

One significant conclusion which can be drawn from eqn. (2.12) is that the relationship between t and θ_i does not depend upon R. Thus, although at first sight it might seem that an enlarged base circle diameter would require a wider

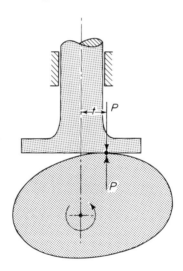

FIGURE 2.12

follower this is not so: the width of the follower is determined by the maximum and minimum values of $dx_o/d\theta_i$. Although the side thrust in this type of follower is due only to friction at the point of contact and is fairly small there may still be a considerable bending moment applied to the stem of the follower. If the centre line of the follower intersects the cam axis, the bending moment is equal to the torque Pt required to turn the cam where P is the contact force.

Disc cam with a rocker follower

The profile of a cam with a rocker follower is set out as indicated in fig. 2.13. The profile of the follower is shown to be flat in the diagram, but the method of setting-out is the same when a roller or any other type of follower is used.

38 Dynamics of Mechanical Systems

There exists a geometrical relationship which enables the precise point of contact between the follower and the cam to be located, and as in the previous section of this chapter the drawing of the cam profile is thereby made much easier.

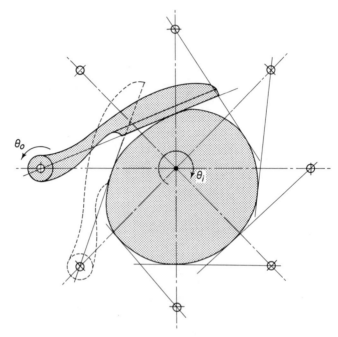

FIGURE 2.13

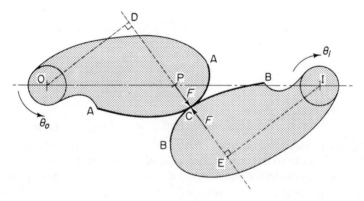

FIGURE 2.14

Let BB, fig. 2.14, be the profile of a cam and AA the profile of the follower. The follower is pivoted at O and the cam at I. DE is the common normal to the two profiles at the point of contact; OD and IE are both perpendicular to DE which intersects OI at P.

SIMPLE MECHANISMS II Displacement analysis

Assume that there is no friction in the mechanism then a driving torque T_i at I and a resisting torque T_o at O cause forces F at the point of contact which lie along the common normal. If the cam rotates through a small angle $\delta\theta_i$ the work $T_i \, \delta\theta_i$ by the input torque must equal the work $T_o \, \delta\theta_o$ by the output torque so that

$$F . \text{IE} \, \delta\theta_i = F . \text{OD} \, \delta\theta_o$$

and in the limit

$$\frac{d\theta_o}{d\theta_i} = \frac{\text{IE}}{\text{OD}}$$

Triangles ODG and ICG are similar so that

$$\frac{d\theta_o}{d\theta_i} = \frac{\text{IP}}{\text{OP}} \tag{2.15}$$

The relationship which this result expresses involves only the geometry of the system and so must be true whatever forces are transmitted. In particular, it is valid even if the initial assumption that there is no friction is not true.

In drawing a cam profile, eqn. (2.15) may be used to determine the position of P on OI for every position of a given follower. The point of contact C with the cam profile is then located for each position by the foot of the normal from P on to the follower profile.

As with other types of cam mechanism, interference sets a lower limit on the size of base circle which can be used. Analytical methods of determining the minimum size of cam can be applied as in the previous examples but they are complicated.

There are two special cases to be considered in relation to eqn. 2.15; (a) the point of contact C coincides with P, (b) P is fixed in position on OI.

2.6 Rolling contact

When the point of contact C lies on OI the relative motion between the cam and follower is one of pure rolling without any sliding. This can be seen as follows.

Let the point of contact on the cam be denoted by C_I and the point of contact on the follower by C_O and imagine C_I and C_O to coincide at P, fig. 2.14. The velocity of C_1 is IP $d\theta_i/dt$ and of C_O is OP $d\theta_o/dt$ both in a direction perpendicular to OI. By eqn. (2.15) these two velocities are equal. But if this is so there is no relative movement between C_I and C_O, and the relative motion can only be rolling. With rolling, as opposed to sliding, contact there is less dissipation of energy in friction and less wear, consequently when it can be achieved conveniently rolling contact is preferred. Furthermore, it is then possible to arrange for a positive drive between the driving cam and the follower, by means of teeth (see p. 55).

When a given transfer relationship is to be generated by a disc cam and follower, and it is specified that only rolling contact is permitted, neither of the contacting profiles is arbitrary. But both profiles are readily calculated.

40 Dynamics of Mechanical Systems

The simplest transfer relationship which can be required is $\theta_o = k\theta_i$. In this case IP/OP $= k$, so that P is a fixed point, and the profiles of both the driver and the follower are circles, with centres at O and I respectively, and touching at P. For the mechanism to be effective it is necessary for the surfaces of the two discs to be pressed hard together. Even then the power which can be transmitted is generally quite small and since the drive depends on friction at the point of contact the transfer relationship cannot be guaranteed.

The following example shows how the profiles are determined when the movements of the input and output shafts are not proportional to each other.

Example

Two parallel shafts I and O carry circular discs evenly graduated as shown in fig. 2.15(a). A cam on shaft I drives a follower on shaft O so that the readings p_i and p_o on the two scales are related by the equation $p_o = \log_{10} p_i$. The profiles of the cam and follower are to be determined.

Let the actual rotations of the input and output shafts be denoted by θ_i and θ_o (radians) respectively. Then,

$$\theta_i = \frac{p_i - 1\cdot 0}{9\cdot 0} \times \pi$$

and

$$\theta_o = \frac{p_o}{1\cdot 0} \times \pi$$

Substituting for p_o and p_i in $p_o = \log_{10} p_i$ we have

$$\theta_o = \pi \log_{10}\left(1 + \frac{9}{\pi}\theta_i\right)$$

$$= \pi \times 0\cdot 434 \log_e\left(1 + \frac{9}{\pi}\theta_i\right)$$

$$= 1\cdot 362 \log_e (1 + 2\cdot 86\theta_i)$$

$$\frac{d\theta_o}{d\theta_i} = \frac{1\cdot 362 \times 2\cdot 86}{1 + 2\cdot 86\theta_i} = \frac{3\cdot 90}{1 + 2\cdot 86\theta_i} = \frac{\text{IP}}{\text{OP}}$$

Denoting the distance OI between centres by h we have

$$\frac{r_i}{h} = \frac{\text{IP}}{\text{OI}} = \frac{\dfrac{3\cdot 90}{1 + 2\cdot 86\theta_i}}{1 + \dfrac{3\cdot 90}{1 + 2\cdot 86\theta_i}} = \frac{3\cdot 90}{4\cdot 90 + 2\cdot 86\theta_i}$$

r_i/h is calculated for suitable values of p_i in the table below. (As $r_i + r_o = h$, $r_o/h = 1 - r_i/h$.)

SIMPLE MECHANISMS II Displacement analysis

p_i	1·0	2	3	4	5	6	7	8	9	10
θ_i	0	20°	40°	60°	80°	100°	120°	140°	160°	180°
r_i/h	0·796	0·662	0·565	0·494	0·438	0·394	0·358	0·328	0·302	0·280
r_o/h	0·204	0·338	0·435	0·506	0·562	0·606	0·642	0·672	0·698	0·720
θ_o	0	54·1°	85·9°	108·4°	126°	140°	152°	162·5°	172°	180°

These profiles are plotted in fig. 2.15(a). Some means must be provided whereby the two profiles are kept in contact, and since the relative motion is

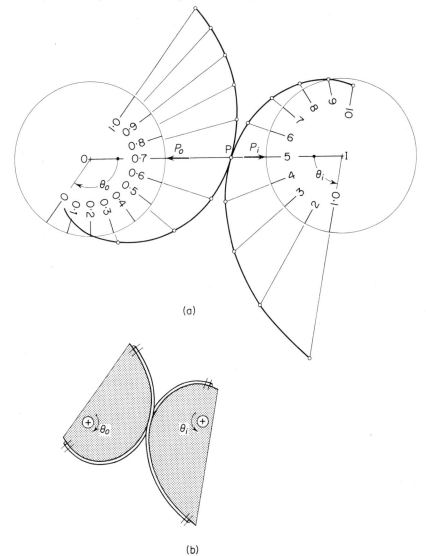

(a)

(b)

FIGURE 2.15

one of pure rolling, contact can conveniently be maintained by using two thin flexible strips as indicated in fig. 2.15(b) provided that both profiles are concave outwards at all points. If heavy forces have to be transmitted mating teeth can be provided (see section 2.15).

2.7 Cam kinematics — general comments

The kinematics of cams have been analysed in this chapter with the object of determining the cam profile required in any particular instance to generate a specified transfer relationship. In general, there is no unique solution to this problem since the size of a cam is determined by the base circle radius which may have any convenient value in excess of a calculable lower limit. This limit is determined mainly by the need to avoid interference, and choice may be further restricted if it is required to place limiting values on the radius of curvature of the profile.

If the cam moves only at moderate speeds there is no restriction on the transfer relationship which it may be designed to generate. With high speed cams the dynamics of the system must be taken into account and this limits the choice of transfer relationship which may be adopted, see exercise 9.11.

2.8 Gear tooth geometry

A gear drive, fig. 2.16, is essentially a succession of cam drives, the cam mechanisms being formed by the pairs of teeth as they mate. During the period in which any one pair of teeth are in contact the transfer relationship is determined by eqn. (2.15),

i.e.,
$$\frac{d\theta_o}{d\theta_i} = \frac{IP}{OP}$$

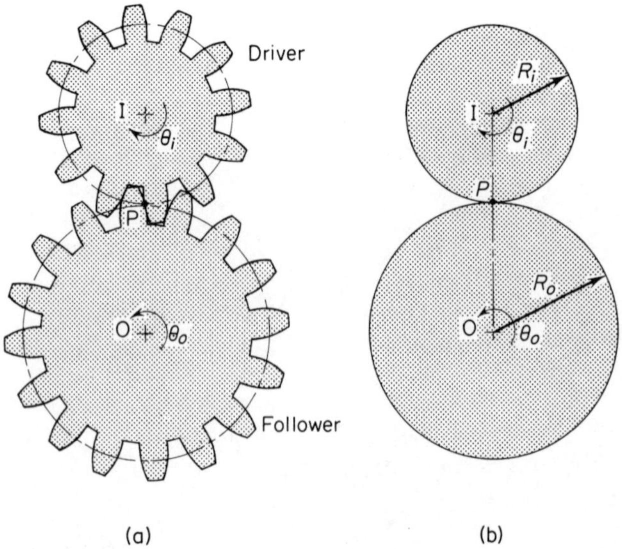

FIGURE 2.16

SIMPLE MECHANISMS II Displacement analysis 43

where P is the intersection between the line of centres OI, and the common normal to the teeth profiles at the point of contact. When P is a fixed point, $d\theta_o/d\theta_i$ is constant, and this condition is required of most of the gear drives which are used in practice.

If the number of teeth on output gear is k times that on the input gear

$$\frac{d\theta_o}{d\theta_i} = k$$

or

$$\frac{d\theta_o}{dt} = k\frac{d\theta_i}{dt}$$

and k is the speed ratio. Since both the input and output gears must each have an integral number of teeth, choice of k is restricted to those values which can be expressed as the ratio of two integers.

The same transfer relationship could be realized by means of a friction drive, fig. 2.16(b) for which

$$R_o\theta_o = R_i\theta_i$$

and

$$k = \frac{d\theta_o}{d\theta_i} = \frac{R_i}{R_o}$$

But we already have shown that

$$\frac{d\theta_o}{d\theta_i} = \frac{\text{IP}}{\text{OP}}$$

Because $OI = R_i + R_o$ it follows that P coincides with the point of contact between the two wheels of the friction drive. P is called the *pitch point* and to locate it in a gear drive imagine the two gear wheels to be replaced by the pair of friction wheels which give the same transfer ratio. In theory we could choose any shape for the tooth profile on one of the gear wheels and so form the teeth on the other wheel that the normal at the point of contact does pass through the pitch point. In practice relatively few gear forms are used and, of the practical possibilities, one particular form is used very much more than the others.

2.9 Involute gearing

The commonest tooth profile at present in use is in the form of an involute. Imagine an inextensible string wrapped tightly round a circular cylinder. If the string is unwrapped and at the same time kept taught the end will describe a curve such as AB in fig. 2.17. A curve so generated is called an involute and the circle which represents the cylinder is called the base circle of the particular involute. It is clear from the method of construction that the string MB is normal to the involute at B and tangential to the base circle at M. Furthermore, MB is the radius of curvature of the involute at B and is equal to the length of the arc AM. Let AB, fig. 2.18, be the profile of a cam. The follower profile CD is also an involute and it is generated from a second base circle. The two profiles are in contact at X. XM is normal to AB and tangential to its base circle at M; XN is normal to CD and tangential to its base circle at N. It follows that MN is the common tangent to the two base circles which must pass through a fixed point P on the line joining the centres of the two base circles. P does not move

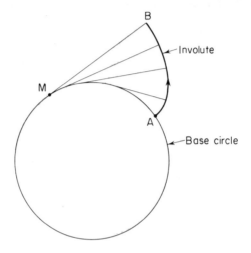

FIGURE 2.17

as the cam and follower rotate so that by eqn. (2.15) $d\theta_o/d\theta_i$ is constant. The same result can be proved independently by the following reasoning. If θ_i is increased by an amount $\Delta\theta_i$ (not necessarily small) then the arc length AM and hence MX both increase by an amount $R_{bi}\,\Delta\theta_i$. Because MN is of fixed length, XN must decrease by the same amount and hence $\Delta\theta_o = R_{bi}\,\Delta\theta_i/R_{bo}$ so that

$$\frac{\Delta\theta_o}{\Delta\theta_i} = \frac{R_{bi}}{R_{bo}} = \frac{\text{IP}}{\text{OP}} \qquad \text{(for triangles ONP and IMP are similar)}$$

The argument just given applies irrespective of the actual distance OI, and it is one of the features of involute gears that even if the centres of the gears are displaced relative to each other the transfer relationship is not altered. To convert the mechanism in fig. 2.18(a) into a practical geared pair a succession of mating profiles must be provided to allow continuous rotation. A second set of profiles, fig. 2.18(b), facing in the opposite directions allow the directions of rotation of the gears to be reversed. The point of contact between members of the second set of profiles moves along the common tangent M'N'.

In order to provide continuity of action, it is essential that contact between a pair of teeth should be established before the contact between the preceding pair is broken. In fig. 2.19 two points of contact X and Y are marked. As θ_i is increased X and Y both move towards N. When X is at E contact is right at the tip of the tooth on the upper wheel and is about to be broken. If the wheels are moved in the opposite direction contact ceases at Y when Y and G are coincident.

SIMPLE MECHANISMS II Displacement analysis 45

Thus, if there is always to be at least one pair of teeth in contact, XY must be less than the length of the *path of contact* EG, which is the distance along the common tangent enclosed by the tip circles of the two gears. Remembering the way in which the involute profiles are generated it can be seen that XY equals

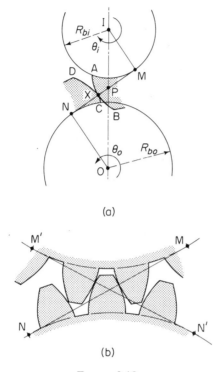

FIGURE 2.18

the arc length p_b. The length p_b is called the *base circle pitch* and it equals the $2\pi R_{bi}/N_i$, where N_i is the number of teeth on the input gear. p_b must be the same for the two sets of teeth, so that

$$p_b = \frac{2\pi R_{bi}}{N_i} = \frac{2\pi R_{bo}}{N_o} \qquad (2.16)$$

and

$$\frac{N_i}{N_o} = \frac{R_{bi}}{R_{bo}} = k \qquad (2.17)$$

The average number of pairs of teeth in contact or *contact ratio* is given by EG/XY. If $1 < EG/XY < 2$, then for part of the time there is contact between only one pair, and for the rest of the time two pairs are in contact. If $2 < EG/XY < 3$ then there are always either two or three pairs in contact.

As the number of pairs of teeth in contact changes, the forces at the individual points of contact alter but, apart from the effects of friction, the total contact force and hence the reaction at the bearings of the shaft carrying the gear wheel is constant, provided that the driving torque is constant. This is because the forces between each pair of teeth must be in the direction of the normal to the

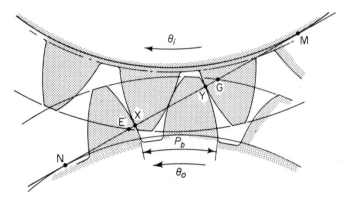

FIGURE 2.19

profiles at the point of contact. Thus all the contact forces are along MN. The total force, and hence the shaft reaction, is $T_i/R_{bi} = T_o/R_{bo}$, where T_i and T_o are the torques applied to the input and output gears respectively. With non-involute gear teeth the shaft reaction fluctuates with a resulting tendency for the system to vibrate. It can be proved that the involute profile is the only one which satisfies both the requirement that $d\theta_o/d\theta_i$ is constant, and that the shaft reaction is constant.[1]

When specifying a particular gear the base circle diameter is not usually one of the dimensions which is quoted. The leading dimension is generally taken to be the *pitch circle diameter*, although it is not possible to pick up a gear wheel and to measure this dimension. The pitch circles of pairs of gear wheels are the two circles which are drawn from the two centres to touch each other at the pitch point P, fig. 2.20. The pitch circle of a gear wheel is not really defined until it is meshed with another gear. Although the teeth of the second gear wheel must have the same base circle pitch as those of the first wheel the tooth thickness at the base circle can be varied with the result that the distance between the centres of the gears changes and with it the pitch circle diameters. This arbitrariness is avoided by specifying the *pressure angle* ψ which is the angle between the tangent to the pitch circles at the point of contact and the common tangent to the base circles. The pitch circle radius R is then determined by the relationship

$$R_b = R \cos \psi \qquad (2.18)$$

In modern practice, ψ is normally 20 degrees. The size of teeth is determined primarily by the distance between successive teeth as measured round the pitch circle. However, as this dimension, known as the *circular pitch*, inevitably involves awkward decimals it is usual to specify the size of gear teeth by their

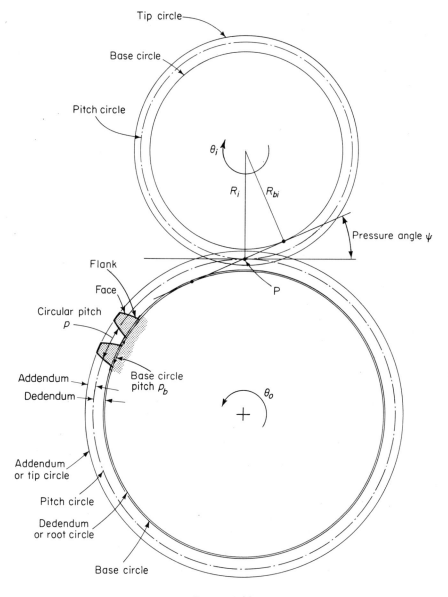

FIGURE 2.20

module m, which is defined as the pitch circle diameter divided by the number of teeth.† The *addendum* is the radial distance that the tooth projects outside the pitch circle and this is commonly equal to m. For standard teeth the

† At the time of writing, the size of gear teeth are more usually specified in Britain in terms of their diametral pitch, which is the number of teeth divided by the pitch circle diameter. With the adoption of metric units this term will, it is to be hoped, become obsolete.

dedendum, which is the radial distance between the roots of the teeth and the pitch circle, is commonly equal to $1 \cdot 25m$. The thickness of the tooth measured by the arc length round the pitch circle is usually half the circular pitch.

2.10 The involute rack

The path of contact is tangential to the base circles of two mating gears and passes through the pitch point. This is true, irrespective of the relative sizes of the two gear wheels considered. If one of the gears has an infinite pitch circle diameter it becomes a rack, and its pitch circle becomes the pitch line. Comparing fig. 2.21 with fig. 2.19, the path of contact is still along MN even though N is removed to infinity. As the rack is moved to the left the point of contact moves down the rack tooth. In each position MN is normal to the profiles of each of the mating surfaces from which it follows that the tooth profile AA in

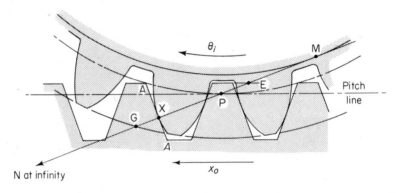

FIGURE 2.21

fig. 2.21 is a straight line. A practical consequence of this is that an accurate involute rack can be made relatively easily and using it as a cutter any involute gear can be generated.

2.11 Interference (involute gears)

Given one gear wheel or rack, the profile of the member with which it is to mate can be determined by the graphical method which was used for cams. In fig. 2.22 the pitch line of a rack is rolled round the pitch circle of a wheel and the teeth of the wheel are generated by drawing the rack teeth in a series of positions. Our experience with cams leads us to expect interference to occur if at any point the radius of curvature of the generated profile drops to zero. As the rack rolls anticlockwise round the wheel in fig. 2.22(a), the point of contact moves up the left-hand face of the rack tooth with a continuous reduction in the radius of curvature of the profile of the wheel tooth at the point of contact. In fig. 2.22(a) the point of contact reaches the tip of the rack tooth before the radius of curvature of profile of the wheel tooth has dropped to zero and no interference occurs even though the tip of the rack tooth penetrates inside the

SIMPLE MECHANISMS II Displacement analysis

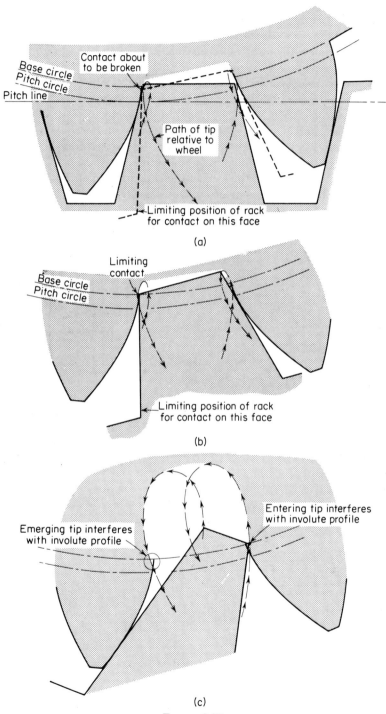

Figure 2.22

base circle of the wheel. In fig. 2.22(b) the rack tooth is somewhat longer and contact is just broken as the point of contact reaches the base circle of the gear wheel. At this point the radius of curvature of the involute is zero, since the length of the tangent which can be drawn from the point of contact to the base circle is zero, and we have a limiting case. In fig. 2.22(c) the rack tooth is longer still, and it can be seen that its tip interferes with the involute portion of the wheel-tooth profile as it emerges from the base circle inside the original involute surface, part of which is removed as a result. Interference is objectional, not only on kinematic grounds but also because the resulting tooth is very much weakened by the removal of material at its root.

Interference is avoided in the case of the rack and pinion by ensuring that contact ceases before the point of contact passes inside the base circle. In

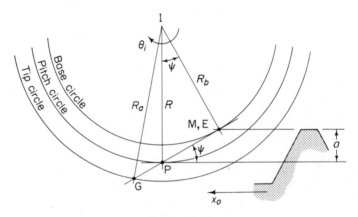

FIGURE 2.23

fig. 2.21 the path of contact is EG and there is no interference because E is outside the base circle of the pinion; there is no question of G being inside the base circle of the rack, as N is at infinity. The limiting case for interference is when E and M coincide as in fig. 2.23, where the addendum a of the rack is as large as possible. To avoid interference we require

$$a \leqslant R - R_b \cos \psi$$

i.e.
$$\leqslant R(1 - \cos^2 \psi) = R \sin^2 \psi \qquad (2.19)$$

If a has the 'standard' value of m then

$$N = \frac{2R}{m} \geqslant \frac{2}{\sin^2 \psi} \qquad (2.20)$$

for
$$\psi = 20°$$

$$N > 17 \cdot 1$$

Thus if the pinion has 18 or less teeth it is necessary to reduce the addendum of the rack in order to avoid interference. Any reduction in the rack addendum a

reduces the length of the path of contact, and there is the possibility that continuity of action will not be maintained. In fig. 2.23, the length of the path of contact is

$$\text{GM} = (R_a^2 - R_b^2)^{\frac{1}{2}} \tag{2.21}$$

and for continuity it is necessary that

$$\text{GM} \geqslant \text{base circle pitch}$$

or
$$(R_a^2 - R_b^2)^{\frac{1}{2}} \geqslant \frac{2\pi R_b}{N} \tag{2.22}$$

When two spur gears mesh there is a possibility of interference on the teeth of both gears. The calculation of maximum and minimum addenda proceeds in a manner similar to that just expounded for a spur gear meshed with a rack. The results are naturally more complex but there is no difference in principle.

Interference sets a geometrical limitation on tooth profiles. For heavily loaded gears the contact stresses impose a lower limit on the permissible radius of curvature of a tooth profile so that the question of interference does not arise. Thus the maximum addendum calculated from the interference standpoint is in general an upper bound rather than a value which can be realized in practice.

When the use of standard tooth proportions leads to interference the dimensions of the tooth must be altered. The simplest modification which can be made is just to reduce the addendum of the offending teeth. If this remedy reduces the length of the path of contact too much, a *corrected* tooth form is used. Having removed the excess addendum from the teeth of the larger gear wheel the addendum of the pinion teeth is increased by an amount sufficient to restore the length of the path of contact to the required value.

2.12 Internal gears

So far we have considered only the case where the pitch circles of two mating gears are external to each other. Almost exactly the same analysis applies when the larger gear wheel has internal teeth, so that instead of the mechanism shown in fig. 2.16 we have that in fig. 2.24. The internal gear teeth are generated from the same base circle as the equivalent external teeth, and the path of contact still lies along the common tangent to the two base circles as shown in fig. 2.25. Interference is avoided if the tip circle radius of the internal gear is restricted as indicated.

One advantage of using an internal gear is that the whole mechanism is much more compact than the equivalent one which uses only external gears. Also, the load-carrying capacity may be greater due to the reduction in surface stresses under a given load which result when two mating convex surfaces are replaced by a pair consisting of one convex and one concave surface. A further consideration which may affect the choice is that θ_o and θ_i are in the same direction for an internal gear whereas with external gears they are in opposite directions.

52 Dynamics of Mechanical Systems

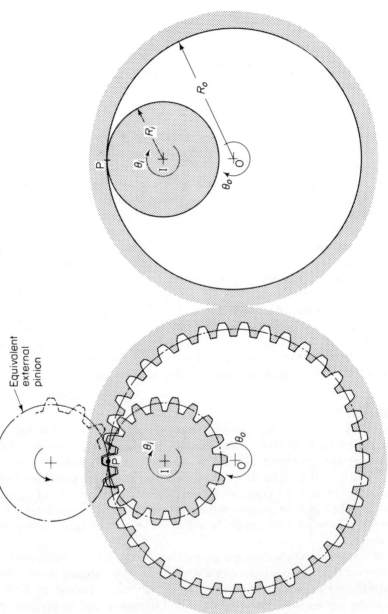

Figure 2.24

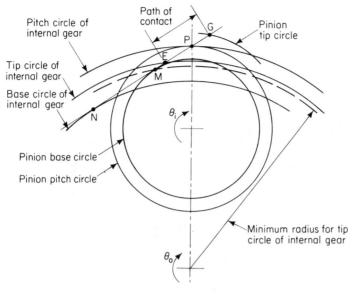

FIGURE 2.25

2.13 Helical gears

When a pair of gears transmit heavy loads, distortion of the teeth affects the geometrical considerations of the preceding paragraphs, and the ideal constant velocity ratio is no longer achieved. If the amount of power transmitted is constant it is possible to modify the profiles of the gear teeth to allow for their distortion, but clearly this is impractical if the gears operate at varying power levels and the teeth must then simply be made stiff enough to reduce distortion to an acceptable level.

As we have already seen, the average number of pairs of teeth in contact is given by the contact ratio, which is the length of the path of contact divided by the base circle pitch, unless the contact ratio happens to be a whole number the actual number of teeth in contact must fluctuate. This means that the load carried by any one tooth must vary during its period of contact due to changes in the number of other teeth which share the total load. If the teeth are not highly stressed this effect may be unimportant, but with heavily loaded gears it is necessary to arrange for the tooth loads to be taken up gradually. As a first step towards this imagine that a gear wheel is split in half by a plane perpendicular to its axis and the two halves are then turned slightly with respect to each other as in figs. 2.26(a) and (b), and the same is done to the mating gear, the load on any one tooth is taken up in two steps instead of one. This results in a smoother and quieter drive. The logical extension of this is to cut the wheel into an infinite number of slices and to displace each by an infinitesimal amount with respect to the next. The generators of the teeth profiles which were originally parallel to the axis of the gear are thus deformed into helices. Such gears are known as *helical gears*.

With helical gears, the load on a tooth has a component parallel to the axis of the shaft on which the gear is mounted, and either a thrust-bearing must be provided to take this axial load, or it must be cancelled by cutting the teeth in the double helical form shown in fig. 2.26(c).

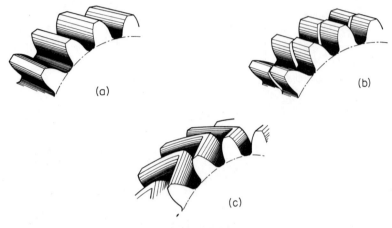

FIGURE 2.26

2.14 Non-involute gears

Helical gears free us from dependence on the involute tooth as the only means of achieving constant velocity ratio and constant bearing reactions. With helical involute gears, as with straight involute gears, contact between teeth is basically one of line contact, with the line sweeping down the whole width of a tooth during its period of contact. Now we do not have to have line contact in order to achieve the basic requirements of constant velocity ratio and constant load. Point contact will do just as well, provided that in axial view the common normal at the point of contact is seen to pass through the pitch point. With circular arc (Novikov) gears contact between teeth occurs, in principle, at a point which moves across the face of a tooth as the gears rotate, so that a transverse section through the gears at the point of contact always shows the mating teeth in exactly the same relationship to each other, as in fig. 2.27. The virtue of this type of gear lies in its load-carrying capacity. This statement may appear paradoxical in view of the fact that the line contact, which is a feature of involute gears, has been abandoned in favour of point contact. In reality, there is no such thing as line or point contact, for both would result in the generation of infinite stresses under load, and deformation of the surfaces in the region of contact causes the load to be spread over an area. The size of this area, and hence the magnitude of the stresses induced, depends on the degree of conformity between the contacting surfaces. Figure 2.27 shows that this is much better in the case of circular arc gears, one surface being convex and the other concave, than it is with external involute gears, for which both surfaces are convex. The resulting reduction in surface stress for a given load means that for the same power the circular arc tooth results in a much more compact gear than the involute tooth.

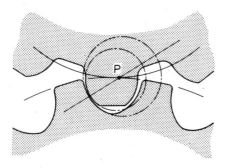

FIGURE 2.27

2.15 Non-circular gears

Gears for which the pitch line are non-circular are used when the transfer relationship is other than $\theta_o = k\theta_i$, and when a cam drive is not acceptable. As the point of contact P lies on the line joining the centres of rotation O and I there is rolling contact between the two cam profiles in fig. 2.28(a). These profiles can therefore be regarded as the pitch lines of two non-circular gears and appropriate tooth profiles can be generated from suitable base curves by precisely the same method as that used to generate conventional involute teeth from a base circle.

Let PT, fig. 2.28(a), be the common tangent to the two pitch curves at the point of contact P. Let MPN be the common normal to the tooth profiles at P. We define $\angle$ TPN as the pressure angle ψ, in conformity with the practice for circular gears.

Now imagine that MP and PN are drawn on the cams so that when the input cam is turned through a small angle $\delta\theta_i$, fig. 2.28(b), the point of contact originally P becomes P_i on the input cam and P_o on the output cam, and the original common normal MPN becomes the two lines P_iN_i and P_oM_o. The new common normal is M'P'N' which intersects P_iN_i and P_oM_o at G and H respectively. As the two gears turn, G and H trace at loci which are the base curves for the gears, fig. 2.28(c). We can now imagine a string wrapped round the two base curves so that it pays off the input gear and on to the output gear as the two rotate. A point fixed to the string traces out one tooth profile A–A on the input gear and the mating profile B–B on the output gear.

It is not immediately clear that the length GH is constant so that the string will stay just taut. This can be proved as follows:

Let $P_iG = p_i$ and $P'G = p_i + \delta p_i$

If $\delta\theta_i$ is infinitesimal then

$$\angle GP_iP' = \psi$$

56 Dynamics of Mechanical Systems

and
$$P'P_i = \delta s_i = \frac{\delta p_i}{\cos \psi}$$

so that
$$\delta p_i = \delta s_i \cos \psi$$

For the output gear
$$\delta p_o = \delta s_o \cos \psi$$

But as the two profiles roll without slip $\delta s_o = \delta s_i$ and
$$(\delta p_i - \delta p_o) = 0$$

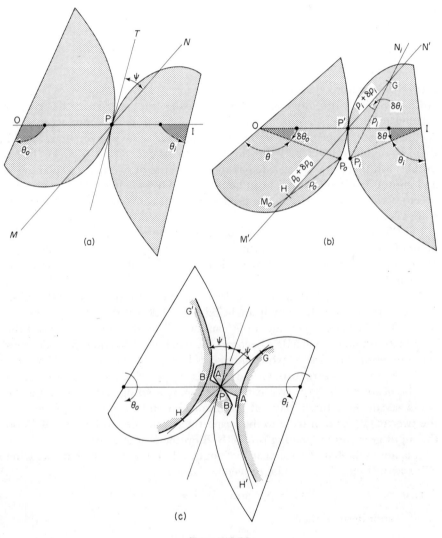

FIGURE 2.28

SIMPLE MECHANISMS II Displacement analysis

It can be seen from the diagram that if δp_o is positive as drawn δp_i must be negative. It follows that:

$$p_i + p_o = \text{constant}$$

To calculate the length PG, which is the radius of curvature of the tooth profile at P we note that

$$p_i \, \delta\theta_i = \delta s_i \sin \psi$$

or
$$p_i = \frac{ds_i}{d\theta_i} \sin \psi \qquad (2.23)$$

But
$$\frac{ds_i}{d\theta_i} = \left[r_i^2 + \left(\frac{dr_i}{d\theta_i} \right)^2 \right]^{\frac{1}{2}} \qquad (2.24)$$

and so can be calculated for all θ_i once r_i is determined.

It should be noted that the opposite faces of a particular tooth are generated from different base curves because the common normal G'H' drawn at an angle ψ on the opposite side to the tangent from GH is not, in general, tangential to the base curves generated by G and H.

2.16 Gear drives between non-parallel shafts

So far we have restricted our attention to gear drives between parallel shafts. Other types of gear must be mentioned but will not be studied in detail; for a fuller analysis the reader should refer to the works listed in the bibliography.

Bevel gearing
When the axes of the two shafts intersect, bevel gears are used. With this type of gearing the concept of an equivalent friction drive applies as for spur gears, the driving surfaces being cones, fig. 2.29(a), instead of cylinders. The apices of the cones coincide at the point of intersection H of the two axes. Imagine two *base cones* fixed within the two *pitch cones* which comprise the equivalent friction drive, the radii of the base cones being proportional to their respective pitch cones. Consider a sheet of flexible inextensible material wrapped round the two base cones so that it unwinds off one and on to the other as the cones turn with the appropriate speed ratio. A line HX drawn on the sheet sweeps out one surface with respect to one base cone and a second surface with respect to the other. The two surfaces touch on XH which lies in the plane NHM tangential to the two base cones. The plane NHM also contains the pitch line PH where the two pitch cones touch. Clearly, if tooth surfaces are cut to the same shapes as the surfaces generated by XH we shall have a gear drive whose action is closely analogous to that of straight involute spur gears, fig. 2.18(a). Note, however, that as MN is not a straight line but part of a circle centred at H, the two involutes which are shown touching at X in fig. 2.29(b) are not quite the same as those in fig. 2.18. In the latter diagram, the involutes lie in a plane, the plane of the paper; in fig. 2.29(b) the curves lie on a spherical surface of radius HX and are described as *spherical involutes*.

58 Dynamics of Mechanical Systems

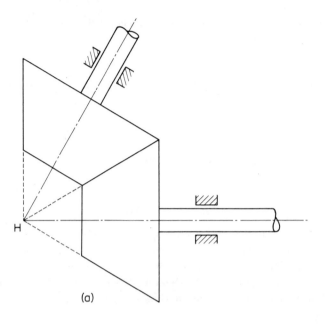

(a)

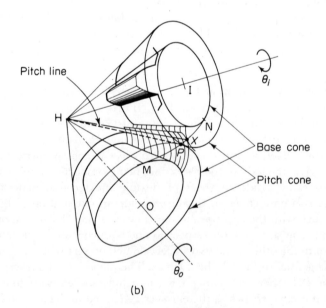

(b)

FIGURE 2.29

If the apex angle of one of the pitch cones is 180 degrees so that the pitch surface is plane the resulting gear is called a *crown wheel* which is the bevel gear equivalent of a rack. The tooth surfaces of a crown wheel with spherical involute teeth are not quite planes. In practice, crown wheels invariably do have teeth with plane surfaces and as a result bevel gear teeth are not of the spherical involute form. They are known as *octoid* teeth. The difference is very small.

Spiral bevel gearing
In fig. 2.29 the tooth surfaces are generated by a straight line, HX, drawn on a flexible sheet which unwinds from one base cone onto another. If a generating line other than a straight line through H is chosen, a spiral bevel gear results. This type has the same advantages over the straight bevel gear as does the helical gear over the straight spur gear.

Skew gearing
Several different types of gearing are used in practice to connect shafts for which the axes are skew, i.e., they are not parallel to each other and they do not intersect. In studying spur and bevel gears we have found it convenient to start by considering an equivalent friction drive. This approach is less helpful in the case of skew gears because it is not possible to design a skew friction drive for which the relative motion at the line of contact is one of pure rolling. Consequently, it is not possible to extend the line of reasoning which we have so far adopted to cover this type of gear. The geometry of skew gearing is a specialized and difficult subject which it would not be appropriate to pursue here.

2.17 Gear trains
Any combination of gears is called a *gear train*. There are three classes of gear train: simple, compound, and epicyclic.

Simple gear trains
A simple gear train is one in which all the gears rotate about fixed axes, and there is only gear wheel on each shaft. Two examples, one using spur gears and the other with bevel gears, are shown in fig. 2.30.

The transfer relationship for such a train of gears is:

$$\theta_o = (-1)^{p-1} \frac{N_1}{N_2} \times \frac{N_2}{N_3} \times \frac{N_3}{N_4} \times \cdots \times \frac{N_{p-1}}{N_p} \theta_i$$

where N_1, N_2, etc. are the numbers of teeth on the wheels and p is the number of wheels.

This expression reduces to

$$\theta_o = (-1)^{p-1}(N_1/N_p)\theta_i \qquad (2.25)$$

so that the transfer factor

$$k = (-1)^{p-1} N_1/N_p$$

is not dependent on the sizes of the intermediate wheels except that the total number of wheels p determines the direction of rotation of the output shaft.

With spur gears the directions of rotation are all viewed from the same end.

In the case of bevel gears, either a clockwise or an anticlockwise rotation is taken as positive when looking out from the point of intersection H of the shafts.

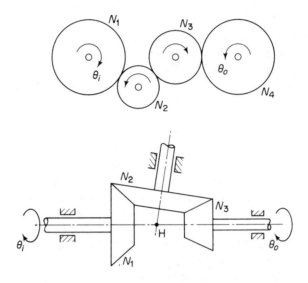

FIGURE 2.30

Compound gear trains
When a pair of gear wheels are made integral with each other so that they turn together on the same shaft the gears are said to be *compounded*. A compound gear train is one which incorporates compounded gears, see fig. 2.31. In any pair of compounded gears, one gear is driven by the preceding gears in the train and the other drives the succeeding gears. If the numbers of teeth in the driving gears are denoted by N_1, N_2, etc., and the numbers of teeth in the driven gears by n_1, n_2, etc., the transfer relationship is

$$\theta_o = \left(\frac{N_1}{n_2}\right)\left(-\frac{N_2}{n_3}\right)\left(-\frac{N_3}{n_4}\right)\cdots\left(-\frac{N_{p-1}}{n_p}\right)\theta_i$$

$$= (-1)^{p-1}\frac{N_1 N_2 N_3 \ldots N_{p-1}}{n_2 n_3 n_4 \ldots n_p}\theta_i \qquad (2.26)$$

With compound gear trains the transfer factor does depend on the number of teeth in gears carried by the intermediate shafts. If any intermediate compounded pair, say the jth pair, are replaced by a single gear wheel then effectively $N_j = n_j$ and these two terms cancel each other out and the size of this particular pair of gear wheels does not affect the ratio, though it does influence the sign of the term $(-1)^{p-1}$.

It is clear from eqns. (2.25) and (2.26) that a gear train can generate a transfer relationship of the form

$$\theta_o = k\theta_i$$

only if k is expressible as the ratio of two integers. The sign of k is irrelevant because we can always change this by adding an uncompounded gear wheel at any convenient point in the train.

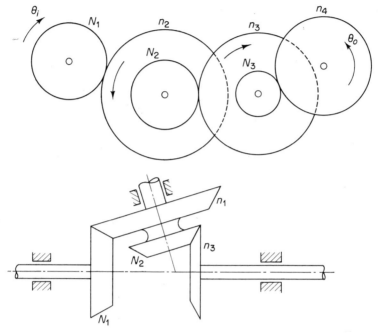

FIGURE 2.31

If k is an irrational quantity, e.g. π or $\sqrt{2}$, it is possible to design a gear train for which the actual constant k' approximates k to any desired degree of accuracy, assuming that no limit is set by manufacturing tolerances, and deformations due to load.

Basically the method is as follows:

the ratio k (or $1/k$ if $k > 1$) may be expressed as a continued fraction thus:

$$k = \cfrac{1}{b_1 + \cfrac{1}{b_2 + \cfrac{1}{b_3 + \text{etc.}}}}, \qquad (2.27)$$

To a first approximation $k \approx k_1 = 1/b_1$. It can be seen that $k_1 > k$. The second approximation

$$k_2 = \cfrac{1}{b_1 + \cfrac{1}{b_2}}$$

is $< k$, the third greater and so on.

62 Dynamics of Mechanical Systems

So that we have

$$k_2 < k_4 < \cdots < k < \cdots < k_3 < k_1 \qquad (2.28)$$

Consider, for example, the design of a gear train for which

$$k = \log_{10} e = 0 \cdot 434\ 294\ldots$$

$$= \frac{434\ 294}{1\ 000\ 000}$$

On dividing through by the numerator we have

$$k = \frac{1}{2 + \dfrac{131\ 412}{434\ 294}}$$

If the remainder in the denominator is ignored we have $k_1 = \frac{1}{2}$ as a first approximation to k. To obtain a second approximation the remainder is divided through by its numerator, thus:

$$k = \frac{1}{2 + \dfrac{1}{3 + \dfrac{40\ 058}{131\ 412}}} \approx \frac{1}{2 + \dfrac{1}{3}} = \frac{3}{7} = k_2$$

On continuing this process we obtain $k_3 = \frac{10}{23}$, $k_4 = \frac{33}{76}$ and so on. The inequality relationship (2.28) is

$$\frac{3}{7} < \frac{33}{76} < \frac{76}{175} < k_8 < k_{10} <$$

$$k_{12} = \frac{217\ 147}{500\ 000} = 0 \cdot 434\ 294$$

$$< k_{11} < k_9 < \frac{271}{624} < \frac{43}{99} < \frac{10}{23} < \frac{1}{2}$$

The proper fractions k_1, k_2, etc. have various properties which can be exploited to enable the optimum gear ratio to be selected and the successive values of k to be calculated more rapidly than is indicated.[1, 2] The most important property is that there is no proper fraction n/d between two successive values of k, $k_i = n_i/d_i$ and $k_{i+1} = n_{i+1}/d_{i+1}$, for which n is smaller than either n_i or n_{i+1}, or for which d is smaller than either d_i or d_{i+1}. So that if k is to be generated by a single pair of gears, and d_{max} is the maximum number of teeth permitted on any wheel, the best approximation which can be achieved is k_i with $d_i < d_{max} < d_{i+1}$. Generally things do not work out in practice as simply as the last statement implies. Suppose for example that in a particular set of circumstances, where $k = \log_{10} e$, tooth loading and space considerations set $d_{max} = 60$. The best approximation to k revealed by the above calculation is

$k_3 = \frac{10}{23} = 0.434\,783$ which differs from k by $+0.112$ per cent. The next approximation $k_4 = \frac{33}{76} = 0.434\,211$, i.e., an error of -0.015 per cent, is not acceptable because the denominator is greater than 60. In accordance with the property already quoted there can be no relevant proper fraction between $\frac{33}{76}$ and $\frac{10}{23}$ for which the numerator is less than 33 and the denominator less than 76. But a fraction which is just outside this range on the $\frac{33}{76}$ bound might still be a better approximation than $\frac{10}{23}$ even though it is not as good as $\frac{33}{76}$. Such a fraction is $\frac{23}{53} = 0.433\,962$ which differs from k by 0.076 per cent. Methods for locating suitable fractions which do not appear in the sequence of contained fractions are given elsewhere[1,2], together with extensions to the design of compound gear trains.

With compound gear trains the problem is that of selecting a fraction which is sufficiently accurate and which can be factorized. In the above sequence

$$k_6 = \frac{76}{175} = \frac{16 \times 19}{28 \times 25}$$

gives an accuracy of 0.002 per cent which is much better than anything which can be achieved by a single pair of gears with $d_{max} = 60$. The appearance within the sequence of continued fractions of a fraction which factorizes so conveniently is quite fortuitous (none of k_7 to k_{12} can be so factorized). In general, the designer must look outside this sequence for a suitable fraction. Again the reader is referred elsewhere for the details of this method.

Epicyclic gear trains

Figure 2.32 shows a simple epicyclic gear train. It differs from the compound gear trains previously considered in that the axis of one of the gears, the planet,

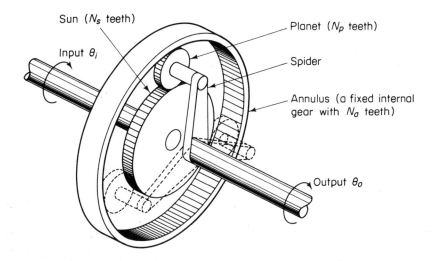

FIGURE 2.32

rotates about the axis of the other gears. In figs. 2.30 and 2.31 the axes of all the gears are fixed. Normally an epicyclic gear has two or three planet gears so that the arms of the spider are at either 180 or 120 degrees to each other. This prevents any side thrust being transmitted to the output shaft and also allows the overall size of the mechanism to be smaller than it would be if a single planet wheel were used. However, except that they impose some restriction on the relative number of teeth in the sun, planet, and annulus, the extra planet wheels have no effect on the kinematics and may be ignored in calculating the gear ratio.

The main advantage which an epicyclic gear train has over an ordinary fixed axis gear train is one of compactness; it is possible to achieve a high input to output gear ratio (i.e., small k) very simply and in a small space. Another feature which may be valuable is that the input and output are generally, but not necessarily, coaxial.

There are a number of methods of determining the transfer factor for an epicyclic gear train. One method will now be given. First, it is as well to observe that we cannot immediately say what the consequences of turning the input shaft through a given angle will be, because we cannot readily visualize what is the effect of revolving the axis of the planet wheel about the axis of the sun wheel. The first step in our analysis is to see what happens if this motion is prevented. Clearly, if we simply fix the output shaft it will not be possible to turn the input, unless the annulus is released from its anchorage and is permitted to turn also. Let us assume that this is done and that the input shaft is turned through an angle ϕ. The annulus turns through an angle of

$$-\frac{N_s}{N_p} \cdot \frac{N_p}{N_a} \phi = -\frac{N_s}{N_a} \phi$$

Now imagine the whole mechanism to be turned bodily, without any relative motion of the gears, through an angle $(N_s/N_a)\phi$. Clearly, the resultant rotation of the annulus is zero. The input will have turned through a total angle

$$\theta_i = \left(1 + \frac{N_s}{N_a}\right)\phi$$

and the output through an angle

$$\theta_o = \frac{N_s}{N_a} \phi$$

Hence

$$\theta_o = \frac{N_s}{N_s + N_a} \theta_i$$

SIMPLE MECHANISMS II Displacement analysis 65

The above calculations are best set out in a table with $\phi = +1$.

Operation	Number of rotations of:			
	Sun (Input)	Planet	Annulus	Output
Fix output shaft and free annulus Rotate input +1 radian	$+1$	$-N_s/N_p$	$-N_s/N_a$	0
Rotate whole $+N_s/N_a$ radians	N_s/N_a	N_s/N_a	N_s/N_a	N_s/N_a
Total	$1 + N_s/N_a$	$N_s/N_a - N_s/N_p$	0	N_s/N_a

For simple epicyclic gear trains this method is quite suitable, but its application to more complex arrangements sometimes requires considerable ingenuity and care (see exercises 2.10 to 2.12 and 4.9 to 4.11). An alternative method which avoids the difficulties is given in section 4.6.

2.18 Linkages

As we have seen, the problem of designing a cam mechanism to generate a given transfer relationship is in general not difficult to solve, the only snag which arises is one of interference. The designing of linkage mechanism is by comparison not so simple. In order to see why this is so consider two mechanisms, fig. 2.33, one a cam with a knife-edge rocker follower and the other a four-bar linkage mechanism. In both cases, the output is required to rotate through $\theta_o^{(1)}$ and $\theta_o^{(2)}$ as the inputs rotate through $\theta_i^{(1)}$ and $\theta_i^{(2)}$ respectively. Let us assume that in each case the centres A and D are specified, but that the lengths AB_o, DC_o and the inclinations of these lines to AD are arbitrary. Now consider the cam mechanism, and let AB_o be arbitrarily specified in direction and length. If the direction of C_oD is also chosen the contact C_oB_o can be drawn in, and the cam profile can be constructed in the manner indicated in the figure. Now in this construction the lengths AB_1 and AB_2, where B_1 and B_2 are points on the cam profile, are both seen to depend upon the initial choice of AB_o, C_oD, etc., but are independent of each other. Indeed any number of positions (θ_i, θ_o) can be specified and in each case the radius AB depends only on the basic set of dimensions.

Now consider the linkage. Again we can arbitrarily specify AB_o in length and direction, and DC_o in direction. By means of a construction to be given below, the length of DC_o and hence C_oB_o can be determined so as to cause DC to move through the angle $\theta_o^{(1)}$ when AB moves through $\theta_i^{(1)}$. The mechanism is now completely specified, and it can be seen that the output angle corresponding to an input $\theta_i^{(2)}$ will not, in general, have the required value $\theta_o^{(2)}$. If the transfer relationship is to be correct for each of the three positions shown, both AB and CD must be treated as unknowns to be determined simultaneously, either by a

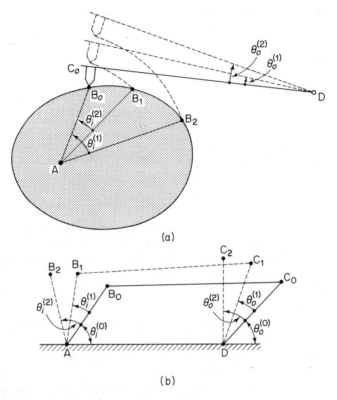

FIGURE 2.33

graphical construction or by calculation. If we treat the angles $\theta_i^{(0)}$ and $\theta_o^{(0)}$ as unknowns we can satisfy up to a maximum of five (θ_i, θ_o) positions, by in effect solving five simultaneous equations.

Thus by comparison with cams, linkage mechanisms are often more difficult to design and there are restrictions on the extent to which any given form of linkage mechanism can generate a given transfer relationship. Nevertheless, linkages form a very important class of mechanism, and for this reason we will now consider how some of the simpler problems are resolved. First it is desirable to state some basic propositions regarding the displacement of rigid bodies.

2.19 Kinematics of rigid body motion in respect of finite displacements

If a body has one point O fixed, any displacement of the body from one given position to another is equivalent to a rotation about some fixed axis through O.

SIMPLE MECHANISMS II Displacement analysis 67

This theorem, due to Euler (Leonard Euler, 1707–1783), is proved in standard textbooks on dynamics.† Euler's theorem, together with its extensions, provides a basis from which methods of designing linkage mechanisms can be evolved. When the body considered is a lamina which moves only in its own plane Euler's theorem states the obvious. It is perhaps less obvious that any displacement of a lamina in its own plane is equivalent to a rotation about a fixed axis perpendicular to the plane, but it can readily be proved.

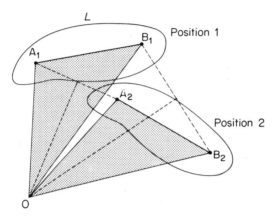

FIGURE 2.34

The lamina L in fig. 2.34 is displaced from position 1 to position 2. Let us identify any two separate points A and B in the lamina so that the line AB drawn on the lamina is displaced from A_1B_1 to A_2B_2. Now let the perpendicular bisector of A_1A_2 and the perpendicular bisector of B_1B_2 intersect at O. Triangles OA_1B_1 and OA_2B_2 are congruent so that the displacement of AB_1 and hence the lamina, from position 1 to position 2 can be achieved by a rotation about the fixed point O.

This extension of Euler's theorem does not apply to the more general case of three-dimensional motion. It can be shown (ref. Ramsey) that any general displacement of a rigid body is equivalent to a translation of the body as a whole, with all parts in the body moving along straight parallel paths, together with a rotation about a fixed axis. For any given displacement the direction of the axis of rotation is determined by the displacement, but its position is not, neither is the direction of the translational motion. The position of the axis of rotation and the direction of the translational motion are, however, interdependent, and only one can be specified arbitrarily. For a particular position of the axis of rotation the translation will be in a direction parallel to the axis. Thus any displacement of a rigid body is equivalent to a screw motion about a unique axis. This is illustrated in fig. 2.35. The upper surface of a square slab

† E.g., EASTHOPE, S. E. *Three-dimensional Mechanics*, Butterworths, 1958, p. 54.
RAMSEY, A. J. *Dynamics: Part II*, chapter III, 2nd edn., C.U.P., 1964.

68 Dynamics of Mechanical Systems

is denoted by the corners ABCD. In fig. 2.35(a), the block is given two consecutive displacements: first a translation $A_1B_1C_1D_1$ to $abcd$, and then a rotation about the axis B_1c to bring the body to its final position $A_2B_2C_2D_2$.

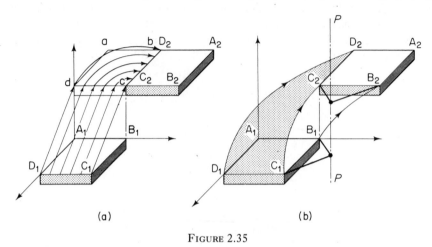

FIGURE 2.35

Figure 2.35(b) shows how the same resultant motion can be achieved by a single screw motion about the axis PP with the points ABCD; and all other points in the body describing helices. A number of facts should be noted: first, that the screw axis PP is parallel to the rotation axis B_1c, second, the combination of translation and rotation in fig. 2.35(a) is not unique, the same end result could have been attained by displacing A straight from A_1 to A_2 and then rotating about an axis passing through A and parallel to PP, third, these alternative translational paths are not parallel to each other. Indeed, apart from the restriction that it may not lie in a direction perpendicular to PP, the direction of translation is arbitrary.

The ratio of the displacement along the screw axis to the angle of rotation is known as the *pitch* of the screw. Pure rotation of a body, as, for example, the motion of a lamina in its own plane, can be regarded as a screw motion with zero pitch, and pure rotation can be regarded either as a screw motion with infinite pitch or a pure rotation about an axis at infinity.

2.20 Design of linkage mechanisms

Spatial (i.e., three-dimensional) motion is not only more difficult to visualize than plane motion but is basically more complex, and it is to be expected that the design of spatial mechanisms is more complicated and less developed than the design of plane mechanisms. We shall concentrate on plane mechanisms, in particular the plane four-bar mechanism. This consists, fig. 2.36, of a fixed member AD, two cranks AB and CD, and a connecting-rod or coupler BC.

The purpose of a four-bar mechanism may be:

1 to achieve a particular correlation between the rotations of the two cranks,

2 to guide the coupler through a series of positions, which may or may not be correlated with rotations of one of the cranks,
3 to cause some point attached to the coupler to trace out a prescribed locus.

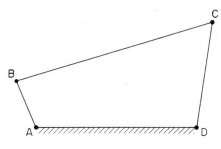

FIGURE 2.36

The extent to which the transfer relationship for a four-bar mechanism can be specified is strictly limited by the small number of parameters available for adjustment. If the specification is simple it is possible to start with an arbitrary assumption of most of the parameters and leave only one or two to be determined. As the specification is tightened the designer is permitted less discretion and more parameters must be left to be determined by whatever method of synthesis has been adopted. The methods themselves become increasingly complicated as the number of conditions which are imposed is increased. As our purpose here is to explore general principles rather than to give a detailed specialist treatment we shall consider only the more elementary problems with which this subject is concerned.

2.21 Synthesis of a plane four-bar linkage for correlated crank displacements

There are three distinct types of four-bar mechanism: the crank and rocker mechanism, in which one crank is able to make complete revolutions while the other reciprocates; the drag link mechanism, in which both cranks rotate continuously in the same direction; and the double rocker mechanism, in which neither crank is able to turn continuously. These different types are shown in fig. 2.37.

The crank and rocker mechanism provides a means of converting a rotation into a reciprocating motion, and the main constraint on its design is likely to be the total angle through which the rocker is required to rotate. This alone is not sufficient to determine the relative lengths of the links so that there is some freedom of choice left to the designer who will also have to take into account the possibility that the mechanism might tend to jam at certain points in the cycle of motion, and the loads which the members will have to carry.

The drag link mechanism is used when it is necessary to turn the output crank in such a way that its velocity varies from a minimum to a maximum and back to a minimum during a complete revolution, assuming that the input

70 Dynamics of Mechanical Systems

crank turns at constant speed. The basic design requirements here are likely to be the maximum and minimum velocity ratios.

The solution of the problems which have just been outlined will not be pursued here (see exercises 2.13 and 2.14). The more interesting problem is

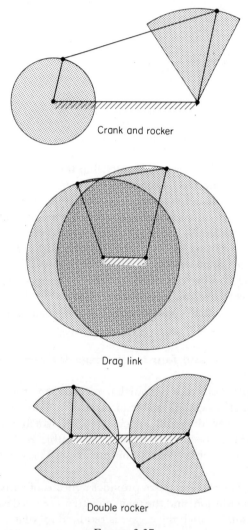

Crank and rocker

Drag link

Double rocker

FIGURE 2.37

that of designing a mechanism to correlate the rotations of the two cranks for a number of discrete displacements. In general, neither crank will be required to rotate through 360 degrees so that for this class of problem the classifications of crank and rocker, drag link, and double rocker are not relevant.

Let us assume that the positions of the hinges A and D of the mechanism in

SIMPLE MECHANISMS II Displacement analysis

fig. 2.38 are specified, together with the length AB of the input crank, the two input angles $\theta_i^{(1)}$ and $\theta_i^{(2)}$, and the corresponding output angles $\theta_o^{(1)}$ and $\theta_o^{(2)}$. The quantities to be determined are the length of the output crank DC and the length of the coupler BC.

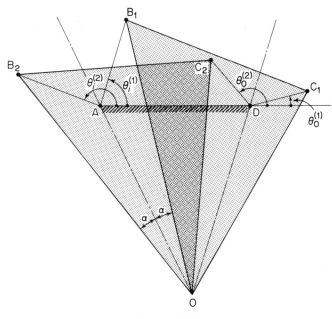

FIGURE 2.38

Let AO bisect the angle between the two given positions of the input crank, and let DO bisect the angle between the two given positions of the output crank. If the length CD is chosen such that the angles which DC_1 and DC_2 subtend at O are made equal to the angle α which AB_1 and AB_2 subtend at O, then triangles OB_1C_1 and OB_2C_2 are congruent, with B_1C_1 equal to B_2C_2. Hence ABCD is the required linkage.

This method is unsatisfactory when the lengths OA and OB are large compared with the lengths of the links. A more convenient method which keeps the construction within an area comparable to that occupied by the actual mechanism uses the *principle of inversion*. Instead of considering the motion relative to the fixed link AD, we suppose link CD to be fixed in its initial position and consider the motion of the other links relative to it, fig. 2.39. Imagine that the mechanism is drawn in the two required positions and that the drawing for the second position is pivoted about D until C_1 and C_2 coincide at C as in fig. 2.39(a) DC is initially of unknown length, but since BC is of fixed length the point C is equidistant from B_1 and B_2, and is therefore located on the perpendicular bisector of B_1B_2.

By rotating the drawing for the second position about D so as to bring A_2

72 Dynamics of Mechanical Systems

back into coincidence with A_1 at A the mechanism now appears, fig. 2.39(b), in its two required configurations.

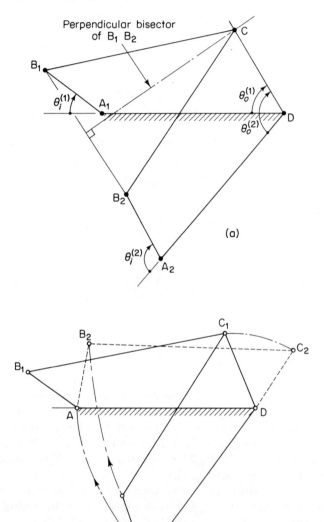

FIGURE 2.39

Graphical methods can be extended to the synthesis of four-bar linkages when more than two pairs of crank positions are specified[3, 4]. Alternatively, this class of problem can be solved by the numerical method which is explained in the next section.

2.22 Freudenstein's method
In fig. 2.40
$$c \cos \phi = a + d \cos \theta_o - b \cos \theta_i$$
$$c \sin \phi = d \sin \theta_o - b \sin \theta_i$$

By squaring and adding these two expressions, and rearranging, we find

$$\frac{a}{b} \cos \theta_o - \frac{a}{d} \cos \theta_i + \left(\frac{a^2 + b^2 + d^2 - c^2}{2bd} \right) = \cos(\theta_o - \theta_i)$$

or
$$R_1 \cos \theta_o - R_2 \cos \theta_i + R_3 = \cos(\theta_o - \theta_i) \qquad (2.29)$$

If three pairs of specified input/output crank angles are specified and the values inserted in this relationship, the resulting three equations enable R_1,

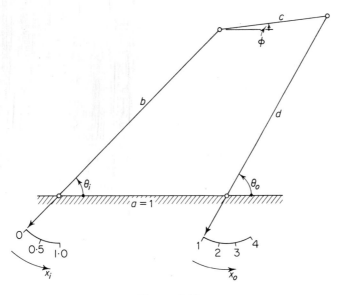

FIGURE 2.40

R_2, and R_3 to be determined, and hence the two crank radii and the length of the coupler, given that the fixed link is of unit length. This is Freudenstein's method, which was developed as a means of designing linkages as function generators in mechanical computers. An example will illustrate how the method is applied.

Let the function to be generated be:

$$x_o = (1 + x_i)^2 \qquad (2.30)$$

where x_i and x_o are proportional to θ_i and θ_o respectively, and let the range over which this function is to be generated by $0 \leqslant x_i \leqslant 1$, so that $1 \leqslant x_o \leqslant 4$. Let the corresponding ranges of the crank angles be $45° \leqslant \theta_i \leqslant 90°$, and

$60° \leq \theta_o \leq 120°$, and the values of x_i and x_o be indicated by means of point moving over uniformly graduated scales, as shown in fig. 2.40.

It is clear that we cannot ensure that the readings x_i and x_o will satisfy eqn. (2.30) at all points in the range of operation, because the true relationship between these readings is controlled by eqn. (2.29). We can, however, ensure that there is accurate correspondence between x_i and x_o at a limited number of points. If these points are well spread over the ranges of x_i and x_o, it is reasonable to suppose that x_o is nowhere considerably in error.

Let us require x_o to be accurately generated for

$$x_i = 0, 0.5, \text{ and } 1.0$$

corresponding to $\quad \theta_i = 45°, 67\cdot5°, \text{ and } 90°$

and $\quad x_o = 1, 2\cdot25, \text{ and } 4$

with $\quad \theta_o = 60°, 85°, \text{ and } 120°$

Substituting these values for θ_i and θ_o in eqn. (2.29) we have

$$R_1 \cos 60 - R_2 \cos 45 + R_3 = \cos 15$$
$$R_1 \cos 85 - R_2 \cos 67\cdot5 + R_3 = \cos 17\cdot5$$
$$R_1 \cos 120 - R_2 \cos 90 + R_3 = \cos 30$$

On solving these we find

$$R_1 = 0\cdot7317; \quad R_2 = 0\cdot8935; \quad R_3 = 1\cdot2319$$

Hence $\quad b = \dfrac{a}{R_1} = 1\cdot369$

$$d = \dfrac{a}{R_2} = 1\cdot120$$

and $\quad c = (a^2 + b^2 + d^2 - 2bdR_3)^{\frac{1}{2}} = 0\cdot587$

In designing this mechanism the initial and final values of θ_i and θ_o were chosen arbitrarily. A more accurate approximation to the desired transfer relationship can be obtained by designing for zero error at more than three *precision points*. This can be done by relaxing some of the arbitrarily imposed constraints on the initial and final positions of the cranks and treating them as unknown quantities. Freudenstein has shown[5] how to synthesize a mechanism using five precision points by prescribing only the total angles through which the input and output cranks are allowed to swing.

None of the above methods of synthesis leads automatically to a practical design. It is always necessary to check that there is no tendency to jam due to two members moving into a position where their axes are colinear.

2.23 Synthesis of a plane four-bar mechanism for specified coupler displacements

(a) Coupler hinges specified

The simplest method of constraining a body in plane motion so that it may be moved from one specified position to another, is to pivot it at the centre of rotation. If the centre of rotation is inaccessible or otherwise unsuitable an alternative is to make the body the coupler of a four-bar mechanism. If the positions of the hinges B and C are located in the body and two positions of BC are specified, as in fig. 2.41(a) it is quite a straightforward matter to locate

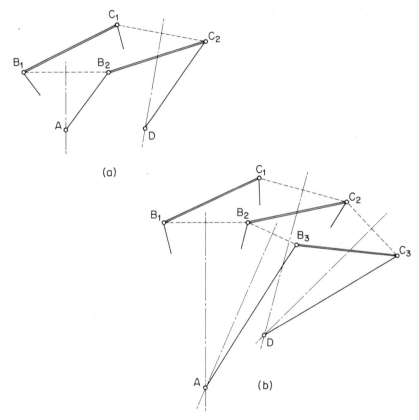

FIGURE 2.41

two suitable crank pivots A and D. Point A must be equidistant from B_1 and B_2 and hence it lies on the perpendicular bisector of B_1B_2, the actual position on this line being arbitrary. Likewise D is an arbitrary point on the perpendicular bisector of C_1C_2.

When a third position of the coupler is given, A must be equidistant from B_1, B_2 and B_3, and is located at the intersection of the perpendicular bisectors of B_1B_2 and B_2B_3 (A is also on the perpendicular bisector of B_1B_3). D is located in a similar manner and the mechanism is then fully determined. When more positions of the coupler are specified the location of the hinges on the coupler are no longer completely arbitrary.

(b) Location of the crank hinges specified
When a coupler is to be guided through two or three specified positions by means of cranks of unspecified lengths rotating about given points, the points of attachment of the cranks to the coupler can be readily determined by inverting the mechanism so that the coupler becomes the fixed member.

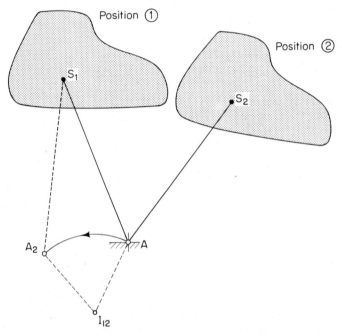

FIGURE 2.42

In fig. 2.42 a lamina which forms the coupler is shown in two positions together with the centre A of one of the cranks. By considering the displacement of any convenient pair of points the centre of rotation I_{12} for this displacement is readily located; the construction lines used to determine I_{12} have been omitted from fig. 2.42.

Let S be a possible point of attachment to the lamina of the crank centred at A. As a result of the displacement, the crank is moved from AS_1 to AS_2. Now imagine that in position 2 the crank is rigidly attached to the lamina, and is disconnected from the pivot at A. If the lamina and the attached crank are rotated about I_{12} until the connecting rod is back in position 1, S moves back to S_1, the end A of the crank moves to A_2, and the crank itself being in the position indicated. Since A_2S_1 must equal AS_1 it follows that S can lie anywhere on the perpendicular bisector of AA_2. The possible points of attachment of the second crank lie on a similar line which can be determined in a like manner. The two lines intersect at I_{12}.

If a third position of the lamina is specified, the position of the rotation axis for the resultant displacement from position 1 to position 3 can be determined, and hence A_3. The point of attachment S is now uniquely located at the intersection of the perpendicular bisectors of AA_2 and AA_3.

When four positions of the lamina are specified the positions of the crank hinges can no longer be chosen arbitrarily. There are an infinite number of points on the moving lamina which for the four given positions lie on circular arcs. The cranks may be attached to the lamina at any pair of such points, and the centres of the two circles thus selected determine the positions of the crank centres. The locus of the crank centres obtained by varying the points of attachment to the lamina is known as Burmester's curve.[8]

2.24 Use of a plane four-bar mechanism as a curve generator

The locus of a point attached to the connecting rod, or coupler, of a four-bar linkage is known as a *coupler curve*. In fig. 2.43, a four-bar linkage is shown

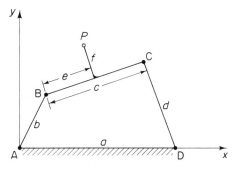

FIGURE 2.43

with xy axes placed so that their origin is at one crank hinge, and the fixed link lies along the x-axis. The equation of the locus of the point P which is rigidly attached to the coupler can be expressed in terms of x and y, and five lengths a, b, c, d, e, f which define the linkage, and is a polynomial equation of the sixth degree. Clearly it will be possible to design a mechanism to guide a point such as P exactly along a prescribed path only if the equation to the path is of this particular form. Fortunately, design requirements are rarely so stringent, and it is usually sufficient to find a mechanism for which the locus of the moving point approximates to the prescribed path. The simplest and most direct method is to consult reference (11). This book is an atlas containing more than 7000 coupler curves drawn to scale, for various combinations of the ratios of the lengths of the links, and positions of the tracing point on the coupler of a crank and rocker mechanism. In order to determine a mechanism to suit his needs the designer simply scans through the atlas until he finds an appropriate

78 Dynamics of Mechanical Systems

one. Since the drag link and double rocker mechanisms are excluded the atlas does not cover all the possibilities, but it seems unlikely that this would in practice often prove to be a restriction.

Figure 2.44 gives a few examples of mechanisms where the kinematic problem is basically that of designing a mechanism to generate a particular type of coupler curve. Although it is occasionally necessary to guide a point along a fairly complicated path, interest tends to centre on those mechanisms for which the complex curve has portions which are either approximately

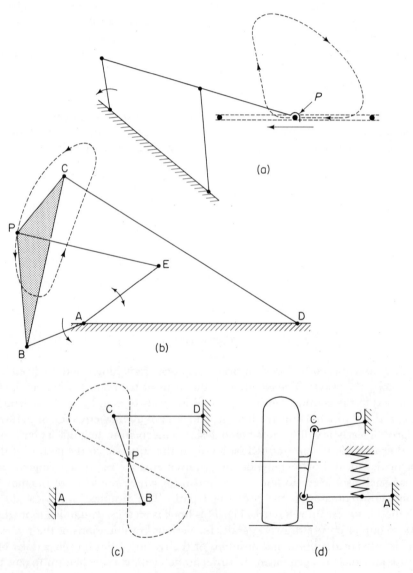

FIGURE 2.44

straight or approximately circular. In fig. 2.44(a) a crank and rocker mechanism is used as an intermittent drive; the coupler curve has a fairly straight portion with the entry and exit well defined. Figure 2.44(b) shows a dwell mechanism. The crank AB rotates at a uniform speed while the output crank AB oscillates, but is stationary for about 40 per cent of the time of one complete oscillation. Such a mechanism can be used to hold a tool in contact with a succession of workpieces for finite periods of time. The principle used here is to select a four-bar linkage for which the coupler curve is approximately circular over an extensive section. The floating link PE is of length equal to the radius of curvature of the circular portion of the coupler curve. The dwell period then lasts as long as P is on the circular portion. The third example, fig. 2.44(c) shows a Watt straight-line mechanism. This is a classical means of guiding a point along a line which is approximately straight by means of a linkage where the use of a simple slider is inappropriate. The mechanism is of the double crank type and the complete coupler curve is in the form of a figure eight. The motion is constrained so that only the straight portion of the coupler curve is used. Figure 2.44(d) shows an alternative form for the Watt linkage as used for the front-wheel suspension of an automobile.[6] The purpose of this linkage is to ensure that as it deflects, due to compression of the suspension springs, the point of contact P between the tyre and the road does not move laterally, and so cause undue wear of the tyre. B and C are ball and socket joints and form the steering pivots.

The straight-line mechanism is an example of a mechanism which is designed to generate a specified coupler curve accurately, but only for a limited range of movement. One method of tackling this problem is to design a mechanism for which the coupler curve passes through a number of points spaced along the path which it is desired to generate. The philosophy of this approach is that if the generation is exact at a succession of finitely spaced points, it is reasonable to expect the generation to be fairly accurate between these points. The geometry of this method is fairly complicated and the reader is referred to specialist books for further details.[7, 8]

If infinitesimal displacements of a mechanism are considered it is possible to calculate the radius of curvature at a particular point on a coupler curve. Conversely, it is possible to design a mechanism to generate a coupler curve which has a specified radius of curvature at a given point.

2.25 Euler-Savary equation

For the purpose of this analysis we regard the plane four-bar linkage as consisting of two planes, F and M, connected by two cranks, AB and CD, fig. 2.45. The moving plane M slides on the fixed plane F with the relative motion depending on the lengths of the cranks, and their points of attachment to the two planes.

Any displacement of plane M is equivalent to a rotation about an axis perpendicular to M. If B moves from B_1 to B_2, and C moves from C_1 to C_2, the rotation axis is located at the intersection of the internal bisectors of angles

80 Dynamics of Mechanical Systems

B_1AB_2 and C_1AC_2 (as in fig. 2.41). Now consider an infinitesimal displacement in which B_1 and B_2 coalesce at B, and C_1 and C_2 coalesce at C, so that the internal bisectors are simply AB and DC respectively. If these two intersect at I, then I is the *instantaneous centre* of rotation of the moving plane.

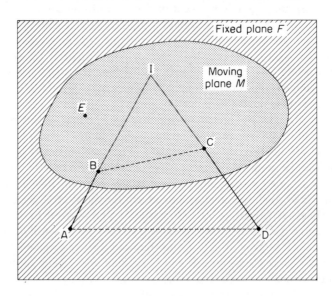

FIGURE 2.45

Although a small displacement of the moving plane from the position given in fig. 2.45 is equivalent to a small rotation about I, this does not mean that I is the centre of curvature of the path traced by some point such as E attached to the moving plane. If the reader finds this statement surprising he should consider the motion of the hinge B whose path is clearly a circle centred on A.

The radius of curvature of the locus of E can be calculated from the Euler-Savary equation which can be derived in the following way:

Let crank AB, fig. 2.46, be deflected through a small angle $\delta\psi$ so that B moves from B_1 to B_2 and the instantaneous centre moves from I_1 to I_2 through a distance δs and point E on the coupler moves from E_1 to E_2 (to simplify the diagram somewhat, the coupler is not drawn in its displaced position). The displacement of the coupler is equivalent to a small rotation about the instantaneous centre I.

The angle through which it rotates is

$$\frac{B_1 B_2}{I_1 B_1} = \frac{AB\ \delta\psi}{I_1 B_1} = \delta\phi \qquad (2.31)$$

and the displacement of E is $I_1 E_1\ \delta\phi$

so that
$$E_1 E_2 = I_1 E_1\ \delta\phi \qquad (2.32)$$

SIMPLE MECHANISMS II Displacement analysis

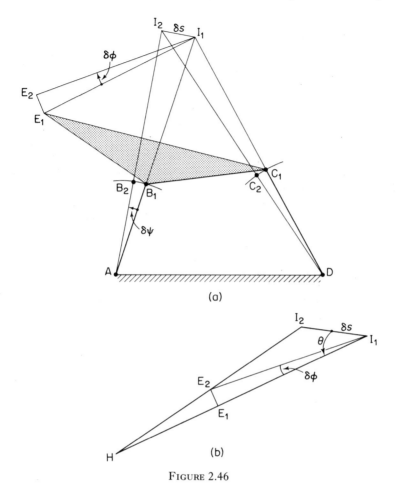

FIGURE 2.46

Since this is a small displacement perpendicular to I_1E_1, it could be achieved by a rotation about any point on I_1E_1. Now a subsequent displacement will be about I_2, the new instantaneous centre. Let this displacement be such that it just cancels out the initial displacement, with E returning from E_2 to E_1 along a line perpendicular to E_2I_2. The centre for this displacement of E can be anywhere on I_2E_2. The centre of curvature H for the displacement E_1E_2 must therefore lie at the intersection of E_1I_1 and E_2I_2, fig. 2.46(b).

If the angle which I_1E_1 makes with I_1I_2 is θ we have, by similar triangles

$$\frac{E_1H}{I_1H} = \frac{E_1E_2}{I_1I_2 \sin \theta}$$

Substituting from eqns. (2.31) and (2.32), and replacing I_1I_2 by δs

$$\frac{E_1H}{I_1H} = \frac{I_1E_1}{\sin \theta} \frac{AB}{IB_1} \frac{\delta\psi}{\delta s}$$

In the limit, as $\delta s \to 0$, we have on rearranging

$$\frac{EH}{IH \cdot IE} \sin \theta = \frac{AB}{IB} \frac{d\psi}{ds}$$

and on substituting $(IH - IE)$ for EH

$$\left\{ \frac{1}{IE} - \frac{1}{IH} \right\} \sin \theta = \frac{AB}{IB} \frac{d\psi}{ds} \qquad (2.33)$$

$d\psi/ds$ is the reciprocal of the rate at which I moves along its locus as ψ varies, and it and the lengths AB and IB are all unaffected by the location of E with respect to BC. Thus for any given configuration of the mechanism the right-hand side of eqn. (2.33) is a constant, which for convenience we will denote by k^{-1}. For certain positions of E the radius of curvature IH is infinite; if such points are denoted by J, eqn. (2.33) becomes

$$IJ = k \sin \theta \qquad (2.34)$$

As θ is measured with respect to a fixed direction, $I_1 I_2$ (which in the limit is the tangent at I_1 to the locus of I), this is an equation in polar coordinates. It is the equation to a circle, of diameter k, known as the *inflection circle*, fig. 2.47.

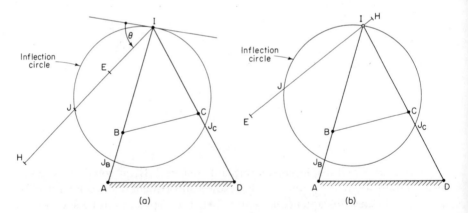

FIGURE 2.47

On eliminating k between eqns. (2.33) and (2.34) we find the following relationship between the lengths IE, IH, and IJ, fig. 2.46(a),

$$\frac{1}{IE} - \frac{1}{IH} = \frac{1}{IJ} \qquad (2.35)$$

This is the Euler-Savary equation. It is a simple exercise to manipulate it into the alternative form

$$EH \cdot EJ = IE^2 \qquad (2.36)$$

SIMPLE MECHANISMS II Displacement analysis   83

The minus sign in eqn. (2.35) arises from the assumption which is implicit in fig. 2.46(b) that H lies on the same side of I as E_1. If IE > IJ, so that E lies outside the inflection circle, IH is negative. This must be taken to mean that H lies on the opposite side of I from E, as in fig. 2.47(b). Equation (2.36) shows that EH and EJ must be of the same sign, because of the squared term on the right-hand side, and hence that J and H always lie on the same side of E.

The inflection circle can be drawn without knowing the direction of $I_1 I_2$ by determining two points in addition to I through which it must pass. J_B and J_C, fig. 2.47, are two such points. Taking J_B as an example; it lies at the intersection of crank AB with the inflection circle, and can be determined immediately from eqn. (2.35) as the centre of curvature of the path of B is known to be at A so that

$$\frac{1}{IJ_B} = \frac{1}{IB} - \frac{1}{IA}$$

The position of J_C can be determined similarly.

The Euler-Savary equation provides a very quick means of determining the radius of curvature at any point on a coupler curve for a given mechanism.

As an example of the use of the Euler-Savary equation as a means of synthesis consider the problem of designing a straight-line mechanism, fig. 2.48.

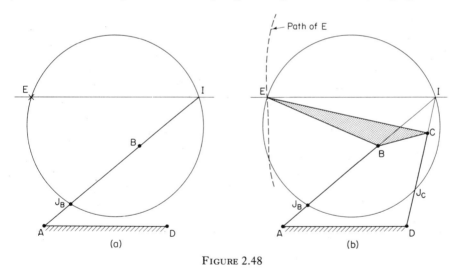

FIGURE 2.48

Let us suppose that the crank centres A and D are specified, and that for small displacements of the crank AB from its given position the mechanism traces a straight line through E perpendicular to AD.

First, it is to be noted that, as the motion at E is to be straight, E must be on the inflection circle. Furthermore, as the motion is perpendicular to AD the instantaneous centre I must be so placed that EI is parallel to AD. But I must also lie on AB, which is specified. The Euler-Savary equation now enables J_B

84   Dynamics of Mechanical Systems

to be located, and the inflection circle can be drawn through E, I, and J_B, as in fig. 2.48(a).

ID is now joined to locate J_C and a further application of the Euler-Savary equation locates C, fig. 2.48(b). The coupler curve is shown and the extent to which it is straight in the vicinity of E can be judged. Whether or not the solution is acceptable depends upon the particular application of the mechanism. As it is given above, the problem has only the one solution which we have obtained. In practice, the specification usually allows more latitude and there is a choice of solutions.

2.26 Roberts' theorem

Having designed a four-bar mechanism which generates a particular coupler curve it may for various reasons be unsuitable: it may be too large, the fixed hinges may be inconveniently placed, it may tend to jam, and so on. The question then arises as to whether there are other four-bar mechanisms which will generate the same curve. A theorem due to Roberts (1875)[9] proves that, in general, there must be two other mechanisms which will generate precisely the same curve.

Linkages which generate a common coupler curve are said to be *cognate*. The two linkages which are cognate to a given linkage are obtained in the following way:

Let ABCD, fig. 2.49, be the given linkage, AD being the fixed link, AB and CD the cranks and BC the connecting-rod. Point E which is rigidly attached to BC, traces the coupler curve. We now add the additional links AF and EFG which are so proportioned that ABEF is a parallelogram and triangle EFG is similar to triangle CBE. Likewise KD and EJK are added so that DCEK is a parallelogram and triangle EJK is similar to triangle BEC. Finally, the links GH and JH are added to form the parallelogram EGHJ. AFGH and DKJH are the two linkages cognate to ABCD.

In order to justify this statement, it is sufficient to show that, given the above method of construction, H is a fixed point for all configurations of the composite linkage. If H is a fixed point we can, for example, attach GH to a fixed hinge at H and throw away linkage HJKD. The point E clearly traces out the same coupler curves irrespective of whether it is attached to ABCD or AFGH.

It can be shown by purely geometrical reasoning that H is a fixed point, but elementary vector analysis provides a much neater proof. (This proof anticipates the introduction to basic vector methods which is given later. The reader who is unfamiliar with vectors should skip this proof until he has studied chapter 4.)

In fig. 2.49, AB, BC, CD, AD, and BE are represented by vectors **a**, **b**, **c**, **d**, and **e** respectively. Our proof hinges on the relationship between vectors **e** and **b**.

In fig. 2.49(b) $\overline{BQ} = \mathbf{b} + \boldsymbol{\Omega} \times \mathbf{b}$, where $\boldsymbol{\Omega}$ is a vector which is normal to the plane of the paper. By changing the magnitude of $\boldsymbol{\Omega}$ we can vary the angle θ at will. We can now define any vector in the direction $\overline{BQ}$ by multiplying by a suitable scale factor k, thus $\overline{BQ} = k(\mathbf{b} + \mathbf{\Omega} \times \mathbf{b})$. In particular, returning to fig. 2.49(a) we can write

$$\mathbf{e} = k(\mathbf{b} + \mathbf{\Omega} \times \mathbf{b}) \qquad (2.37)$$

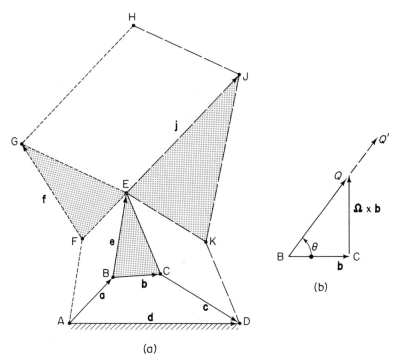

FIGURE 2.49

In triangle FEG we have $\overline{FE} = \mathbf{a}$, and $\overline{FG} = \mathbf{f}$. Furthermore, since triangle FEG is similar to BCE, $\mathbf{f} = k(\mathbf{a} + \mathbf{\Omega} \times \mathbf{a})$, and likewise $\mathbf{j} = k(\mathbf{c} + \mathbf{\Omega} \times \mathbf{c})$.

Point H is located by the vector $\overline{AH}$ which is given by

$$\overline{AH} = \overline{AF} + \overline{FG} + \overline{GH}$$
$$= \mathbf{e} + \mathbf{f} + \mathbf{j}$$

On substituting from eqn. (2.37), etc., we have

$$\overline{AH} = k(\mathbf{a} + \mathbf{b} + \mathbf{c} + \mathbf{\Omega} \times (\mathbf{a} + \mathbf{b} + \mathbf{c}))$$
$$= k(\mathbf{d} + \mathbf{\Omega} \times \mathbf{d}) \qquad (2.38)$$

But $\overline{AD}$ is a fixed vector so that $\overline{AH}$ is also a fixed vector and H is a fixed point. Furthermore, by comparison of eqns. (2.37) and (2.38) we see that triangle AHD is similar to triangle BEC.

2.27 Symmetrical coupler curves

If the two cranks of a four-bar mechanism are of equal length it is obvious that a point attached to the perpendicular bisector of the coupler must trace a symmetrical path. The Tchebichev mechanism in fig. 2.50 is an example of

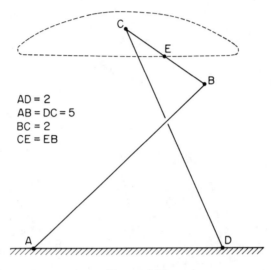

AD = 2
AB = DC = 5
BC = 2
CE = EB

FIGURE 2.50

such a mechanism. It is also a classical example of a straight-line mechanism, for during the phase of motion for which the cranks are crossed, the coupler curve approximates very closely to a straight line.

It is, however, not essential for a mechanism itself to be symmetrical in order for it to generate a symmetrical coupler curve. In the four-bar mechanism which is shown in fig. 2.51(a), the two cranks are of unequal length and point E is not on the perpendicular bisector of the line BC which passes through the two moving hinges. Nevertheless, the coupler curve is symmetrical about a line through the fixed hinge A and making an angle of $\frac{1}{2} \angle$ EDC with AD. The coupler curve has this property because the mechanism is so dimensioned that AB = BC = BE. To prove the property we draw in the cognate mechanisms as shown in fig. 2.51(b). It can be seen that linkage AFGH is the mirror image of linkage ABCD although at the given instant, the configurations are not mirror images. It follows that the locus of E which is common to both linkages must be symmetrical about the bisector of $\angle$ HAD. As triangle ADH is similar to triangle DEC, the angle which the axis of symmetry makes with AD is half the angle EBC.

An application of the symmetrical coupler curve which has been suggested[10] in the design of dwell mechanisms is shown in fig. 2.52. The link PQ rotates once for every two rotations of the driving link FA with two periods during which PQ is virtually stationary.

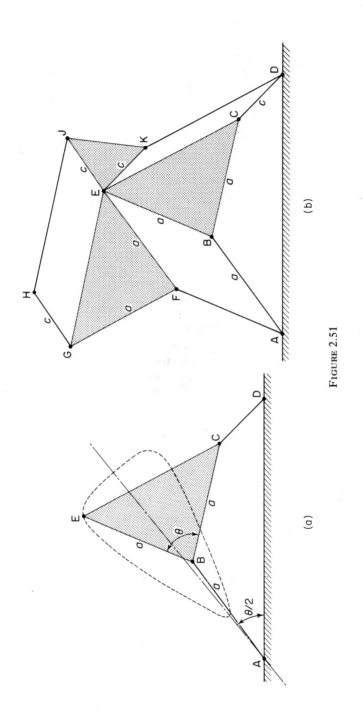

Figure 2.51

88 Dynamics of Mechanical Systems

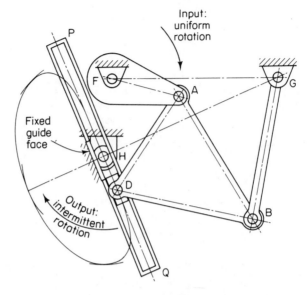

Figure 2.52

2.28 Classification of four-bar mechanisms

Reference has already been made to the fact that a mechanism whose main dimensions have been determined by the methods described in this chapter will not necessarily operate in a satisfactory manner. In particular, the mechanism may tend to jam in certain phases of its motion. Although a full assessment of this possibility involves consideration of how the forces are transmitted and the effect of friction, it is possible to arrive at a qualitative assessment of the possibility that a mechanism may jam from its geometrical properties alone.

It has been mentioned that a four-bar mechanism may fall into one of three categories: the drag link mechanism, in which both cranks are capable of rotating through 360 degrees; a crank and rocker mechanism, in which one crank, usually the driver, can rotate through 360 degrees; and the double crank mechanism in which neither crank can rotate fully. The category into which a mechanism falls depends on the relative lengths of the links, the order in which they are joined, and which one is fixed.

It is possible to imagine that by changing the length of the coupler, say, of a crank and rocker mechanism we can transform it into a double crank mechanism. If the change is small the operation of the crank and rocker mechanism is unlikely to be satisfactory, for at a certain point the crank will only just be able to continue rotating in the same direction, and if there is friction it is likely to cause the mechanism to jam at this point. In other words, the satisfactory operation of a crank and rocker mechanism is likely to depend on its being quite definitely a crank and rocker and not close to being something else. This brings us to the question of how we can formally distinguish between

SIMPLE MECHANISMS II Displacement analysis 89

the different classes. The criteria to be satisfied will now be derived, with particular reference to the crank and rocker mechanism.

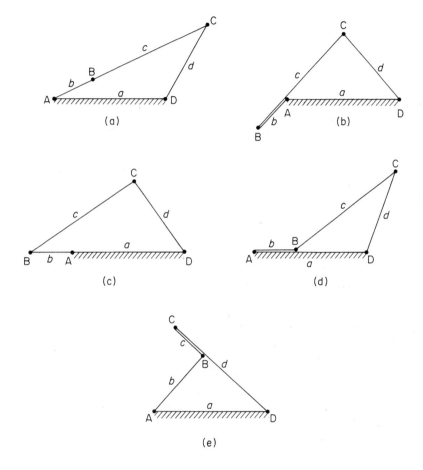

FIGURE 2.53

First, we note that for it to be possible to connect all the hinges no single member may be longer than the total length of the other three members. In the nomenclature of fig. 2.53 this requirement leads to the following inequalities:

$$a < b + c + d \quad \text{(i)}$$
$$b < c + d + a \quad \text{(ii)}$$
$$c < d + a + b \quad \text{(iii)}$$
$$d < a + b + c \quad \text{(iv)}$$

Let us assume that AB is the crank, and that CD is the rocker. Figures 2.53(a) and (b) shows the rocker in its two extreme positions. The existence of two such positions is determined by the two inequalities

$$b + c < a + d \qquad \text{(v)}$$

and
$$a < (c - b) + d$$

or
$$a + b < c + d \qquad \text{(vi)}$$

It will be noted that the last condition ensures that crank b can rotate through its extreme outer position, fig. 2.53(c). Finally, if crank b is to rotate through its inner extreme position as depicted in fig. 2.53(d), then

$$c > b + d - a \qquad \text{(vii)}$$

or
$$b + d < a + c \qquad \text{(viii)}$$

The same criterion may be deduced by arguing that b cannot revolve completely if the configuration of fig. 2.53(e) is possible.

The above inequalities can all be depicted graphically; consider first condition (iv)

$$d < a + b + c$$

Figure 2.54 shows the relationship

$$d = a + b + c$$

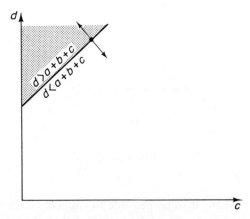

FIGURE 2.54

plotted with respect to c, d axes. The equation is represented by a straight line at 45 degrees to the d-axis, and cutting it at the point $d = a + b$. For any point in the c, d plane to the left of this line

$$d > a + b + c$$

and for any point to the right

$$d < a + b + c$$

As it is the latter inequality which must be satisfied the area to the left of the line is shaded to indicate that it is excluded from further consideration. By applying similar reasoning to the other criteria we conclude that for a mechanism to be a crank and rocker, with AB as crank, it must be represented by a point which lies in the central unshaded strip in fig. 2.55. No mechanisms

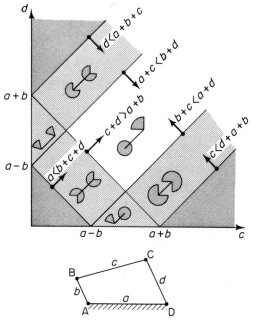

FIGURE 2.55

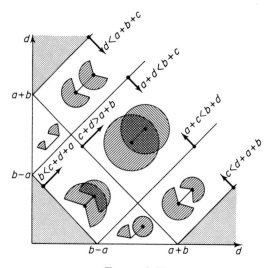

FIGURE 2.56

92 Dynamics of Mechanical Systems

are possible in the heavily shaded regions, and the small diagrams in each of the lightly shaded regions show the type of mechanism represented by points therein. It may be noted that a crank and rocker for which CD is the rocker is represented by the triangular region adjacent to the c-axis.

The drag link mechanism is automatically excluded from fig. 2.55 because it is assumed therein that $b < a$, whereas in the drag link mechanism it is essential for b to be greater than a. The corresponding diagram for $b > a$ is given in fig. 2.56. The two diagrams are coalesced into a single working diagram in fig. 2.57.

For satisfactory operation a particular crank and rocker mechanism must be represented by a point which is well inside the boundaries of the permissible

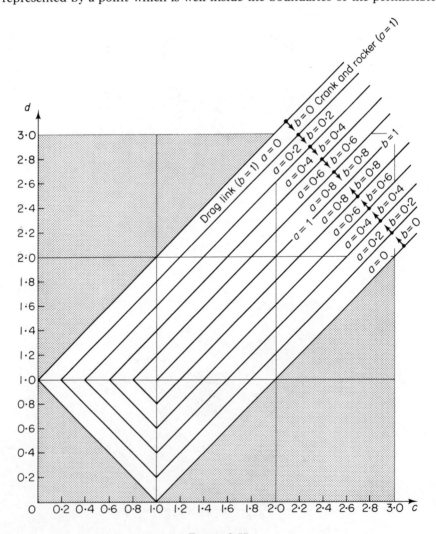

FIGURE 2.57

region. This statement is necessarily vague because by unsatisfactory operation we mean a tendency to jam as a result of friction, and we have as yet made no formal investigation into friction effects. The effect of friction will be considered in the next chapter, where a modified form of fig. 2.57 is developed.

REFERENCES

1. BEGGS, J. S. *Mechanism*, McGraw-Hill, 1955.
2. MERRITT, H. E. *Gears*, Pitman, 1942.
3. HUNT, K. H. *Mechanisms and Motion*, E.U.P., 1959.
4. SVOBODA, A. *Computing Mechanisms and Linkages*, McGraw-Hill, 1948.
5. FREUDENSTEIN, F. 'Approximate Synthesis of Four-bar Linkages', *Trans. A.S.M.E.* **77**, pp. 853–61, 1955.
6. MORRISON, J. L. M. and CROSSLAND, B. *An Introduction to Mechanics of Machines*, Longmans, 1964.
7. BEYER, R. *The Kinematic Synthesis of Mechanisms*, Chapman & Hall, 1963.
8. HIRSCHHORN, J. *Kinematics and Dynamics of Plane Mechanisms*, McGraw-Hill, 1962.
9. ROBERTS, S. 'On Three-bar Motion in Plane Space', *Proc. London Math. Soc.*, 1875, p. 14. See also: Cayley, A. 'On Three-bar Motion', *ibid.*, p. 136.
10. HUNT, K. H., FINK, N., and NAGER, J. 'Linkage Geneva Mechanisms: A Design Study in Mechanism Geometry', *Proc. I.M.E.* **174**, No. 21, 1960.
11. HRONES, J. A. and NELSON, G. L. *Analysis of a Four-bar Linkage*, Wiley, 1951.

3

FORCE RELATIONSHIPS IN MECHANISMS
I Transmitted forces and friction effects

3.1 Introduction
The point has been made in the last chapter that a mechanism which has been designed to meet a kinematic specification will not necessarily operate successfully. It may tend to jam due to friction, the forces in the members may turn out to be greater than what can readily be coped with, and in the case of high-speed mechanisms there may be intolerable inertia effects. In this chapter, we will study the methods of calculating the forces which act on the members of a mechanism as a result of the transmitted power, taking into account friction effects. Inertia forces will be considered in a later chapter.

3.2 Forces in mechanisms without friction
Simple statics
The most direct way of obtaining the forces which act in the various members of a mechanism given the force (or torque) applied to the input member, is to trace the progress of the load through the mechanism, establishing the conditions of equilibrium for each member. It is necessary that the reader should dismiss from his mind any thought that this is the same problem as that of analysing a pin-jointed framework of the type commonly considered in the theory of structures. The essential feature of pin-jointed structures is that the members are considered to act only in tension or compression; with mechanisms this is not the case, so that whilst it is possible to show the relevant forces unambiguously on a line diagram of a complete pin-jointed framework, as in fig. 3.1(a), it is not so easy to do this for a mechanism where the members are subjected to shearing loads as well as end loads, such as that in fig. 3.1(b). The best technique is to sketch the mechanism in 'exploded' form as in fig. 3.1(c), and to show the forces which act on each member, rather than, as in fig. 3.1(a), the forces which act at each joint.

The problem is to know where to start: clearly we can draw a triangle of forces to determine the relationship between the gas force P, the cylinder reaction R, and the connecting-rod force F which act on the gudgeon-pin which connects the piston and connecting-rod, provided we know the direction of F. This can be determined only by considering the equilibrium of the connecting-rod. Since the only forces which act on the connecting-rod are the

FORCE RELATIONSHIPS IN MECHANISMS I 95

forces at each end those forces must be equal and opposite to each other, that is to say F must lie along the axis of AB. The triangle of forces at A can now be drawn and F determined. The crank BC is in equilibrium under the force F applied at B in the direction AB, a bearing reaction at C equal and parallel to the force at B, and a torque $T = F.d$ applied to the crankshaft where d is the perpendicular distance between the force at B and the reaction at C.

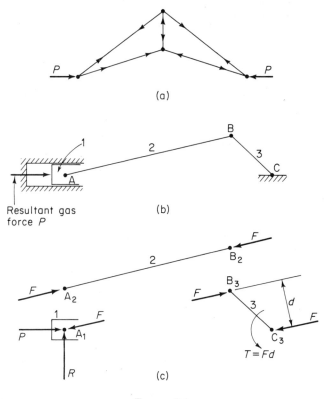

FIGURE 3.1

It will be noted that as a matter of convenience the members have been numbered so that in the exploded diagram point A becomes A_1 and A_2. The identification of a member by a number as well as the letters at its ends is found to be a convenience later on.

Other examples of this type of analysis will appear later in different contexts. It is sufficient for the moment to note that the method of determining the forces in a mechanism is the most direct and, if the forces are required in all the members of a mechanism, it is generally the most convenient method to use. If, on the other hand, one is interested in determining the force only in a particular member it is necessary to have a method which circumvents the task of working right through the mechanism.

96 Dynamics of Mechanical Systems

Virtual work

The principle of virtual work states that, if a system is in equilibrium under a set of forces, no net work is done in a small displacement of the system from its equilibrium position.

It follows that if the piston in fig. 3.1(b) is displaced a small amount Δx to the right, so causing a small clockwise rotation $\Delta\theta$ of the crank, the work $P\,\Delta x$ done by the force P at the piston must equal the work $T\,\Delta\theta$ required to overcome the torque T at the crank. Thus

$$P\,\Delta x - T\,\Delta\theta = 0 \tag{3.1}$$

The external reactions F and R do no work and, in this displacement, the internal forces do no net work. For example, the work done by the force F acting at B_2 is counterbalanced by the work required to overcome the equal and opposite force at B_3.

The relationship between Δx and $\Delta\theta$ may be obtained by graphical construction, fig. 3.2, or by calculation, though calculation is only rarely easy or advisable, even for mechanisms as simple as that in fig. 3.1.

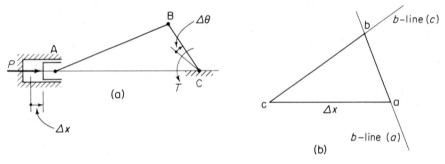

FIGURE 3.2

We use the fact that displacement is a vector quantity, i.e., it has magnitude and direction, and may be combined with other displacements according to the parallelogram rule. The displacement of the piston A relative to the fixed cranks centre C is Δx to the right, and may be represented to a suitable scale by line ca in fig. 3.2(b). The point a is to the right of c because the displacement of A relative to c is to the right. A corollary of Euler's theorem (page 66) is that any displacement of a rigid body is equivalent to a translation of any designated point in the body plus a rotation about that point. If we regard A as the designated point on the rigid body AB, the displacement of B is the displacement of A plus an additional displacement due to the rotation of AB. This additional displacement must, as AB is rigid, be perpendicular to AB. So returning to the displacement diagram, fig. 3.2(b), we can add a line through a perpendicular to AB and label it *b-line (a)*; the displacement of b is represented by some point on this line, as yet unknown because we do not know the angle $\Delta\phi_{AB}$ through which AB has rotated. When the displacement diagram has been completed we can measure ab and then determine $\Delta\phi_{AB} = ab/AB$. For the moment we must seek some other way of locating b, and we turn our attention to BC. As C is a fixed point the displacement of B relative to C must be perpendicular to BC, accordingly we add the line labelled *b-line (c)*, the point where this intersects the other *b*-line locates b. It merely remains to measure cb and so determine $\Delta\theta = bc/BC$.

In this example we have applied the method of virtual work in a rather special way in that the displacements used have been real displacements, and so eqn. (3.1) is a real work equation. The point is only worth mentioning, with the dismissal that it is a mere playing with words, because some students worry about it. Nevertheless the essence of the method of virtual work is that it is not restricted to real displacements.

Let us suppose that we wish to determine the reaction R in fig. 3.1 in terms of the crank torque T. We can do this without introducing the force P by imagining that the constraints on the piston are such as to allow a small displacement Δy (a virtual displacement) of the piston in the direction perpendicular to P. The only work done by the external forces at the piston is then $R\,\Delta y$. As before the forces F do no net work and the equation of virtual work is $R\,\Delta y - T\,\Delta\theta = 0$, where $\Delta\theta$ is now the rotation of the crank due to the virtual displacement Δy of the piston. The relationship between Δy and $\Delta\theta$ is determined graphically by drawing a displacement diagram, fig. 3.3.

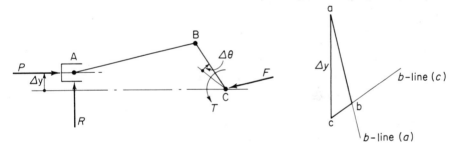

FIGURE 3.3

Internal forces are calculated by considering a virtual distortion of the relevant member. In the case of the connecting-rod force F we imagine that the connecting-rod extends by an amount Δz, say. If we wish to obtain F in terms of the piston force P we prevent crank torque T from doing work by keeping the crank fixed, and allowing the piston to move. To obtain F in terms of T we fix the piston so that P does no work. The displacement diagrams for both these cases are given in fig. 3.4. It should be noted that the relative displacement of the two ends A and B of the connecting-rod is now the sum of two vectors, a displacement perpendicular to AB due to its rotation, as before, together with the displacement Δz parallel to AB.

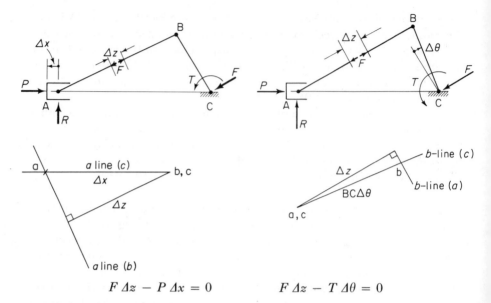

FIGURE 3.4

3.3 Friction

Of all the physical properties which we have to use in calculating the behaviour of mechanical systems the coefficient of friction is probably the one which can be specified with the least amount of precision. Fortunately the friction forces which operate between properly lubricated surfaces are generally so small that they can be neglected, and in those mechanisms which are designed to jam as a result of friction it is usually possible to make generous allowance for variations in the coefficient of friction. In intermediate cases, where friction forces are not negligible in relation to other forces but are insufficient to cause jamming, it must be recognized that it is difficult to arrive at anything more than a reasonable estimate of the forces involved.

In order to define the *coefficient of friction* we consider a small body of weight W resting on a horizontal plane and subjected to a force P_n applied normal to the plane. The force applied to the body by the plane, that is, the reaction R_n of the plane, is clearly normal to the surface and given by

$$R_n = P_n + W$$

If a vanishingly small force P_t is now applied to the body in a direction tangential to the plane, the body will be held in equilibrium by the development of an equal and opposite force R_t due to friction between the plane and the body. Up to the point where the body starts to slip, increasing the applied tangential force gives rise to a corresponding increase in the friction force. The coefficient of static friction μ is defined as the value of limiting value of the friction force R_t, just before slipping occurs, divided by the normal reaction R_n.

As slipping starts, the friction force falls so that the friction force and the applied tangential force are no longer in equilibrium and the body accelerates. The sliding friction force divided by the reaction is known as the coefficient of sliding friction.

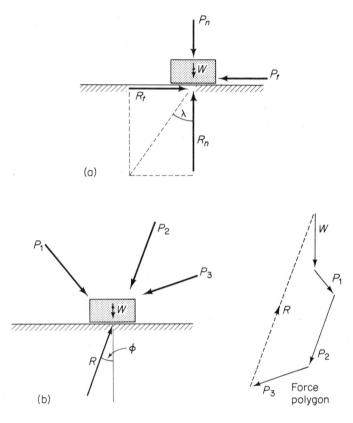

FIGURE 3.5

The angle λ between the resultant R of R_n and R_t and the direction normal to the surface is given by the expression

$$\tan \lambda = \frac{R_t}{R_n} = \mu$$

If a number of forces P_1, P_2, etc. act on the body, as in fig. 3.5(b), and the plane reaction required to put the block in equilibrium is R, the necessary and sufficient condition for slip not to occur is that the angle ϕ between R and the normal to the surface must be less than λ. Slipping will start if ϕ is greater than $\tan^{-1} \mu_{\text{static}}$, and once started will continue if ϕ is greater than $\tan^{-1} \mu_{\text{sliding}}$.

This reasoning applies whether the surfaces are flat or curved at the point of contact. The plain journal bearing (fig. 3.6) is a particular case which deserves individual mention. In fig. 3.6, clearance between the spindle and the bearing has been exaggerated to show the point of contact clearly. If the spindle were subjected simply to a load P acting through the centre O of the spindle, the point of contact between the spindle and the bearing would be at B. With the application of a torque T, in the direction shown, the spindle would roll to the right. Provided that T were not too large the spindle would take up the

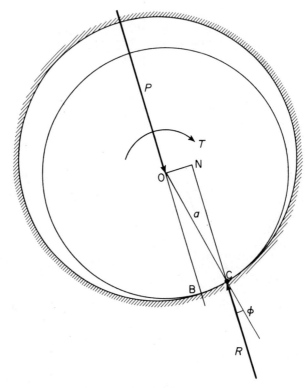

FIGURE 3.6

equilibrium position shown with the point of contact at C. As the load P, the bearing reaction R and the torque T are the only forces which act in the spindle, R must be equal to P and have its line of action CN parallel to the direction P. If ON is the common perpendicular between P and R, the torque T must equal $P \cdot ON$. As T is increased, ON increases proportionally, until the angle ϕ between the reaction R and the common normal OC to the surfaces at the point of contact equals the friction angle λ. Any attempt to increase T further causes slip at the point of contact and the spindle to accelerate. In the limiting case $ON = a \sin \lambda$, where a is the radius of the spindle. If the coefficient of friction is small, λ is small, and we can conveniently use the approximation

$$\sin \lambda \approx \tan \lambda = \mu$$

FORCE RELATIONSHIPS IN MECHANISMS I 101

and so obtain
$$ON = \mu a$$
$$T = \mu a P$$

There are two ways of looking at the forces which act on the spindle. Either, we can consider that rotation of a spindle which sustains a load P requires the application of a torque $\mu a P$ to overcome friction, or we can combine P and T and say, that, in order to overcome friction, the line of action of the force P applied to a spindle must be displaced by an amount μa. This displacement naturally brings P into the position where it passes through C to counterbalance R directly. Since the distance ON is fixed it is convenient to think of P as being tangential to a circle with its centre at O and of radius μa, or more precisely $a \sin \lambda$. This circle is known as the *friction circle*.

To see how the concept of the friction circle is used, let us consider the bell-crank lever in fig. 3.7 where the horizontal force P supports the weight W.

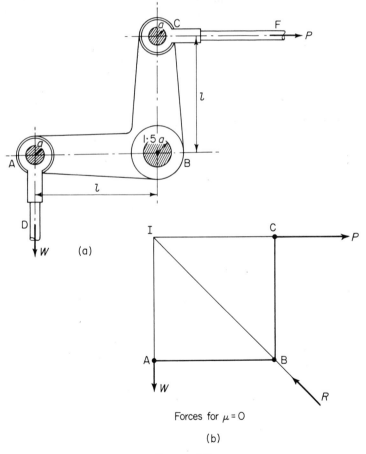

FIGURE 3.7

102 Dynamics of Mechanical Systems

The arms AB and BC of the bell-crank lever are each of length l between bearing centres and are at right angles to each other. The bearing radius is a at A and C and $1{\cdot}5a$ at B. We will now determine the limiting values of P:

(i) to raise W,
(ii) to prevent W from falling,

assuming that the coefficient of friction has the same value μ at each of the bearings. It is wise first to consider the forces which operate if μ is zero.

As only three forces are applied to the bell-crank lever, P, W and the reaction R at B, the reaction R must pass through the point of intersection I of P and W. By drawing the triangle of forces, or by taking moments and resolving forces, we find that

$$W = P \quad \text{and} \quad R = P\sqrt{2}$$

These results hold whether W is being raised or lowered.

With friction, the system of forces is changed because P, W, and R no longer act through the centres of the respective bearings, but move through lines of action so as to be tangential to the relevant friction circles. The problem is to determine, for each case, which way the force shifts. The general line of reasoning is as follows: friction always tends to prevent relative motion of the contacting elements, so that with the introduction of friction the force applied by one element to another must move its line of action in such a way that it tends to oppose the relative motion. We will now apply this argument to each of the bearings B, A, and C in turn for the case when W is being raised.

Figure 3.8(a) shows the bell-crank lever displaced slightly in the clockwise direction, so raising W. Now let us consider the effect of friction on the reaction R at B. Relative to the fixed bearing at B the bell-crank lever moves in the clockwise direction. Due to friction, the force applied by the bearing to the lever must shift so as to oppose this motion of the lever, i.e., it must move so as to produce an anticlockwise torque. It therefore shifts as indicated in fig. 3.8(b), where the friction circles are shown much exaggerated in size.

At A, the displacement of the lever relative to member AD is clockwise, consequently the force moves its line of action so as to apply an anticlockwise torque to the lever, and at C likewise.

The forces P, W, and R must intersect at a single point I', as shown in fig. 3.8(c). In order to determine the relationship between P and W we may take moments about any convenient point on the line of action of R; we will choose the point H where R intersects AB produced. It is not possible to derive an exact expression for BH in terms of μ, l, and a, and even if these quantities are given numerical values, the calculation of BH is a fairly difficult exercise. However, provided that μ is small the radii of the friction circles will be small compared with the length l of the arms of the bell-crank lever, and an approximate value for BH can be obtained without much difficulty. It must be recognized that the size of friction circles have been much exaggerated in fig. 3.8(c), and that in reality, with small μ, I' is located very close to the point I shown in fig. 3.7(b), and H is really very close to B. Hence the triangle JBH is, for all practical purposes, similar to triangle IAB. It follows that BH ≈ 1·5 $\mu a \sqrt{2}$ so that, taking moments about H, we have

$$P(l - \mu a) \approx W(l + 1\cdot 5\mu a\sqrt{2} + \mu a)$$

Approximating again

$$P \approx W(1 + 1\cdot 5\sqrt{2}\,\mu a/l + \mu a/l)(1 + \mu a/l)$$

and again

$$P \approx W[1 + (1\cdot 5\sqrt{2} + 2)\mu a/l]$$

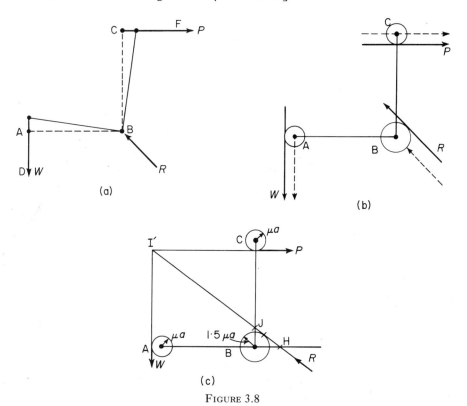

FIGURE 3.8

If W is lowered, the direction of rotation of the bell-crank lever is reversed and all the forces move the opposite sides of their friction circles. It is left as an exercise to the reader to show that, in this case,

$$P \approx W[1 - (1\cdot 5\sqrt{2} + 2)\mu a/l]$$

3.4 Use of virtual work with friction

There are basically two ways in which the principle of virtual work can be applied in determining the force relationship in a mechanism when friction is present, but both are subject to certain limitations.

One technique which may be used is to give the mechanism a set of virtual displacements in which the friction forces do no work. By way of illustration let us consider the situation depicted in fig. 3.9, where a ladder stands on rough ground with its top against a rough vertical wall. The problem is to determine

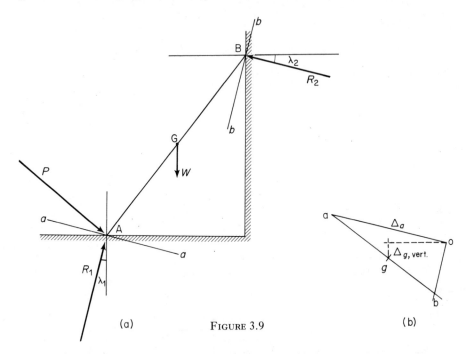

FIGURE 3.9

the least force P needed to prevent the foot of the ladder slipping away from the wall. The ladder is assumed to be in a vertical plane.

If the ladder is on the point of slipping away from the wall, the reactions R_1 and R_2 at the foot and top respectively of the ladder will be inclined at their appropriate angles of friction λ_1 and λ_2 to the normals at the surfaces. The ladder is in equilibrium under the action of the forces W, P, R_1 and R_2. We can determine P without also calculating R_1 and R_2 if we allow the ladder to have a vertical displacement in which R_1 and R_2 do no work. In such a displacement, the foot of the ladder moves along aa which is perpendicular to R_1, and the top moves along bb which is perpendicular to R_2. A displacement diagram can be drawn for an arbitrary displacement $\varDelta_a$ of the foot of the ladder along aa to determine the consequent movement of the centre of gravity G of the ladder. The equation of virtual work is then

$$W \times \text{(vertical displacement of G)} - \varDelta_a \times \text{(component of } P \text{ along } aa\text{)} = 0$$

Clearly P has its least possible value when it is directed along aa, so that

$$P_{\min} = W \frac{\varDelta_{g,\,\text{vert.}}}{\varDelta_a}$$

If it appears from the displacement diagram that G rises in the arbitrary displacement then P is negative, so that instead of a force being required to prevent slip one is needed to promote it.

The method just used has only limited application to mechanisms. The limitations may be seen in relation to the slider-crank mechanism, fig. 3.10. If

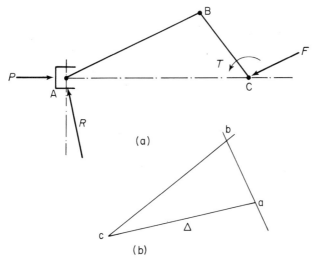

FIGURE 3.10

friction occurs between the piston and cylinder we can derive the relationship between the piston force and crank torque by allowing a virtual displacement of the piston $\varDelta$ perpendicular to the reaction R, which is as shown in fig. 3.10(a) when P is overcoming T. Figure 3.10(b) shows the displacement diagram. We cannot, however, use the same method to allow for friction at the hinge P, for example, because to prevent the friction torque doing work we must consider a displacement in which there is no rotation of AB relative to BC. Given this constraint it is no longer possible to find a set of virtual displacements in which only P and T do work. The same argument holds for hinge A. We can, however, allow for friction at the main crank bearing C by allowing a virtual rotation of the crank about the actual point of contact between the crankshaft and its bearing, instead of one about its centre.

The crux of the problem as just described lies in finding a set of virtual displacements in which only one unknown force does work. If there are too many unknown forces it may be impossible to eliminate them all simultaneously from the equation of virtual work. One way out of this difficulty is to relax somewhat, and be content with using approximate values for the forces in question. To illustrate this approach we will use the example of the bell-crank lever which we studied earlier; the relevant diagrams are repeated in fig. 3.11, which shows the mechanism, the forces when there is no friction, and the displacement diagram for a small rotation about hinge B with P raising the load W.

The effect of friction is to displace the lines of action of the forces from the positions shown in fig. 3.11(b) and consequently to change their values somewhat. If the friction effects are small, the changes in the magnitudes of the forces will be small, so that to a fair degree of approximation we can say that

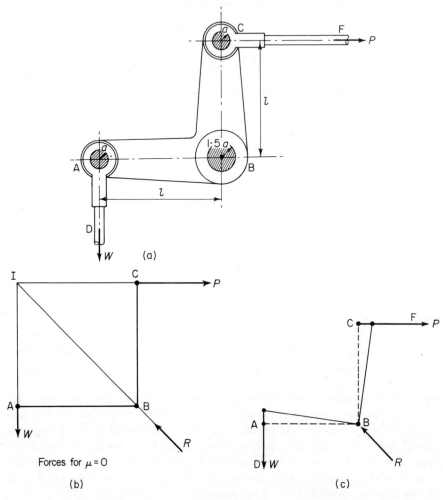

FIGURE 3.11

the friction torque generated at a joint is the force which operates when there is no friction, multiplied by the radius of the friction circle. This torque multiplied by the relative rotation at the joint gives the energy dissipated. The forces in the problem are P at C, $R = P\sqrt{2}$ at B, and $W = P$ at A. The relative rotation happens to be the same $\delta\theta$ at each joint, μ is also the same so the energy dissipated is

$$P\mu a\ \delta\theta \text{ at C}, \quad P\sqrt{2} \times 1{\cdot}5\mu a\ \delta\theta \text{ at B, and } P\mu a\ \delta\theta \text{ at A}$$

The work equation is therefore

$$Wl\,\delta\theta = Pl\,\delta\theta - P\mu a\,\delta\theta - 1\cdot 5\sqrt{2}P a\mu\,\delta\theta - P\mu a\,\delta\theta$$

Hence
$$W \approx P\{1 - (2 + 1\cdot 5\sqrt{2})\mu a/l\}$$

In this equation, the effect of friction has been over-emphasized; due to the friction losses, the hinge forces become progressively less than the assumed values as joints further and further from the input are considered. We have, for example, assumed the torque at A to be $\mu a P$, whereas it has the lesser value $\mu a W$. If we now express all the hinge forces in terms of W rather than P we shall underplay the friction effects. The work equation becomes

$$l\,\delta\theta \approx Pl\,\delta\theta - W\mu a\,\delta\theta - 1\cdot 5\sqrt{2}W a\mu\,\delta\theta - W\mu a\,\delta\theta$$

Hence
$$W \approx P\{1 + (2 + 1\cdot 5\sqrt{2})\mu a/l\}^{-1}$$

By first over-emphasizing and then under-emphasizing friction we have established lower and upper bounds on the value of W, i.e.,

$$P\{1 - (2 + 1\cdot 5\sqrt{2})\mu a/l\} < W < P\{1 + (2 + 1\cdot 5\sqrt{2})\mu a/l\}^{-1}$$

The closeness of the two limits depends upon the value of the factor $\mu a/l$. It will be noted that to a first degree approximation in $\mu a/l$ the two limits are equal to each other, and that the relationship between P and W is then the same as we obtained previously.

If friction effects are not small this method leads to a pair of limits which are so wide apart as to be useless. A further point to notice is that the calculation does not reveal the conditions under which the mechanism may jam. Provided that these limitations are accepted, this method is of general validity and is proved formally in the appendix to this chapter.

3.5 Friction effects in particular mechanisms
Cams

As it is often difficult to ensure good lubrication between a cam and its follower, and between the follower and its guides, it is important to make generous allowance for the effects of friction in the design of cams.

The forces which act at the tip of a knife-edged follower, fig. 3.12, may be resolved into P along the axis of the follower and Q perpendicular to it. If, in the given configuration, the follower is rising, the driving force R from the cam is inclined at the friction angle λ to the common normal at the point of contact in the direction shown. On resolving forces we find

$$Q = P \sec(\phi - \lambda)$$

and
$$R = P \operatorname{cosec}(\phi - \lambda)$$

The importance of keeping ϕ as large as possible is clear. If ϕ is allowed to decrease until it approximates to λ the total reaction R and the transverse

108 Dynamics of Mechanical Systems

force Q both become large compared with the transmitted force P. As a result, the contact force between the cam and follower, the bending moment on the follower, and the contact forces between the follower and its guides, all become very much larger than would be expected if friction were neglected. The possibility of the mechanism jamming is also to be considered.

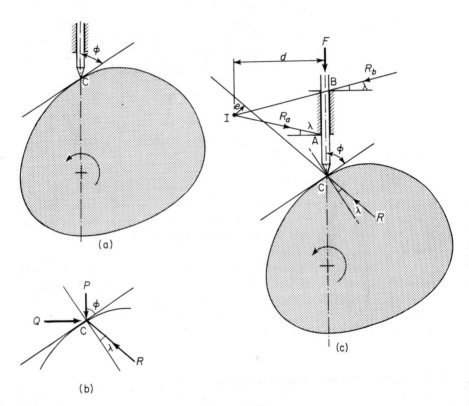

FIGURE 3.12

The set of forces which acts on the follower as it is lifted by the cam is shown in fig. 3.12, with the reaction at each point of contact inclined at λ to the common normal. The reactions R_A and R_B between the follower and its guides intersect at I, so that the relationship between the contact force R and the output force R from the follower obtained by taking moments about I is $Fd = Re$. As ϕ increases, the offset e decreases, and may become zero. When this happens, the output force F is zero and the mechanism has jammed. The remedy for this situation is: (a) to reduce ϕ by increasing the radius of the base circle of the cam; (b) to move point I to the left by increasing the length AB of the guide; or (c) to replace the knife-edged follower by a roller follower, and thereby cause the reaction between the follower and the cam to be directed more closely along the normal at the point of contact.

FORCE RELATIONSHIPS IN MECHANISMS I 109

Slider-crank mechanism

The tendency for the piston of a single-cylinder engine to stick in the dead-centre position is well known. Figure 3.13 shows the system of forces operating when the crank is a little way past the dead-centre position, in fact in the

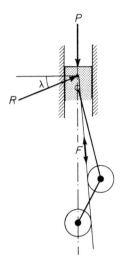

FIGURE 3.13

position where P is just on the point of being able to overcome the friction effects. If the crank is closer to the dead-centre position the lines of thrust for the connecting-rod will cut one or other of the friction circles at the ends of the crank. At whichever bearing this occurs the angle between the normal and the reaction at the point of contact will be less than the friction angle, so that there can be no slip.

Four-bar mechanism

The four-bar mechanism tends to jam at certain configurations just as a slider-crank mechanism tends to jam near the dead-centre position. In order to determine the configurations in which jamming might be expected to occur we will examine the changes which occur in the pattern of forces as the driving crank AB, fig. 3.14, makes one complete revolution. Starting with the position shown in fig. 3.14(a), the forces change in the following manner:

1 In fig. 3.14(a), where a driving torque Q_i is applied at A to overcome the output torque Q_o applied at D, the force in the connecting-rod BC is compressive. (The reader should verify that each of the forces is drawn tangentially to the correct side of each of the friction circles, the direction of rotation of AB being clockwise.)

2 As AB is moved through the position in which it is in line with BC, the input torque drops to zero and then picks up again. Assuming that the

110 Dynamics of Mechanical Systems

direction of Q_o is always to oppose the rotation of CD it reverses as the direction of rotation of CD reverses and the force in the connecting-rod changes from compression to tension. As the directions of the relative motions at joints B and C do not change at this stage, the line of action of the force in the connecting-rod changes to the position shown in fig. 3.14(b), so that at each joint the torque continues to resist the relative motion.

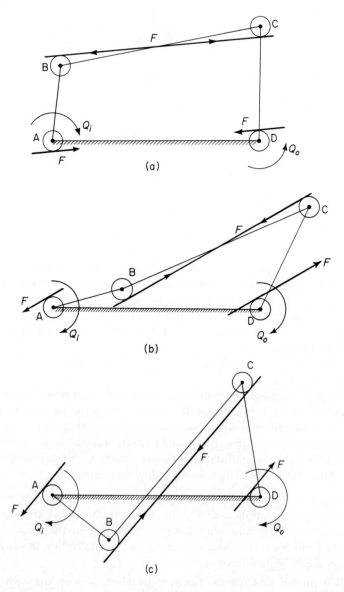

FIGURE 3.14

FORCE RELATIONSHIPS IN MECHANISMS I 111

3 In fig. 3.14(c), the angle BCD has clearly increased from what it was in fig. 3.14(b), so that at some intermediate stage the direction of rotation of BC relative to CD must have reversed, with the consequent change in the line of action of the force at C from one side of the friction circle to the other.

The reader is invited to trace the subsequent course of events as AB rotates on round to the initial position in fig. 3.14(a).

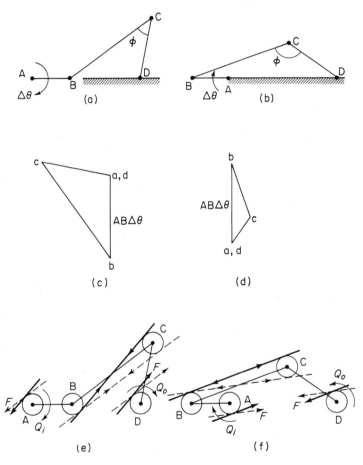

FIGURE 3.15

The tendency for the mechanism to jam as a result of friction is greatest at the instants when the direction of rotation of link BC relative to CD changes. This happens when crank AB is colinear with the fixed link AD, as in figs. 3.15(a) and 3.15(b). The proof of this statement follows from a consideration of the displacement diagrams for a small rotation $\Delta\theta$ of the crank AB from the given position, figs. 3.15(c) and 3.15(d) respectively. The angle through which

BC rotates is bc/BC, and the angle through which CD rotates is cd/CD, anti-clockwise for the configuration of fig. 3.15(a), and clockwise for that in fig. 3.15(b). The change in the angle ϕ is in each case

$$\Delta\phi = bc/BC - cd/CD$$

In the displacement diagram, ab is perpendicular to AB (i.e., bd is perpendicular to BD, bc is perpendicular to BC, and cd is perpendicular to CD. It follows that the triangles dbc and DBC are similar, that $\Delta\phi = 0$, and hence that $d\phi/d\theta = 0$. The angles ϕ_1 and ϕ_2 in figs. 3.15(a) and 3.15(b) respectively are therefore the minimum and maximum values of ϕ. The changes, as in the pattern of the forces at these instants, are shown for the two cases in figs. 3.15(e) and 3.15(f). In both cases, the forces indicated by full lines obtain just before the instant considered and dotted line just after.

If the links are so proportional that in the extreme positions BC lies very close to AD, with ϕ_1 tending to zero, and/or ϕ_2 tending to 180 degrees, the possibility has to be considered that the line of thrust in BC which is shown dotted may intersect the friction circle at D. If this happens, the mechanism jams.

The minimum angle, ϕ_{min}, between the axis of the connecting-rod BC and the rocker CD is known as the *transmission angle*, and it is generally accepted that for satisfactory performance in relation to the tendency to jam the transmission angle or its supplement should not be less than 40 degrees for low-speed mechanisms. For high-speed mechanisms, which we have defined earlier as mechanisms for which the inertia forces are significant, there is probably no simple criterion. Using the nomenclature of the last chapter and putting $AB = b$, $BC = c$, $CD = d$, and $AD = a$, the following relationships must hold if $40° \leqslant \phi \leqslant 140°$,

$$(a - b)^2 \geqslant c^2 + d^2 - 2cd \cos 40$$
$$(a + b)^2 \leqslant c^2 + d^2 + 2cd \cos 40$$

The reader may verify that the same inequalities must hold for the drag link mechanisms.

In chapter 2 we saw that if AB is the driving crank a four-bar mechanism will be either a crank and rocker mechanism ($b < a$), or a drag link mechanism ($b > a$), provided that the length of the links are such as to define a point in the unshaded rectangular region of the chart in fig. 3.16. A crank and rocker mechanism is also defined by a point in the unshaded triangular region, but in this case with CD as the crank. We will assume, however, that in the case of a crank and rocker mechanism AB is always the crank, and concentrate on the unshaded rectangular region of the chart in fig. 3.16. For all points on the boundary of this region the transmission angle ϕ is zero and unless steps are taken to assist the mechanism through its dead-centre position jamming is inevitable. Friction will cause jamming for mechanisms in the unshaded

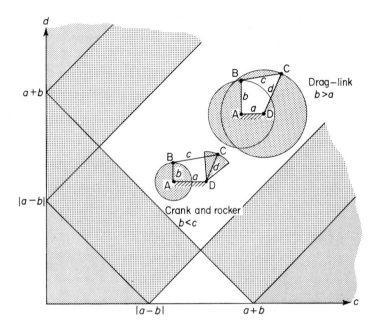

FIGURE 3.16

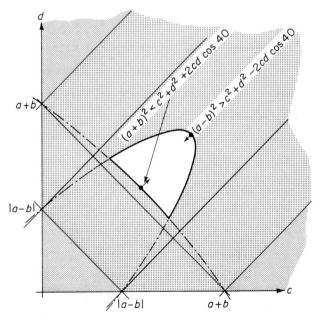

FIGURE 3.17

region if the points which define them are too close to the boundary. If we accept 40 degrees as being the minimum allowable transmission angle the boundaries of the permissible regions are defined by the equations

$$c^2 + d^2 - 2cd \cos 40 = (a - b)^2$$
$$c^2 + d^2 + 2cd \cos 40 = (a + b)^2$$

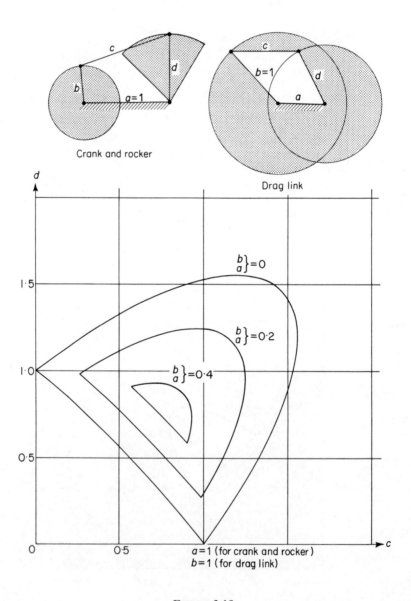

FIGURE 3.18

These are the equations to two ellipses shown in fig. 3.17, and the permissible region is the unshaded area between them. A chart suitable for design purposes can be constructed as in fig. 3.18. In the case of the crank and rocker mechanism we take $a = 1$, and the given contours are for the stated values of b. For a drag link mechanism a and b are interchanged.

APPENDIX 3.1

Calculation of upper and lower bounds on the efficiency of a mechanism with friction

We will consider a mechanism with a single input and a single output. Let the displacement of the input member and the force applied to it be Δx_i and P_i, and let the corresponding quantities for the output member be Δx_o and P_o, the displacements being small.

If there is no friction the work equation gives

$$P_i \cdot \Delta x_i = P_o \cdot \Delta x_o$$

Hence $$P_o = kP_i$$

where $$k = \Delta x_i / \Delta x_o$$

At a typical joint in the mechanism there is, at the point of contact between the two members,

1 a normal force $\quad P_r = \alpha P_i = \alpha P_o / k$

2 a tangential relative displacement

$$\Delta x_r = \beta\, \Delta x_i = \beta k\, \Delta x_o \tag{A3.1}$$

where α and β are constants which depend on the configuration of the mechanism.

(In the case of a hinge of radius a_r, the displacement is $\Delta x_r = a_r\, \Delta\theta_r$, where $\Delta\theta_r$ is the relative angular displacement of the two members.)

Assume now, that friction is introduced and that the same input force is applied. The normal force P'_r at the typical joint will be less than it was before the introduction of friction and the output force P'_o will be relatively even less, so that

$$P'_r < \alpha P_i \tag{A3.2}$$

and $$\alpha P'_o / k < P'_r \tag{A3.3}$$

The work equation is now

$$P_i \cdot \Delta x_i - \sum \mu_r P'_r \cdot \Delta x_r = P'_o \cdot \Delta x_o \tag{A3.4}$$

but the original displacement relationship

$$\Delta x_r = \beta\, \Delta x_i = \beta k\, \Delta x_o$$

still holds.

On substituting for P'_r in eqn. (A3.4) using the inequality (A3.2), and eliminating Δx_r by means of eqn. (A3.1), we find

$$P_i \Delta x_i - \sum \mu_r \alpha P_i \beta \Delta x_i < P'_o \Delta x_o$$

and hence
$$(1 - \sum \mu_r \alpha \beta) P_i \Delta x_i < P'_o \Delta x_o \qquad (A3.5)$$

Similarly, we have

$$P_i \Delta x_i > \left(P'_o + \mu_r \frac{\alpha}{k} P'_o k \beta \Delta x_o \right)$$

and
$$P_i \Delta x_i > (1 + \sum \mu_r \alpha \beta) P'_o \Delta x_o \qquad (A3.6)$$

Equations (A3.5) and (A3.6) yield respectively lower and upper bounds on the efficiency $\eta = P'_o \Delta x_o / P_i \Delta x_i$:

$$(1 - \sum \mu_r \alpha \beta) < \eta < (1 + \sum \mu_r \alpha \beta)^{-1} \qquad (A3.7)$$

4

VELOCITIES AND ACCELERATIONS

4.1 Introduction
In high-speed mechanisms, the inertia forces due to the accelerations of the moving parts may be comparable with, or even much greater than, the forces due to transmitted power. The two main effects of inertia force are, first, a change in the relationship between the input and output forces; and second, additional stresses in the members of the mechanism. A machine which operates quite satisfactorily at low speeds may literally shake itself to pieces if run at high speed.

We have already observed in earlier chapters that it is rarely possible to deduce simple analytical expressions for the displacements in linkage mechanisms. If a displacement equation is complicated, differentiation to obtain velocity and acceleration equations is likely to lead to extremely cumbersome expressions. To avoid this difficulty, we will employ extensions of the vector diagram method which we have hitherto used in displacement analysis. A few exceptional mechanisms for which algebraic analysis is feasible will be dealt with later.

4.2 Vector representation of velocity and acceleration
A vector is a quantity which has magnitude and direction and which can be compounded in accordance with the parallelogram rule. The movement of a particle from O to A, fig. 4.1 is defined both in direction and magnitude by the

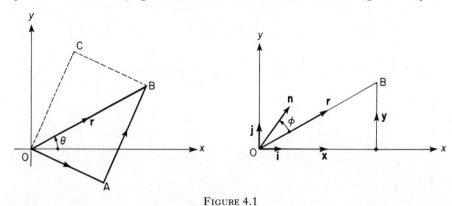

FIGURE 4.1

line OA. A subsequent movement to B is defined by the line AB. The resultant displacement is represented in magnitude and direction by the line OB. But OB is a diagonal of the parallelogram OABC, so that the two separate displacements OA, AB, can be said to be combined by the parallelogram rule to give the resultant displacement OB. Displacements then have all the properties of a vector quantity. The position of a point measured with respect to an origin is defined by the vector which represents the displacement of a particle from the origin to that point so that OB can be said to be position vector **r** of the point B with respect to the origin O.

Assuming for a moment that **r** lies in the place of the paper, its magnitude and direction may be specified by the length OB and the angle θ which OB makes with a datum. Such a specification is obviously directly in accord with the definition of a vector quantity as a quantity which has both magnitude and direction. The additional requirement, that a vector quantity is one which may be compounded according to the parallelogram law, allows an alternative specification to be made; for, if a quantity can be compounded according to the parallelogram law it may, conversely, be resolved into components, and we can then specify the vector in terms of its components. The choice of directions into which we make the resolution is quite arbitrary, but usually it is most convenient to resolve in the directions of cartesian axes Oxy so that

$$\mathbf{r} = \mathbf{x} + \mathbf{y}$$

where **x** and **y** are vectors in the x and y directions. If we define **i** and **j** as vectors of unit length, generally referred to simply as *unit vectors*, in the x and y directions respectively, we can write

$$\mathbf{r} = x\mathbf{i} + y\mathbf{j}$$

where x and y are *scalar* quantities which specify the magnitudes of the components in the direction of the unit vectors. If we discard the assumption that **r** lies in the only plane Oxy, i.e., the plane of the paper, we have

$$\mathbf{r} = x\mathbf{i} + y\mathbf{j} + z\mathbf{k} \tag{4.1}$$

where **k** is a unit vector in the Oz direction.

The component of a vector in a direction other than one of the coordinate directions is obtained by resolution. Thus the component of **r** in the direction of the unit vector **n** is $r \cos \phi . \mathbf{n}$. In vector terminology the magnitude $r \cos \phi$ of this component is referred to in the *scalar product* of **r** and **n** and is written as **r.n** the order of the two terms being of no significance, so that **r.n** = **n.r**. This is a rather special example of a scalar product in that one of the vectors is a unit vector; in general, the scalar product of two vectors **a** and **b** is defined as $ab \cos \phi$, where a and b are their magnitudes, and ϕ is the angle between them. If two vectors are parallel their scalar product is simply the product of their magnitudes; the scalar product of two parallel and equal vectors is the square of their magnitude, e.g., $\mathbf{a}.\mathbf{a} = a^2$. The scalar product of two vectors which are perpendicular to each other is zero.

The following results should be noted:

$$\left.\begin{array}{l}\mathbf{i}.\mathbf{i} = \mathbf{j}.\mathbf{j} = \mathbf{k}.\mathbf{k} = 1 \\ \mathbf{i}.\mathbf{j} = \mathbf{j}.\mathbf{k} = \mathbf{k}.\mathbf{i} = 0\end{array}\right\} \quad (4.2)$$

If $\quad \mathbf{a} = (a_x\mathbf{i} + a_y\mathbf{j} + a_z\mathbf{k})$ and $\mathbf{b} = (b_x\mathbf{i} + b_y\mathbf{j} + b_z\mathbf{k})$

we have $\quad \mathbf{a}.\mathbf{b} = a_xb_x + a_yb_y + a_zb_z \quad (4.3)$

As it is sometimes comforting to verify the obvious we will note in passing that putting $\mathbf{a} = \mathbf{b} = \mathbf{r}$, as defined in eqn. (4.1), we have

$$\mathbf{r}.\mathbf{r} = r^2 = x^2 + y^2 + z^2$$

The velocity of B, fig. 4.1, is the rate of change of position of B and is obtained by differentiating eqn. (4.1) with respect to time. Provided that axes $Oxyz$ are fixed in direction, $\mathbf{i}$, $\mathbf{j}$, and $\mathbf{k}$ do not change with time, so that $d\mathbf{i}/dt = 0$, etc. Hence

$$\frac{d\mathbf{r}}{dt} = \frac{dx}{dt}\mathbf{i} + \frac{dy}{dt}\mathbf{j} + \frac{dz}{dt}\mathbf{k}$$

$$= v_x\mathbf{i} + v_y\mathbf{j} + v_z\mathbf{k} \quad (4.4)$$

It is immediately clear that velocity is a vector quantity because all the terms on the right-hand side are vector quantities; v_x, v_y, and v_z are the components of the velocity $\mathbf{v} = d\mathbf{r}/dt$ in the coordinate directions.

A second differentiation with respect to time shows that acceleration is, likewise, a vector quantity.

$$\frac{d\mathbf{v}}{dt} = \frac{d^2\mathbf{r}}{dt^2} = \frac{d^2x}{dt^2}\mathbf{i} + \frac{d^2y}{dt^2}\mathbf{j} + \frac{d^2z}{dt^2}\mathbf{k}$$

$$= \frac{dv_x}{dt}\mathbf{i} + \frac{dv_y}{dt}\mathbf{j} + \frac{dv_z}{dt}\mathbf{k}$$

$$= a_x\mathbf{i} + a_y\mathbf{j} + a_z\mathbf{k} \quad (4.5)$$

Superficially it may appear that eqns. (4.4) and (4.5) provide a simple and convenient way of determining velocities and accelerations. In fact, a direct application of these equations often leads to results which obscure rather than reveal the essentials features of the motion. Let us take as a simple example the case where the point B, fig. 4.2, moves with constant speed along a circular

VELOCITIES AND ACCELERATIONS

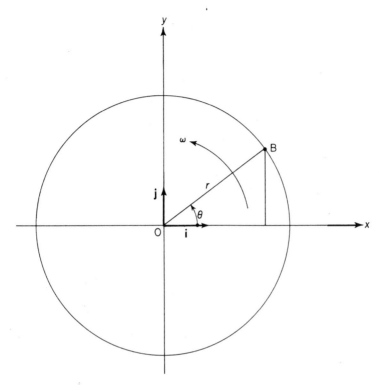

FIGURE 4.2

path in the Oxy plane with O as centre, i.e., r is constant and $d\theta/dt$ has a constant value ω. In vector form, the position of B is defined by

$$\mathbf{r} = r\cos\theta\,\mathbf{i} + r\sin\theta\,\mathbf{j}$$

If we differentiate with respect to time the velocity is

$$\mathbf{v} \equiv \frac{d\mathbf{r}}{dt} = -r\sin\theta\frac{d\theta}{dt}\mathbf{i} + r\cos\theta\frac{d\theta}{dt}\mathbf{j}$$

$$= -r\omega\sin\theta\,\mathbf{i} + r\omega\cos\theta\,\mathbf{j}$$

Differentiating again

$$\mathbf{a} = \frac{d^2\mathbf{r}}{dt^2} = -r\omega^2\cos\theta\,\mathbf{i} - r\omega^2\sin\theta\,\mathbf{j}$$

These equations are surely an unnecessarily elaborate way of saying that the velocity is $r\omega$ along the tangent at B and that the acceleration is $r\omega^2$ towards the centre O. We can obtain expressions for the velocity and acceleration which conjure up immediately the correct physical picture if we use a slightly more sophisticated approach.

122 Dynamics of Mechanical Systems

We will define two unit vectors, **i** along OB and rotating with it, and **j** perpendicular to it is shown in fig. 4.3, so that

$$\mathbf{r} = r\mathbf{i}$$

To obtain the velocity of B we differentiate and have

$$\frac{d\mathbf{r}}{dt} = \frac{dr}{dt}\mathbf{i} + r\frac{d\mathbf{i}}{dt} \tag{4.6}$$

As r is constant, $dr/dt = 0$. The unit vector **i** is (by definition) constant in length, but as it is changing in direction, $d\mathbf{i}/dt \neq 0$.

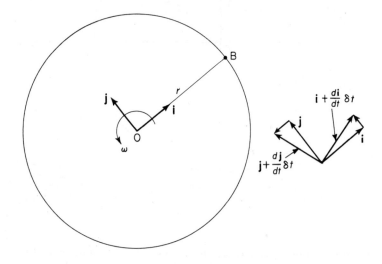

FIGURE 4.3

In a small time interval δt the vectors **i** and **j** rotate through an angle $\omega\,\delta t$. The unit vector **i** is thus changed by an amount $\omega\,\delta t$ perpendicular to **i**, i.e., in the **j** direction, so that

$$\frac{d\mathbf{i}}{dt}\delta t = \omega\,\delta t\,\mathbf{j}$$

and

$$\frac{d\mathbf{i}}{dt} = \omega\mathbf{j}$$

In the case of the **j** vector, a change of sign is involved as can be seen by inspection of fig. 4.3, so that

$$\frac{d\mathbf{j}}{dt} = -\omega\mathbf{i}$$

Substituting for $d\mathbf{i}/dt$ in eqn. (4.6) we have

$$\frac{d\mathbf{r}}{dt} = r\omega\,\mathbf{j}$$

that is the velocity of B is of magnitude $r\omega$ and is directed along the tangent at B.

The acceleration of B is

$$\frac{d^2\mathbf{r}}{dt^2} = r\omega\,\frac{d\mathbf{j}}{dt}$$

$$= -r\omega^2\mathbf{i}$$

that is, it is of magnitude $r\omega^2$ and is directed in the radial direction from B towards O.

The description which we now have of the motion is very much clearer than the one which we derived first. Instead of projecting its motion directly on to a fixed background (the fixed Oxy set of axes) we have viewed B against a moving system of axes (the rotating $O\mathbf{ij}$ axes), and taken separate account of the effect of the motion of the moving system.

4.3 Relative velocity (plane motion)

When we state that the velocity of a moving point A is defined by the vector $\mathbf{v_A}$ we imply the existence of a frame of reference against which the magnitude of $\mathbf{v_A}$ and its direction can be measured. The velocity is then relative to the frame of reference. If the frame of reference is fixed $\mathbf{v_A}$ is the absolute velocity of A. The need for the existence of a frame of reference, whether or not formal reference is made to it, gets us into some difficulty when more than one moving point is involved: the difference between the absolute velocities of two points is generally referred to as the velocity of one point relative to the other, thus the velocity of B relative to A is defined as

$$_A\mathbf{v_B} = \mathbf{v_B} - \mathbf{v_A} \tag{4.7}$$

and the velocity of A relative to B as

$$_B\mathbf{v_A} = \mathbf{v_A} - \mathbf{v_B} = -{_A\mathbf{v_B}}$$

Strictly, it is inconsistent to refer to the velocity of one point relative to another, because a point by itself does not have the sense of direction which is fundamental to a vector quantity. Be this as it may, it will be convenient to ignore this inconsistency. In general, we will wish to determine $\mathbf{v_B}$ from a knowledge of $\mathbf{v_A}$ and $_A\mathbf{v_B}$ so that eqn. (4.7) is generally written as

$$\underset{\text{(velocity of B)}}{\mathbf{v_B}} = \underset{\text{(velocity of A)}}{\mathbf{v_A}} + \underset{\text{(velocity of B relative to A)}}{_A\mathbf{v_B}} \tag{4.8}$$

Relative motion of points of a rigid body

Before eqn. (4.8) can be applied to the analysis of the motion of mechanisms, it is necessary to have a physical interpretation of the term relative velocity when it is applied to two points which are fixed to a rigid body.

We will deal first with the special, but nevertheless very common, case of plane motion.

Figure 4.4 shows two points A and B fixed to a plane body which rotates with angular velocity ω about an axis perpendicular to its plane. The position

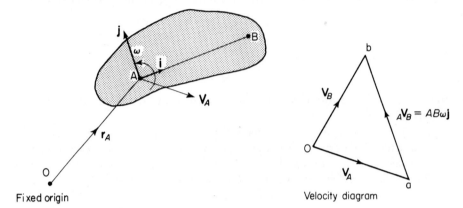

FIGURE 4.4

of A is defined relative to the fixed origin O by the vector $\mathbf{r}_A$. The position of B relative to A is defined by the vector $AB\mathbf{i}$, where $\mathbf{i}$ is a unit vector in the AB direction. The position of B relative to the origin is thus given by

$$\mathbf{r}_B = \mathbf{r}_A + AB\mathbf{i}$$

To obtain the velocity of B we differentiate this expression with respect to time. Since A and B are fixed to the moving plane the length AB is constant and the vector $\mathbf{i}$ rotates with the plane at a rate ω. It follows that

$$\frac{d\mathbf{r}_B}{dt} = \frac{d\mathbf{r}_A}{dt} + AB\frac{d\mathbf{i}}{dt}$$

or
$$\mathbf{v}_B = \mathbf{v}_A + AB\omega\mathbf{j} \tag{4.9}$$

where $\mathbf{j}$ is a unit vector perpendicular to $\mathbf{i}$ as shown in fig. 4.6.

Comparing eqn. (4.9) with eqn. (4.8) we see that

$$AB\omega\mathbf{j} = {}_A\mathbf{v}_B = \text{velocity of B relative to A} \tag{4.10}$$

The velocity of B relative to A is therefore perpendicular to AB and arises solely because of the angular velocity ω. Any relative velocity in the direction AB would imply a change in the length AB and this is precluded because the body to which A and B are fixed is rigid.

If in a specific example $\mathbf{v}_A$ and ω are known, eqn. (4.10) may be evaluated to determine $\mathbf{v}_B$ by drawing to scale a velocity diagram as in fig. 4.4.

As an example of the application of this analysis let us consider the problem of determining the angular velocity of crank CD of the mechanism in fig. 4.5, given the angular velocity of the crank AB. The velocity of B is perpendicular to AB and of magnitude $AB\omega$. As A is a fixed point we can conveniently think of it as the origin relative to which all velocities will be measured, and the

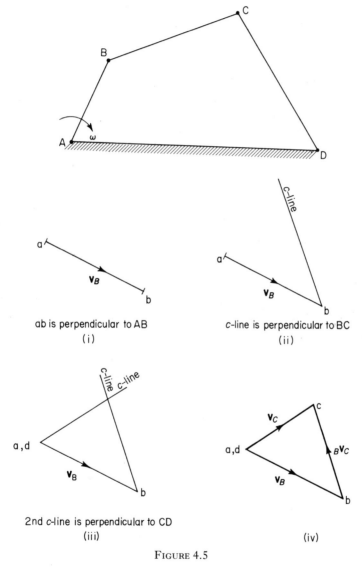

FIGURE 4.5

velocity of B is therefore represented by the vector *ab* drawn to a convenient scale in fig. 4.5(i).

The velocity of C is the velocity of B plus the velocity of C relative to B. The velocity of C relative to B is $BC\omega_{BC}$ perpendicular to BC where ω_{BC} is the angular velocity of BC (*note*: it does not matter whether we call this ω_{BC} or

126 Dynamics of Mechanical Systems

ω_{CB}). Point C in the velocity diagram must therefore lie on a line through B and perpendicular to BC. The position of C on this line is unknown because ω_{BC} is unknown, so we simply attach the label '*c-line*' to this vector as in fig. 4.5(ii), and look for some other means of fixing the point C: we consider the motion of C relative to D.

First we must observe that D is a fixed point, like A. As two fixed points can have no relative velocity A and D are represented in the velocity diagram by a single point, fig. 4.5(iii). The velocity of C relative to D is unknown in magnitude but must be in a direction perpendicular to CD. We can therefore add a second *c-line* to the velocity diagram. Point c is located at the intersection of the two *c-lines*. The vectors representing $\mathbf{v_C}$ and $_B\mathbf{v_C}$ can now be lined in, and the completed velocity diagram appears as in fig. 4.5(iv).

The magnitude of $\mathbf{v_C}$ is now obtained by measurement from the diagram and, as $v_C = CD\omega_{CD}$, the angular velocity of the rocker can be calculated. If it is required, the angular velocity of the connecting-rod can be obtained from $_Bv_C = BC\omega_{BC}$. To obtain the sense of this rotation we note that the velocity of C relative to B is in a direction which is roughly from the bottom to the top of the page. Thus, when C is viewed from point B it appears to move towards the top of the paper and hence in an anticlockwise sense about B. We conclude that ω_{BC} is in the anticlockwise sense.

Absolute velocity of a point whose motion is known relative to a moving body
We will now consider how to obtain the absolute velocity of a point which moves in a specified way relative to a body which is itself in motion. In fig. 4.6, the small block B slides in a slot in the moving plane P. We can determine the absolute velocity of the block if we know the velocity of some fixed point such as A on the plane, the angular velocity ω of the plane, the position of the block in relation to A, and the speed with which it slides along the slot. Understanding of the analysis is helped if we imagine a set of axes $Ax'y'$ drawn on the plane P. The directions of these axes are quite arbitrary, they merely provide a background against which the motion of the block relative to the plane can be recognized by virtue of change in the $x'y'$ coordinates. The position of B relative to O is therefore

$$\mathbf{r_B} = \mathbf{r_A} + {}_A\mathbf{r_B}$$
$$= \mathbf{r_A} + x'\mathbf{i} + y'\mathbf{j}$$

Differentiating with respect to time

$$\mathbf{v_B} = \frac{d\mathbf{r_B}}{dt} = \frac{d\mathbf{r_A}}{dt} + \frac{dx'}{dt}\mathbf{i} + x'\frac{d\mathbf{i}}{dt} + \frac{dy'}{dt}\mathbf{j} + y'\frac{d\mathbf{j}}{dt}$$

Now $\quad \dfrac{d\mathbf{i}}{dt} = \omega\mathbf{j}\quad$ and $\quad\dfrac{d\mathbf{j}}{dt} = -\omega\mathbf{i}$

so that $\quad \mathbf{v_B} = \mathbf{v_A} + \omega(x'\mathbf{j} - y'\mathbf{i}) + \left(\dfrac{dx'}{dt}\mathbf{i} + \dfrac{dy'}{dt}\mathbf{j}\right) \quad$ (4.11)

The second of the terms on the right-hand side of this equation becomes recognizable if we express it in terms of a different pair of unit vectors, **n** in the direction AB and **t** perpendicular to AB is shown in fig. 4.6(b). The magnitudes

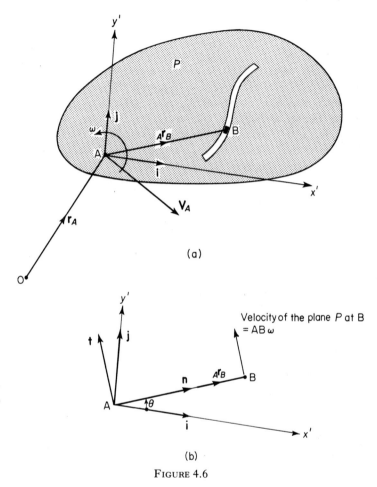

FIGURE 4.6

of the components of **i** in the **n** and **t** directions are $1.\cos\theta$ and $-1.\sin\theta$ respectively, hence

$$\mathbf{i} = \cos\theta.\mathbf{n} - \sin\theta.\mathbf{t}$$

Similarly,

$$\mathbf{j} = \sin\theta.\mathbf{n} + \cos\theta.\mathbf{t}$$

The coordinates $x'y'$ of B are

$$x' = AB\cos\theta$$
$$y' = AB\sin\theta$$

128 Dynamics of Mechanical Systems

Isolating the second term on the right-hand side of eqn. (4.11), we have on substituting for **i**, **j**, x', and y'

$$\omega(x'\mathbf{j} - y'\mathbf{i}) = \omega AB(\sin\theta\cos\theta.\mathbf{n} + \cos^2\theta.\mathbf{t})$$
$$- \omega AB(\sin\theta\cos\theta.\mathbf{n} - \sin^2\theta.\mathbf{t})$$
$$= \omega AB\mathbf{t}$$

But this is the velocity relative to A of a point fixed to the plane and coincident with B. Thus the first two terms on the right-hand side give the velocity of the plane at the point where B is situated. The third term we recognize as the rate of change of the position of B relative to the $Ax'y'$ axes or, more generally as these axes are arbitrary, the velocity of B relative to the plane. The direction of this velocity must be tangential to the centre line of the slot at B. Equation (4.11) therefore becomes

$$\mathbf{v_B} \quad = \quad \mathbf{v_A} \quad + \quad \omega AB\mathbf{t} \quad + \quad _P\mathbf{v_B} \qquad (4.12)$$

| velocity of B | velocity of A | velocity relative to A of the plane at B | velocity of B relative to plane |

velocity of plane at B

where **t** is a unit vector in the direction perpendicular to AB obtained by rotating $_A\mathbf{r_B}$ through 90 degrees in the same sense as ω.

As an example of the application of eqn. (4.12) we will draw the velocity diagram for a cam and follower mechanism, fig. 4.7. The tip of the knife-edged

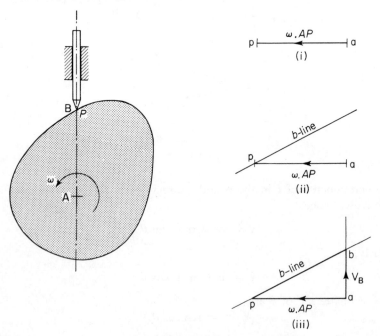

FIGURE 4.7

follower is labelled B and the point of contact on the cam is P. The velocity diagram is constructed in three steps as shown in the figure:

1. the velocity of P is ωAP in the direction shown,
2. the velocity of B relative to P must be in the direction of the tangent at P so that we can draw the first *b-line*,
3. the second *b-line* follows from the fact that the velocity of B must be parallel to the follower guides.

Equation 4.12 is a sufficient basis for the construction of the velocity diagram for any plane mechanism, although there are a number of specialized techniques which can be used to advantage in dealing with certain standard situations and particular types of mechanism. These extensions of the basic method are studied in the next section.

4.4 Velocity diagrams for plane mechanisms
The image theorem

Let us suppose that, having drawn the velocity diagram for a mechanism, we wish to find the velocity of a point rigidly attached to one of the links, such as point E in fig. 4.8. ABCD is the four-bar linkage shown in fig. 4.5, for which we have already constructed the velocity diagram reproduced in fig. 8.4(i).† To locate E in the velocity diagram we proceed as shown in figs. 4.8(ii) and (iii). First, we observe that as E is rigidly attached to BC the line BE is inextensible so that the velocity E relative to B must be perpendicular to BE, hence one *e-line*. By a similar reasoning, we determine a second *e-line* perpendicular to EC. The intersection of the two *e-lines* locates *e* in the velocity diagram.

As *cb*, *be*, and *ec*, are perpendicular to CB, BE, and EC respectively, it follows that triangle *bec* in the velocity diagram is similar to triangle BEC in the mechanism. This result is generalized in the *velocity image theorem*:

If $a, b, c, d, e, \ldots$ are points in a velocity diagram which represent the velocities of points A, B, C, D, E, ... on a rigid link, the two figures *abcde*... and ABCDE... are geometrically similar, and in the same sense.

Obviously, it is easier to locate E in fig. 4.8 directly, as we have done, rather than by applying the image theorem. Furthermore, we avoid the possibility of error due to placing *e* on the wrong side of *bc*, i.e., so that whilst BEC is clockwise *bec* is anticlockwise. If, however, E is very close to BC, the direct method would be unsatisfactory and in the limit when E lies on BC inapplicable. In the latter case, the image theorem gives us the ratio BE:EC into which *bc* must be divided to locate *e*.

We can use the converse of the image theorem to determine the location on a moving link of the point which has a specified velocity. If, for example, we wish to determine E on link BC such that it has the velocity represented by *ae*,

† Note that in reproducing the velocity diagram only the labels at the terminal points of the vectors have been retained. While arrow heads and the other labels in fig. 4.5 are helpful in explaining the construction of the diagram, they are not essential.

130 Dynamics of Mechanical Systems

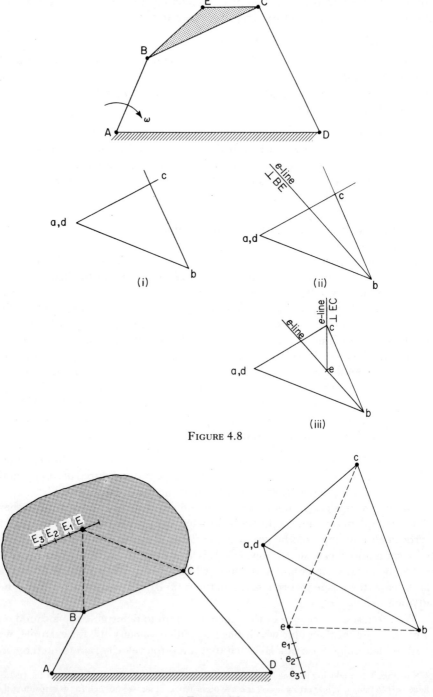

FIGURE 4.8

FIGURE 4.9

VELOCITIES AND ACCELERATIONS 131

fig. 4.9, we first complete the triangle *ecb* and then locate E by drawing lines through B and C perpendicular to *eb* and *cb* respectively.

By varying the magnitude of $\mathbf{v_E}$, but keeping its direction constant, we can construct the locus of the points on the moving plane BC which have their velocity in the specified direction. As $e, e_1, e_2, \ldots$ is a straight line $E, E_1, E_2, \ldots$ must, by the image theorem, also be a straight line.

The instantaneous centre

One point attached to the link BC, fig. 4.10, which is of particular interest is the *instantaneous centre*, which we have already encountered in chapter 2, and is the point I in a moving plane which is instantaneously at rest. As I has zero velocity *i* must be located at the origin of the velocity diagram, fig. 4.10, and is located by drawing perpendiculars to *bi* and *ci* through B and C. As *ab* is perpendicular to AB it follows that I lies on AB produced, and likewise on DC produced. The same conclusion was reached in chapter 2 by considering small displacements of the mechanism.

The instantaneous centre provides a means of determining the velocity relationships of mechanisms without drawing the velocity diagram. Taking, for example, the mechanism of fig. 4.10, we have

$$v_B = \omega_i AB = \omega_{BC} IB$$

hence
$$\omega_{BC} = \omega_i \frac{AB}{IB}$$

Further
$$v_C = v_{BC} IC = \omega_o CD$$

hence
$$\omega_o = \omega_i \frac{AB \cdot IC}{IB \cdot CD}$$

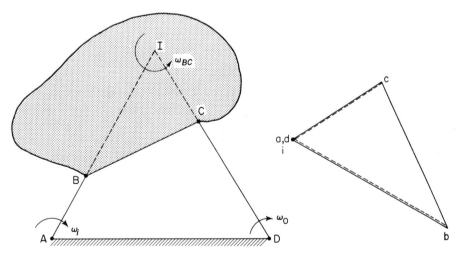

FIGURE 4.10

132 Dynamics of Mechanical Systems

The instantaneous centre may also be of assistance in constructing a velocity diagram. Take, for instance, the mechanism in fig. 4.11; ABCD is a four-bar linkage whose position is controlled by the jack AE. If the velocities of the links are to be determined in terms of a given velocity of the piston in the jack some ingenuity is required in drawing the velocity diagram. The drawing is quite straightforward up to the stage shown in fig. 4.11(i). The piston velocity

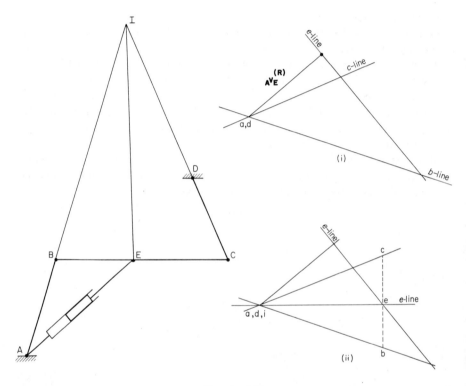

FIGURE 4.11

is represented by the vector $_A\mathbf{v_E}^{(R)}$, where the superscript $(\mathbf{R})$ denotes a component of the velocity of E along the radius AE; there is also a second component of the velocity of E normal to AE because the jack is free to rotate bodily about A. The velocity of E is therefore represented by some unknown point on e-line. The velocities of B and C are likewise known in direction but not in magnitude, hence the b- and c-lines. By the image theorem the velocities of link BEC will be represented by a line bec in the velocity diagram perpendicular to BEC and with e dividing bc in the same proportions as E divides BC. The problem is to determine the location of this line.

Provided that the instantaneous centre I of BC is accessible we can quickly determine the location of e by observing first that since I is stationary it must be coincident with a, d in the velocity diagram, and second that $\mathbf{v_E}$ is perpendicular to IE. These two facts give us a second *e-line*, fig. 4.11(ii), and so the position of e is fixed. Finally, the points b and c are determined by drawing *bec* perpendicular to BEC. A check on the accuracy of drawing is obtained verifying that e divides bc in the same ratio that E divides BC.

A similar method is to start by ignoring the known jack velocity, and assuming instead that crank AB has unit angular velocity. The reader may verify that the drawing of the complete velocity diagram proceeds in a quite straightforward manner; just by dealing with the four-bar mechanism ABCD, locating E by the image theorem, and then completing the diagram with the two components of $_A\mathbf{v_E}$. The resulting jack velocity $_A\mathbf{v_E}^{(R)}$ is measured, and all velocities are then scaled up to give the correct jack velocity.

The main disadvantage of methods based on the use of the instantaneous centre, which applies equally to that just presented, is that they cannot readily be extended to deal with accelerations. Alternative methods of solving the problem just considered, to which this objection does not apply, will be studied in section 4.9 on acceleration. A further difficulty with the use of the instantaneous centre can be visualized by imagining AB and CD in fig. 4.11 to be more nearly parallel so as to make I inaccessible. Obviously, simple auxiliary constructions can be devised to get round this snag but then the main virtue of the method, namely its simplicity and economy in drawing, is lost.

4.5 Velocity relationships in space mechanisms

Plane motion of a body, such as that involved with the mechanisms which we have just studied, is a special case of the rotation of a solid body about a fixed axis, as shown in fig. 4.12. In plane motion we ignore any thickness which the body may have parallel to the axis of rotation and, in effect, consider it to be squashed into a lamina whose plane is perpendicular to the axis or rotation. Any vectors which relate to this motion are then considered to lie in this plane and we close our eyes to the fact that the angular velocity might be represented by a vector whose length is proportional to the magnitude of the angular velocity and is directed along the axis of rotation in accordance with the right-hand screw rule. The use of this fact in the study of three-dimensional motion must depend on whether an angular velocity vector can be manipulated according to the same rules as other vectors which we have used; in particular, whether angular velocities can be added in accordance with the parallelogram rule.

Let a point B fixed in the body be located in relation to a fixed point O on the axis of rotation by the vector $\mathbf{r}$, fig. 4.12, and let N be the foot of the perpendicular from B to the axis of rotation. The velocity of B is then $\omega \cdot \mathrm{BN}$ in a direction perpendicular to both the axis of rotation and BN, that is perpendicular to the triangle OBN and hence perpendicular to $\mathbf{r}$. To sum up, $\mathbf{v_B}$ has magnitude $\omega r \sin \phi$ where ϕ is the angle between $\mathbf{r}$ and $\boldsymbol{\omega}$. The sense of this velocity is positive in the direction for which a rotation from $\boldsymbol{\omega}$ to $\mathbf{r}$ through the angle ϕ is clockwise, fig. 4.4(b). $\mathbf{v_B}$ is said to be the *vector product* (or *cross*

134 Dynamics of Mechanical Systems

product) of $\boldsymbol{\omega}$ and **r**. It is written as $\boldsymbol{\omega} \times \mathbf{r}$ and spoken of as $\boldsymbol{\omega}$ cross **r**, so that we have

$$\mathbf{v_B} = \boldsymbol{\omega} \times \mathbf{r} \qquad (4.13)$$

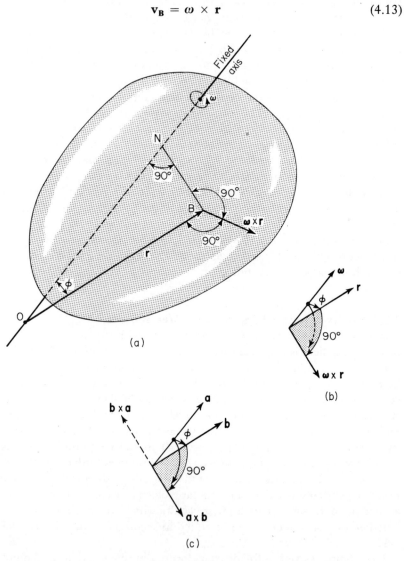

FIGURE 4.12

As the cross product arises in other contexts we will redefine it in general terms. The cross product $\mathbf{a} \times \mathbf{b}$ is a vector of magnitude $ab \sin \phi$, where ϕ is the angle between **a** and **b**, mutually perpendicular to **a** and **b**, and in the direction for which a rotation from **a** to **b** through the angle ϕ is clockwise, fig. 4.12(c).

It follows from this definition that

$$\mathbf{b} \times \mathbf{a} = -\mathbf{a} \times \mathbf{b} \tag{4.14}$$

It is to be noted that if $\mathbf{a}$ and $\mathbf{b}$ are parallel to each other their cross product is zero, because the angle ϕ is zero; in particular, $\mathbf{a} \times \mathbf{a} = 0$. Conversely, it can be concluded that if $\mathbf{a} \times \mathbf{b} = 0$, then $\mathbf{a}$ and $\mathbf{b}$ must be parallel to each other.

In order to show that angular velocity vectors add according to the parallelogram law, let us consider the system in fig. 4.13, whereby a body is given two

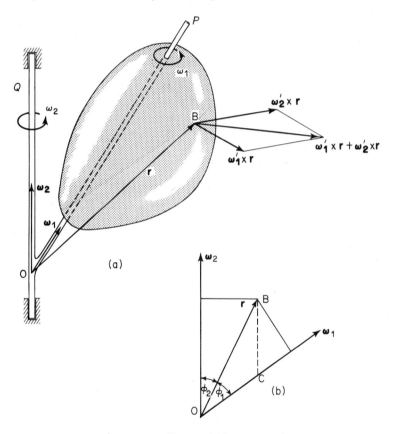

FIGURE 4.13

angular velocities simultaneously, ω_1 about OP and ω_2 about OQ. These two rotations impart independently velocities $\omega_1 \times \mathbf{r}$ and $\omega_2 \times \mathbf{r}$ to a typical point B of the body. The two velocities add in accordance with the parallelogram rule to give the resultant velocity of B. If this velocity happens to be zero then B must lie on the resultant axis of rotation. As $\omega_1 \times \mathbf{r}$ is perpendicular to both ω_1 and $\mathbf{r}$, and as $\omega_2 \times \mathbf{r}$ is perpendicular to both ω_2 and $\mathbf{r}$, it is clear that for $\mathbf{v_b}$ to be zero $\mathbf{r}$ must be coplanar with ω_1 and ω_2, for only then is it possible for $\omega_1 \times \mathbf{r}$ and $\omega_2 \times \mathbf{r}$ to be in the same direction, which is essential if they are to cancel out. Hence ω_1, ω_2, and their resultant $(\omega_1 + \omega_2)$, are coplanar. Given that B is coplanar with ω_1 and ω_2, $\mathbf{v_B}$ is zero if in fig. 4.13(b)

$$\omega_1 r \sin \phi_1 = \omega_2 r \sin \phi_2$$

that is, if

$$\frac{\omega_1}{\omega_2} = \frac{\sin \phi_2}{\sin \phi_1}$$

Let there be drawn from B a line parallel to ω_2 to intersect ω_1 at C, thus making $\angle OBC = \phi_2$. We then have

$$\frac{OC}{OB} = \frac{\sin \phi_2}{\sin \phi_1} = \frac{\omega_1}{\omega_2}$$

so that OC and OB are proportional to ω_1 and ω_2 respectively. It follows that ω_1 and ω_2 add according to the parallelogram rule.

Relative velocities
With a three-dimensional system, eqn. (4.12) becomes

$$\mathbf{v_B} = \mathbf{v_A} + \omega \times {}_A\mathbf{r_B} + {}_p\mathbf{v_B} \tag{4.15}$$

where ${}_p\mathbf{v_B}$ is the velocity of **B** relative to the moving body. In the case of plane mechanisms all the vectors on the right-hand side of this equation lie in the plane of the mechanism and a graphical representation then is simple. With three-dimensional mechanisms the terms on the right-hand side may not be coplanar, and a proper graphical representation requires the projection of the components of the vectors onto two planes. When the relative motion of the points fixed to a rigid link is considered, eqn. (4.15) becomes

$$\mathbf{v_B} = \mathbf{v_A} + \omega \times {}_A\mathbf{r_B} \tag{4.16}$$

The graphical representation of this equation is of course a plane triangle of velocities. Unfortunately, with a space mechanism the plane of this will not, in general, be the same as the plane of the velocity triangles for other links in the mechanism so that we are still forced to employ two projections. Whilst this provides quite a feasible method of analysing the velocity relationships for plane mechanisms,[1] its complexity is such as to make it unsuitable for further study here. We will consider instead a straightforward analytical application of eqn. (4.16).

We will take as our first example the mechanism in fig. 4.14.† Two rods are constrained by guides to move parallel to the axes Ox and Oy as shown. Their ends A and B are connected by means of spherical hinges to the ends of a rigid

---

† This example is taken from reference 1 where it is used to demonstrate the graphical method of determining velocity relationships.

VELOCITIES AND ACCELERATIONS

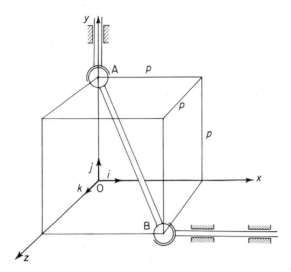

Figure 4.14

push-rod AB. At the instant considered A and B are situated at diagonally opposite corners of a cube with sides of length p. We have therefore

$$\mathbf{r}_A = p\mathbf{j}; \quad \mathbf{r}_B = p\mathbf{i} + p\mathbf{k}$$
$$\mathbf{v}_A = v_A\mathbf{j}; \quad \mathbf{v}_B = v_B\mathbf{i}$$
$$_A\mathbf{r}_B = \mathbf{r}_B - \mathbf{r}_A = p\mathbf{i} + p\mathbf{k} - p\mathbf{j}$$
$$_A\mathbf{v}_B = \mathbf{v}_B - \mathbf{v}_A = v_B\mathbf{i} - v_A\mathbf{j}$$

As AB is inextensible, the velocity of B relative to A must be perpendicular to AB so that

$$_A\mathbf{v}_B \cdot {_A\mathbf{r}_B} = 0$$

or

$$(v_B\mathbf{i} - v_A\mathbf{j}) \cdot (p\mathbf{i} + p\mathbf{k} - p\mathbf{j}) = 0$$

Remembering that $\mathbf{i} \cdot \mathbf{i} = 1$, and $\mathbf{i} \cdot \mathbf{j} = 0$, etc., we have on expanding this product

$$v_B p + v_A p = 0$$

whence

$$v_B = -v_A$$

Knowing $\mathbf{v}_A$ and $\mathbf{v}_B$, the angular velocity ω of the rod is obtained by solving eqn. (4.16), which may be written

$$\omega \times {_A\mathbf{r}_B} = (\mathbf{v}_B - \mathbf{v}_A) = {_A\mathbf{v}_B} \tag{4.17}$$

138 Dynamics of Mechanical Systems

This equation tells us that ω and $_A\mathbf{r}_B$ are both perpendicular to $_A\mathbf{v}_B$. It follows that ω can be expressed in terms of two components, one in the direction $_A\mathbf{r}_B$ and the other in a direction perpendicular to both $_A\mathbf{r}_B$ and $_A\mathbf{v}_B$, i.e., in the direction $_A\mathbf{r}_B \times {}_A\mathbf{v}_B$, so that

$$\omega = \alpha {}_A\mathbf{r}_B + \beta {}_A\mathbf{r}_B \times {}_A\mathbf{v}_B$$

where α and β are constants.

On substituting for ω in eqn. (4.17) we have

$$_A\mathbf{v}_B = \alpha {}_A\mathbf{r}_B \times {}_A\mathbf{r}_B + \beta ({}_A\mathbf{r}_B \times {}_A\mathbf{v}_B) \times {}_A\mathbf{r}_B$$

As the cross product of vector with itself is zero, $_A\mathbf{r}_B \times {}_A\mathbf{r}_B = 0$.

To evaluate $({}_A\mathbf{r}_B \times {}_A\mathbf{v}_B) \times {}_A\mathbf{r}_B$ we refer to fig. 4.15, and note first that as $_A\mathbf{r}_B$ and $_A\mathbf{v}_B$ are mutually perpendicular the magnitude of $({}_A\mathbf{r}_B \times {}_A\mathbf{v}_B)$ is

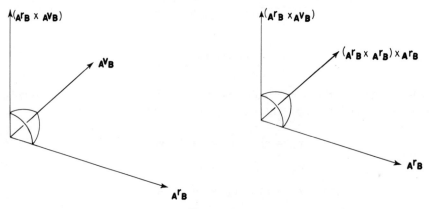

FIGURE 4.15

simply $AB \cdot {}_A v_B$. Again, as $({}_A\mathbf{r}_B \times {}_A\mathbf{v}_B)$ and $_A\mathbf{r}_B$ are perpendicular, the magnitude of $({}_A\mathbf{r}_B \times {}_A\mathbf{v}_B) \times {}_A\mathbf{r}_B$ is $AB^2 {}_A\mathbf{v}_B$ and is in the direction of $_A\mathbf{v}_B$. Thus we have

$$_A\mathbf{v}_B = \beta \cdot AB^2 {}_A\mathbf{v}_B$$

hence
$$\beta = 1/AB^2$$

and
$$\omega = \alpha {}_A\mathbf{r}_B + ({}_A\mathbf{r}_B \times {}_A\mathbf{v}_B)/AB^2 \qquad (4.18)$$

The arbitrary component $\alpha {}_A\mathbf{r}_B$ represents a rotation about the axis of the connecting-rod. In the given configuration the second component is

$$(p\mathbf{i} + p\mathbf{k} - p\mathbf{j}) \times (v_B \mathbf{i} - v_A \mathbf{j})/AB^2$$

On substituting
$$v_B = -v_A, \quad AB^2 = 3p^2$$

and (from fig. 4.14)
$$\mathbf{i} \times \mathbf{j} = \mathbf{k}, \quad \mathbf{j} \times \mathbf{k} = \mathbf{i}, \quad \mathbf{k} \times \mathbf{i} = \mathbf{j}$$

we have
$$\omega = \alpha {}_A\mathbf{r}_B + v_A(\mathbf{i} - \mathbf{j} - 2\mathbf{k})/3p$$

The same type of calculation can be applied to a four-bar mechanism consisting of two cranks which rotate about fixed skew hinges, and a coupler which is attached to the ends of the cranks by spherical hinges, fig. 4.16. We

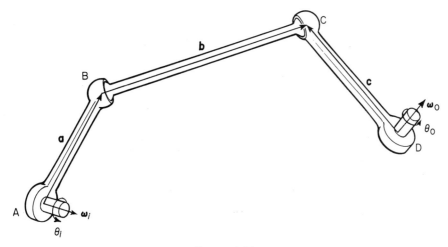

FIGURE 4.16

saw in chapter 1 that such a mechanism has two degrees of freedom, one of which being a rotation of the coupler about its longitudinal axis. There is therefore a one to one relationship between the rotation θ_i of the input crank and the rotation θ_o of the output. A common example of this mechanism is provided by its use to connect the accelerator linkage of a motor car to the carburettor. The relationship between the input and output velocities can be obtained as follows:

$$\mathbf{v}_B = \omega_i \times \mathbf{a}, \quad \mathbf{v}_C = \omega_o \times \mathbf{d}$$

and

$$_B\mathbf{v}_C = \omega_o \times \mathbf{d} - \omega_i \times \mathbf{a}$$

As the velocity of B relative to C must be perpendicular to D

$$(\omega_o \times \mathbf{d} - \omega_i \times \mathbf{a}).\mathbf{b} = 0 \qquad (4.18a)$$

This equation can readily be solved for ω_o.

4.6 Velocity diagrams for epicyclic gear trains

While it is possible to draw conventional velocity diagrams for epicyclic gears little is gained by doing so. Figure 4.17 shows a simple epicyclic gear consisting of a sun gear, driven at an angular velocity ω_S, a planet gear, and annulus driven at an angular velocity ω. If A is the point of contact between the pitch circles of the annulus and the planet, and S is the point of contact between the pitch circles of the planet and sun, a conventional velocity diagram for the mechanism is as shown in fig. 4.17(a). It can be seen that if P is the centre of the planet, $v_A = \frac{1}{2}(v_A + v_S)$, and the angular velocity of the output shaft can readily be calculated, but the diagram is hardly necessary to deduce this result and it does not give a particularly helpful pictorial representation of the velocity relationships.

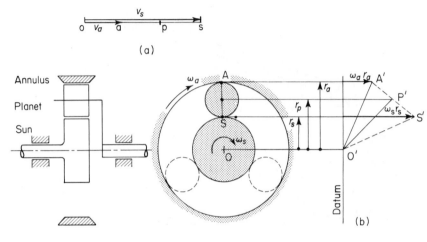

FIGURE 4.17

An alternative form of velocity diagram is given in fig. 4.17(b). This diagram shows the distribution of velocities along the common diameter of the three gears. We start by drawing the vector $\mathbf{v_S}$ of length $\omega_S r_S$ and terminating at S'. The line O'S' gives the distribution of velocities along the diameter of the sun wheel, and its slope relative to the datum is proportional to the angular velocity, ω_S of the sun. O'A' serves the same purpose in respect of the annulus. Since the velocities of the planet wheel at A and S must be the same as $\mathbf{v_A}$ and $\mathbf{v_S}$ respectively, A'S' represents the distribution of velocity along the diameter AS of the planet wheel. Its centre P has velocity $\frac{1}{2}(\omega_A r_A + \omega_S r_S)$ and $\omega_o = \frac{1}{2}(\omega_A r_A + \omega_S r_S)/r_P$.

This method of determining the velocity relationships for epicyclic gear trains is, in general, easier to apply than the tubular method given in an earlier chapter (page 64) and errors are less likely. Furthermore, it provides a greater insight: it is for example immediately clear that, because O'P' lies between O'A' and O'S', the angular velocity of the output shaft must be intermediate between the velocities of the sun and annulus.

It should be noted that with this type of velocity diagram, we can readily determine a general expression for the velocity relationship from a sketch, this is in contrast to the velocity diagrams which we have studied in relation to linkage mechanisms and which, in general, must be drawn accurately to scale. It is possible to draw the velocity diagram for an epicyclic gear to scale but, as well as being unnecessary, it is unsatisfactory if the overall speed ratio is high. This may very well be the case because a most useful feature of epicyclic gear trains is that they provide a compact means of obtaining a high velocity ratio.

VELOCITIES AND ACCELERATIONS 141

Figure 4.18 shows a compound epicyclic gear train where the two compounded planet wheels replace the single wheel of the previous example, and a second sun wheel replaces the fixed annulus. Let us start by assuming that the

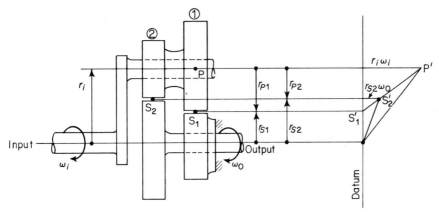

FIGURE 4.18

input shaft rotates with an angular velocity ω_i so that the speed of a point P on the axis of the planet wheels is $\omega_i r_i$. Because sun wheel No. 1 is fixed, the velocity distribution along a vertical diameter of the planet No. 1 is represented in the velocity diagram by $S_1'P'$. However, as the two planet wheels comprise a single body, $S_1'P'$ must also apply to planet No. 2, and S_2' lies on $S_1'P$ as shown. The more nearly that we make r_{S2} equal to r_{S1} the closer S_2' approaches S_1' and the smaller is the speed of the output shaft. Thus, by making the radii of the two sun wheels approximately equal, we can achieve a very considerable reduction in speed between the input and output shafts. As the mechanism appears in fig. 4.18, the input and output shafts turn in the same direction; if r_{S2} is less than r_{S1} the directions are opposite to each other.

We see from the velocity diagram that

$$\frac{r_{S2}\omega_o}{r_i\omega_i} = \frac{r_{S2} - r_{S1}}{r_{P1}}$$

hence
$$\frac{\omega_i}{\omega_o} = \frac{r_{S2} r_{P1}}{r_i(r_{S2} - r_{S1})}$$

$$= \frac{r_{S2} r_{P1}}{r_{S2}(r_{S1} + r_{P1}) - r_{S1}(r_{S2} + r_{P2})}$$

$$= \left(1 - \frac{r_{S1} r_{P2}}{r_{S2} r_{P1}}\right)^{-1}$$

Alternatively, as
$$\frac{r_{S1}}{r_{P1}} = \frac{n_{S1}}{n_{P1}} \quad \text{and} \quad \frac{r_{S2}}{r_{P2}} = \frac{n_{S2}}{n_{P2}}$$

where n denotes number of teeth

$$\frac{\omega_i}{\omega_o} = \left(1 - \frac{n_{S1} n_{P2}}{n_{S2} n_{P1}}\right)^{-1}$$

4.7 Relationship between velocity and displacement diagrams

The reader will perhaps already have noted that the velocity diagram for, say, a four-bar mechanism with a given input velocity ω_i, is identical with the displacement diagram for the same configuration when the input is displaced through an infinitesimal angle $\delta\theta_i$. If we divide all the vectors of the displacement diagram by δt, the time interval during which the displacement takes place, the displacement vectors become mean velocity vectors. In the limit as $\delta\theta_i$ and $\delta\theta_o$ tend to zero the mean values tend to the instantaneous values of the velocities when the input velocity is $\omega_i = \mathrm{d}\theta_i/\mathrm{d}t$. It is perfectly valid to use this argument as the sole basis for velocity diagrams; we have preferred the approach of this chapter because it enables us to take space mechanisms in our stride and also because it is easier to extend to acceleration diagrams.

4.8 Relative accelerations (plane motion)

We will find it convenient to develop our argument in a way similar to that adopted for velocities, dealing first with the relative motion of two points fixed in a body which is moving in a plane, then the absolute motion of a point which has specified motion relative to a body which is itself in plane motion. The more general case of motion in three dimensions will be left to a later section.

Figure 4.19 shows a rigid body whose position relative to a fixed origin is defined by the position vector $\mathbf{r}_A$, the point A fixed in the body, and the direction $\mathbf{i}$ of the line AB in the body, B being a second point fixed in the body. The motion, assumed to be wholly in the plane of the paper, is defined instantaneously by $\mathbf{v}_A$ and the angular velocity ω about an axis perpendicular to the plane of the paper. We have from eqn. (4.9)

$$\mathbf{v}_B = \mathbf{v}_A + \mathrm{AB} \cdot \omega \mathbf{j}$$

This equation is represented graphically by the velocity diagram. On differentiating with respect to time we obtain

$$\frac{\mathrm{d}\mathbf{v}_B}{\mathrm{d}t} = \frac{\mathrm{d}\mathbf{v}_A}{\mathrm{d}t} + \mathrm{AB} \cdot \frac{\mathrm{d}\omega}{\mathrm{d}t} \mathbf{j} + \mathrm{AB} \cdot \omega \frac{\mathrm{d}\mathbf{j}}{\mathrm{d}t}$$

or

$$\underset{\text{acceleration of B}}{\mathbf{a}_B} = \underset{\text{acceleration of A}}{\mathbf{a}_A} + \underset{\substack{\text{tangential acceleration} \\ \text{of B relative to A}}}{\mathrm{AB} \cdot \dot{\omega} \mathbf{j}} - \underset{\substack{\text{radial acceleration} \\ \text{of B relative to A}}}{\mathrm{AB} \cdot \omega^2 \mathbf{i}} \quad (4.19)$$

The acceleration diagram which represents this equation is shown in fig. 4.19. The important thing to notice is that the acceleration of B relative to A has two components, a radial component which is identified by the superscript

VELOCITIES AND ACCELERATIONS

(R), and a tangential component identified by the superscript (T). So that whilst we can say that because A is rigid there can be no velocity of B relative to A in the direction of AB, it does not follow that there is no acceleration of

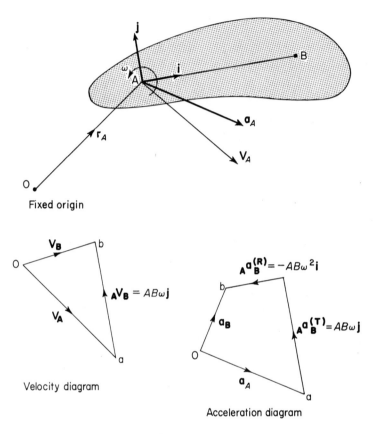

FIGURE 4.19

B relative to A in the direction AB. If A is a fixed point so that it has zero velocity and acceleration, and ω is constant, then

$$\mathbf{a_B} = -AB \cdot \omega^2 \mathbf{i}$$

which we recognize as the familiar centripetal acceleration of a point towards the centre of the circle round which it moves with constant angular velocity. In view of what is to follow it is as well to remind ourselves of the physical basis for this result, which the mathematics has produced so neatly. In fig. 4.20 a particle B moves with constant angular velocity ω about A as centre. At time t_1 the particle is at B, and has a velocity $AB \cdot \omega$ perpendicular to AB, so that the component of B's velocity in the direction AB, is zero. At time t_2 the speed of B is still $AB \cdot \omega$ but is now directed along the tangent at B_2. Since the

144 Dynamics of Mechanical Systems

velocity of B is no longer perpendicular to AB, it has a component in the direction AB. Thus in the time interval t_1 to t_2 the particle has acquired a velocity component in the direction B_1A and hence it has an acceleration in this

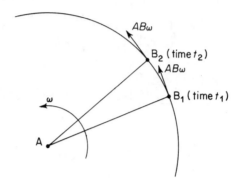

FIGURE 4.20

direction. Coming a little bit closer to the mathematics, we can say that in this example acceleration of B is due to the changing direction of its velocity rather than a change in the magnitude of the velocity.

To illustrate the application of eqn. (4.19) to a particular mechanism we will draw the acceleration diagram for the four-bar mechanism shown in fig. 4.20 where the input crank AB rotates with constant angular velocity ω. This is the same mechanism which we chose to explain the construction of a simple velocity diagram (fig. 4.5), which is re-drawn in fig. 4.21 without further comment. We start the acceleration diagram (i) by drawing the vector $\mathbf{a_B}$ which represents the acceleration of B which is $AB.\omega^2$ in the direction BA. There is no component in the direction perpendicular to BA because the link is rotating with constant angular velocity. The acceleration of C relative to B (ii), however, has two components, a radial component $_B\mathbf{a_C}^{(R)}$ parallel to CB and a tangential component $_B\mathbf{a_C}^{(T)}$ perpendicular to CB. The radial component $_B\mathbf{a_C}^{(R)}$ whose magnitude $_Bv_C^2/BC$ can be calculated because $_Bv_C$ can be determined from the velocity diagram. As we are considering the acceleration of C relative to B, this vector must be drawn from b in the acceleration diagram in the direction CB. The tangential component of $_B\mathbf{a_C}$ is of unknown magnitude so that we draw the c-line perpendicular to $_B\mathbf{a_C}^{(R)}$, and can proceed no further (ii). It may be noted that the point at which $_B\mathbf{a_C}^{(R)}$ meets the c-line does not represent any particular point in the mechanism and so is left unlabelled. To locate c we must now consider the acceleration of C relative to D. This is made up of two components (iii), the known $_D\mathbf{a_C}^{(R)}$ and the unknown $_D\mathbf{a_C}^{(T)}$ which together define the second c-line. The point of intersection of the two c-lines fixes the point c and the acceleration diagram is complete (iv). The angular accelerations of BC and CD are respectively $_Ba_C^{(T)}/BC$ and $_Da_C^{(T)}/CD$, both in the anticlockwise sense.

VELOCITIES AND ACCELERATIONS 145

Unfortunately, eqn. (4.19) is not a sufficient basis for the construction of acceleration diagrams for all types of mechanism. In this particular example we have needed to know the acceleration relationship between two points on a

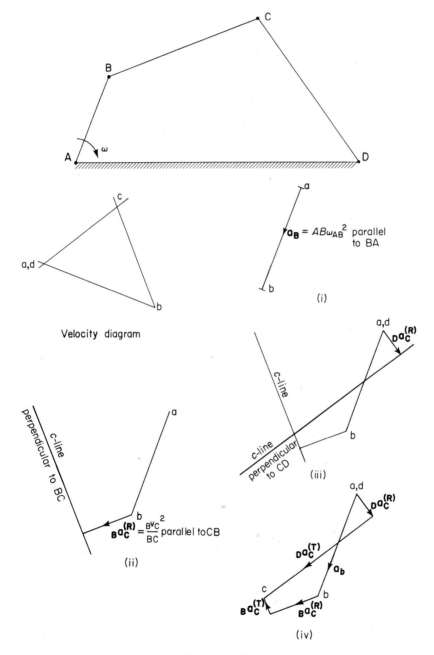

FIGURE 4.21

146 Dynamics of Mechanical Systems

moving body which are a fixed distance apart, and that is all. We shall now have to consider how to obtain the absolute acceleration of a point which moves in a specified way relative to a body which is itself in motion. We have already

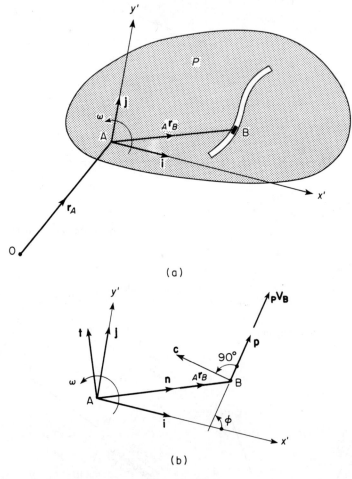

FIGURE 4.22

deduced the velocity relationship in these circumstances; referring to fig. 4.22 we have deduced [eqn. (4.12)] that

$$\mathbf{v_B} = \mathbf{v_A} + \omega AB\mathbf{t} + {_P}\mathbf{v_B}$$

Whilst it is possible to deduce $\mathbf{a_B}$ by direct differentiation of this expression, the argument can be more readily understood by going back a stage to eqn. (4.11) which is

$$\mathbf{v_B} = \mathbf{v_A} + \omega(x'\mathbf{j} - y'\mathbf{i}) + \left(\frac{dx'}{dt}\mathbf{i} + \frac{dy'}{dt}\mathbf{j}\right)$$

On differentiating this equation we find that

$$\mathbf{a_B} = \mathbf{a_A} - \omega^2(x'\mathbf{i} + y'\mathbf{j}) + \frac{d\omega}{dt}(x'\mathbf{j} - y'\mathbf{i})$$

$$+ \left(\frac{d^2x'}{dt^2}\mathbf{i} + \frac{d^2y'}{dt^2}\mathbf{j}\right) + 2\omega\left(\frac{dx'}{dt}\mathbf{j} - \frac{dy'}{dt}\mathbf{i}\right)$$

Now let us define two unit vectors $\mathbf{p}$ in the direction of $_P\mathbf{v_B}$ and $\mathbf{c}$ perpendicular to $\mathbf{p}$ such that the 90 degrees rotation from $\mathbf{p}$ to $\mathbf{c}$ is in the same sense as ω (both anticlockwise in fig. 4.22(b)). If ϕ is the angle between $\mathbf{p}$ and the x'-axis we have

$$\mathbf{i} = \mathbf{p}\cos\phi - \mathbf{c}\sin\phi$$

$$\mathbf{j} = \mathbf{p}\sin\phi + \mathbf{c}\cos\phi$$

and

$$\frac{dx'}{dt} = {}_Pv_B\cos\phi; \quad \frac{dy'}{dt} = {}_Pv_B\sin\phi$$

Substituting these expressions in the last term on the right-hand side of $\mathbf{a_B}$, we have

$$2\omega\left(\frac{dx'}{dt}\mathbf{j} - \frac{dy'}{dt}\mathbf{i}\right) = 2\omega\,{}_Pv_B\mathbf{c}$$

The relationship between the pair of unit vectors $\mathbf{i}$ and $\mathbf{j}$ and the pair $\mathbf{n}$, $\mathbf{t}$ have already been explained in the process of deducing eqn. (4.12). Expressing all the terms in their most convenient form we find

$$\mathbf{a_B} = \mathbf{a_A} - \underbrace{AB\omega^2\mathbf{n} + AB\dot{\omega}\mathbf{t}}_{\text{acceleration relative to}} + \underbrace{{}_P\ddot{\mathbf{r}}_B + 2\omega\,{}_Pv_B\mathbf{c}}_{\text{acceleration of B relative}} \quad (4.20)$$

acceleration of B   acceleration of A   acceleration relative to A of the plane at B   acceleration of B relative to the plane P

If B is fixed to the plane so that $_P\mathbf{v_B}$ and $_P\ddot{\mathbf{r}}_B$ are zero, and if $\mathbf{i}$ and $\mathbf{j}$ are chosen to coincide with $\mathbf{n}$ and $\mathbf{t}$ respectively, eqn. (4.20) is identical to eqn. (4.19). The motion of B relative to the plane is seen to give rise to two components of acceleration. The first of these $_P\ddot{\mathbf{r}}_B$ is often loosely referred to as the acceleration of B relative to the plane. It is the acceleration which would be deduced by an observer riding on the plane using a ruler and stop watch as his means of measurement. If s is the distance which B has moved along its path on the plane then $_P\ddot{\mathbf{r}}_B$ is the vector sum of $\ddot{s}$ tangential to the path and $\dot{s}^2/\rho$ normal to the path and directed towards the centre of curvature, ρ being the radius of curvature of the path at B. Now, if we define the acceleration of B relative to the plane as the vector difference between the absolute acceleration of B and the absolute acceleration of the point on the plane with which B is instantaneously coincident, we must consider the term $2\omega\,{}_Pv_B\mathbf{c}$ to be a component of the relative acceleration. This component is known as the *Coriolis*

148 Dynamics of Mechanical Systems

component of acceleration and arises from the interaction between the velocity $_p v_B$ of relative to the plane and the angular velocity ω of the plane.

The physical basis for the Coriolis component of acceleration is perhaps not immediately obvious, but it can be explained by reference to a simple example. Figure 4.23 shows a rod OP rotating about O with constant angular velocity ω.

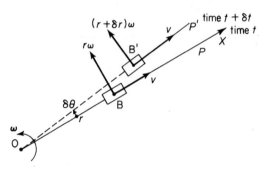

FIGURE 4.23

A block B slides along OP with a constant velocity v relative to the rod. During a time interval δt the block moves from B to B' while the rod turns through an angle $\delta\theta = \omega\,\delta t$. Suppose that at time t a line OX is drawn on a fixed plane to show the direction of the rod at that instant. The absolute velocity of B can then be said to be composed of two components, v along OX and $r\omega$ perpendicular to OX. After the time interval δt components of the velocity of the block are

$$v \cos \delta\theta - (r + \delta r)\omega \sin \delta\theta \quad \text{along OX}$$

and

$$v \sin \delta\theta + (r + \delta r)\omega \cos \delta\theta \quad \text{perpendicular to OX}$$

Putting

$$\sin \delta\theta = \delta\theta = \omega\,\delta t, \quad \cos \delta\theta = 1$$

and

$$\delta r = v\,\delta t$$

we find on neglecting second order terms that the two components are

$$v - r\omega^2\,\delta t \quad \text{along OX}$$

and

$$r\omega + 2v\omega\,\delta t \quad \text{perpendicular to OX}$$

The total acceleration of B is therefore composed of $-r\omega^2$ along the rod and $2r\omega$ perpendicular to the rod. The first of these terms represents the acceleration of the point on the rod with which B is coincident and hence $2v\omega$ is the acceleration of the block relative to the rod. Thus the Coriolis component of acceleration can be seen to arise from two effects: first, the changing magnitude of the tangential component $r\omega$ of the velocity of the block; second, the changing direction of the radial component v.

VELOCITIES AND ACCELERATIONS

As an example of the application of eqn. (4.20) we will determine the acceleration of the follower for the cam mechanism in fig. 4.24. The successive

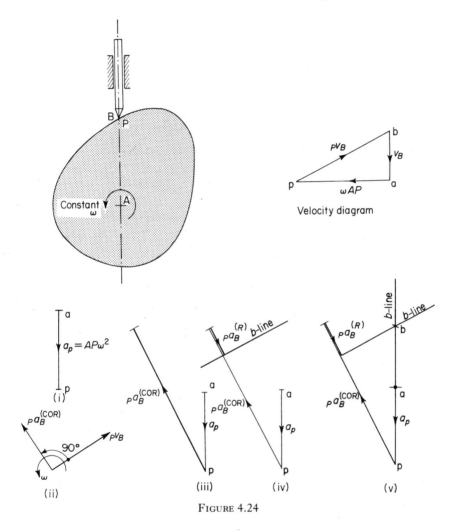

FIGURE 4.24

steps in the construction of the velocity diagram are given in fig. 4.7; the acceleration diagram is constructed as follows:

1. A is a fixed point so that a is the origin of the acceleration diagram. Since ω is constant the acceleration of P relative to A is simply $AP\omega^2$ in the direction from P to A.
2. The terms on the right-hand side of eqn. (4.20) may be added in any order, so as a matter of convenience we take next the Coriolis component $2\omega_P v_B$ in a direction perpendicular to $_P\mathbf{v}_B$. The velocity diagram shows the direction of $_P\mathbf{v}_B$ so that the direction of Coriolis component $_P\mathbf{a}_B^{(Cor)}$ is deduced as

150 Dynamics of Mechanical Systems

shown in fig. 4.24(ii), and added to $\mathbf{a_P}$ to bring the acceleration diagram to the stage shown in fig. 4.24(iii).

3 The term $_P\mathbf{\ddot{r}_B}$ consists of two components, first $_P\mathbf{a_B}^{(R)}$ of magnitude $_Pv_B^2/\rho$ directed from P towards the centre of curvature of the profile at P, and second a component $_P\mathbf{a_B}^{(T)}$ of unknown magnitude tangential to the profile at P. Putting in $_P\mathbf{a_B}^{(R)}$ first we are left with b located somewhere on the *b-line* shown in fig. 4.24(iv).

4 But the acceleration of B must be in the direction of the follower guides, so that a second *b*-line drawn from a parallel to the follower guide enables b to be located and the acceleration diagram is complete, fig. 4.24(v).

Equation (4.20) is a sufficient basis for the construction of the acceleration diagram for any plane mechanisms. One or two useful extensions of the theory and their application to particular problems are studied in the next section.

4.9 Acceleration diagrams for plane mechanisms

This section repeats the pattern of section 4.4, which dealt with some elaborations of the basic method of drawing velocity diagrams, and introduces a few ideas which are of extensions of the acceleration analysis presented in the preceding section.

The image theorem

The relationship between the acceleration of one point on a link and the accelerations of any other two points on the same link is determined by an image theorem similar to that already deduced for velocity. Suppose that the accelerations $\mathbf{a_A}$ and $\mathbf{a_B}$, fig. 4.25, are known. The acceleration of B relative to A can be split into the two components, $_A\mathbf{a_B}^{(R)}$ of magnitude $\omega^2 AB$ and $_A\mathbf{a_B}^{(T)}$ of magnitude αAB, where ω and α are respectively the angular velocity and angular acceleration of the plane to which AB is attached. The acceleration of B relative to A is of magnitude $AB(\alpha^2 + \omega^4)^{\frac{1}{2}}$, and the angle between this

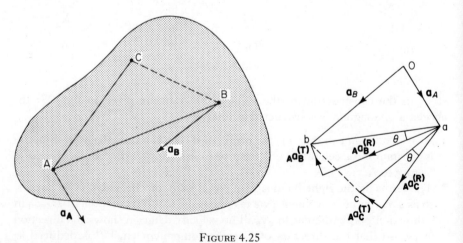

FIGURE 4.25

acceleration and the direction BA is $\theta = \tan^{-1}({}_A a_B{}^{(T)}/{}_A a_B{}^{(R)}) = \tan^{-1}(\alpha/\omega^2)$. Note that θ is independent of the length AB and its direction on the plane. It follows that $|{}_A \mathbf{a}_C| = AC(\alpha^2 + \omega^4)^{\frac{1}{2}}$ and that the angle between ${}_A \mathbf{a}_C$ and CA is also θ. Hence $\angle bac = \angle BAC$, so that triangles BAC and bac are similar. This result is generalized in the *acceleration image theorem*:

If $a, b, c, d, e, \ldots$ are points in an acceleration diagram which represent the accelerations of points A, B, C, D, E, ... on a rigid link, the two figures $a, b, c, d, e, \ldots$ and A, B, C, D, E, ... are geometrically similar and in the same sense.

The instantaneous centre for accelerations
By application of the image theorem it is possible to determine the point on a moving body which has zero acceleration, i.e., the instantaneous centre of accelerations. In general, very little purpose is served by knowing the position of this point so that the matter will not be pursued.

The three-line construction
The basic techniques as described in sections 4.3 and 4.8, together with the image theorems, enable us to draw velocity and acceleration diagrams for most plane mechanisms. However, as we saw in section 4.4 in relation to the mechanism depicted in fig. 4.11 and redrawn in fig. 4.26 difficulties can arise which require special treatment. The particular difficulty which arises in constructing the velocity diagram for this mechanism is illustrated in fig. 4.26(i). Starting with the given jack velocity ${}_A \mathbf{v}_E{}^{(R)}$ the velocity diagram is readily taken to the stage shown. The problem now is to fit a line bec in the diagram which is perpendicular to BEC and divided up in the same ratio. The method given in section 4.4, which utilized the instantaneous centre of BC, cannot be generalized to provide a means of solving the similar problem which arises in constructing the acceleration diagram, and a different approach is needed. One method is as follows.

A point c_1 on the *c-line* is selected arbitrarily. The line $c_1 b_1$ is then drawn perpendicular to BC and e_1 is located on $c_1 b_1$ so that it divides $c_1 b_1$ in the same ratio that E divides CB. If e_1 happens to lie on the *e-line* then we will accidentally have selected the correct position for ceb. If this does not happen we can try again, $c_2 e_2 b_2$; and then again, $c_3 e_3 b_3$. The locus $e_1 e_2 e_3 \ldots$ is then drawn in and the point where it intersects the *e-line* locates e. In fact, it is not necessary to draw $c_2 b_2$, etc. because it is clear from the diagram that the locus $e_1 e_2 e_3$ is a straight line passing through a, d so that e is located at the intersection of $e_1 a$ with the *e-line*.

In order to determine the acceleration of E relative to A it is convenient to visualize the cylinder of the jack being extended right up to the point E, the point on the axis of the cylinder with which E is momentarily coincident being

152 Dynamics of Mechanical Systems

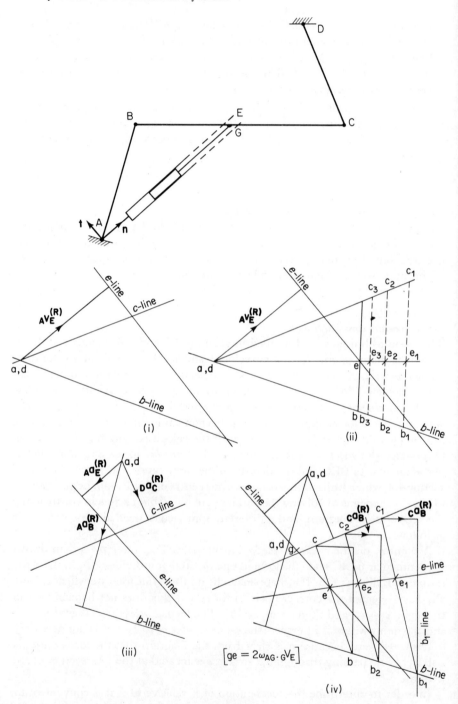

$$[ge = 2\omega_{AG} \cdot {}_GV_E]$$

Figure 4.26

labelled G. Then in terms of the two unit vectors **n** and **t** we have by applying eqn. (4.20)

$$\mathbf{a_E} = \underbrace{\mathbf{a_A}}_{\substack{\text{acceleration} \\ \text{of E}}} \underbrace{\vphantom{\mathbf{a_A}}}_{\substack{\text{acceleration} \\ \text{of A}}} \underbrace{- AG\omega_{AG}^2 \mathbf{n} + AG\dot{\omega}_{AG}\mathbf{t}}_{\substack{\text{acceleration of G} \\ \text{relative to A}}} + \underbrace{{}_G\ddot{\mathbf{r}}_E + 2\omega_{AG}\,{}_G v_E \mathbf{t}}_{\substack{\text{acceleration of E} \\ \text{relative to G}}}$$

Since A is a fixed point $\mathbf{a_A}$ is zero. If we assume for the sake of simplicity that the jack velocity is constant, then ${}_G\ddot{\mathbf{r}}_E = 0$ so that

$$\mathbf{a_E} = -AG\omega_{AG}^2\mathbf{n} + (AG\dot{\omega}_{AG} + 2\omega_{AG}\,{}_G v_E)\mathbf{t}$$

As $AG \equiv AE$ and $\omega_{AG} = \omega_{AE} = {}_A v_E/AE$, the first term on the right-hand side of the equation, which is really ${}_A\mathbf{a_E}^{(R)}$, is known in magnitude and direction. The second and third terms are both perpendicular to AE, the second being of unknown magnitude, and the third (Coriolis) of known magnitude. The acceleration diagram is therefore started by drawing ${}_A\mathbf{a_E}^{(R)}$, magnitude ${}_A v_E^2/AE$, as shown, and the *e-line* perpendicular to it. The radial component of the acceleration of B relative to A, and the radial component of the acceleration of C relative to D, can both be calculated (${}_A\mathbf{a_B}^{(R)}$ and ${}_D\mathbf{a_C}^{(R)}$) and entered on the acceleration diagram. The *b-line* and *c-line* are then added to bring the acceleration diagram to the stage shown in fig. 4.26(iii). The line *ceb* must now be fitted in. As with the velocity diagram we start with an arbitrary selection of a point c_1 on the *c-line*: The radial component of the acceleration of B relative to C, can be calculated and used to locate the b_1-*line*. The intersection of the b_1- and *b-lines* determines b_1. Point *e* is then obtained by dividing c_1b_1 in the appropriate ratio. If this procedure is repeated for other freely selected positions of *c* the locus of *e* is obtained and found to be a straight line. The intersection of the two *e-lines* now determines the location of *e*, and the diagram is readily completed. The main disadvantage of this three-line construction is that the diagrams are confused by the addition of extraneous lines. A more elegant solution of the same problem is provided by the auxiliary point method.

Auxiliary point method
Referring again to the mechanism in fig. 4.26 we proceed as before, taking the construction of the velocity diagram up to the stage where the *e-*, *b-*, and *c-lines*, fig. 4.27(i) have been determined. We now consider B, E, and C to be points attached to a rigid plane P which extends indefinitely. Now, instead of concentrating on points B, E, and C (whose associated points in the velocity diagram we are trying to find), we look for some other point on this plane whose associated point can be readily located in the velocity diagram. Such a point is X, which is located at the intersection of AE produced with DC. It is important to emphasize that X is considered to be a point on the plane P, and not a point on the link CD.

Now $$\mathbf{v_X} = \mathbf{v_E} + {}_E\mathbf{v_X}$$

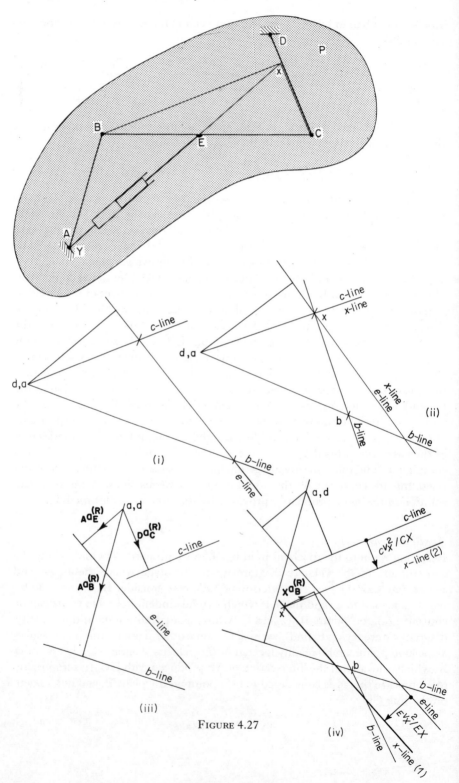

FIGURE 4.27

The location of e in the velocity diagram is not known, but we do know that it must lie on the *e-line* which is perpendicular to AE. The velocity of X relative to E is of unknown magnitude, but it must lie on a line drawn from e and perpendicular to AEX. It follows that the *e-line* in the velocity diagram is also an *x-line*. A similar argument to the effect that, since $\mathbf{v_C}$ and $_C\mathbf{v_X}$ are perpendicular to CD and CX respectively, the *c-line* is also an *x-line*, enables us to locate x at the intersection of the *c-* and *e-lines*, fig. 4.27(ii).

A ray from x perpendicular to BX now gives us a second *b-line* which intersects the first at the point b. From b a ray drawn perpendicular to BEC intersects the *e-* and *c-lines* at e and c respectively.

Provided that the drawing is accurate, it will be found that e divides bc in the proper ratio.

The same auxiliary point X enables the acceleration diagram to be completed. Figure 4.27(iii) shows the diagram at the state when the *e-*, *b-*, and *c-lines* have been determined.

$$\mathbf{a_X} = \mathbf{a_E} + {}_E\mathbf{a_X}$$
Now
$$= \mathbf{a_E} + {}_E\mathbf{a_X}^{(T)} + {}_E\mathbf{a_X}^{(R)}$$

If we were to add the vectors on the right-hand side of this equation in the order that they are given, we would start at e, located somewhere on the *e-line*, add the vector $_E\mathbf{a_X}^{(T)}$ which is in a direction perpendicular to EX and so takes us to some other point on the *e-line*, and finally we would add $_E\mathbf{a_X}^{(R)}$ in a direction perpendicular to the *e-line*. From this it follows that x must lie on a line parallel to the *e-line* and offset from it by an amount $_Ev_X^2/EX$. As $_E\mathbf{a_X}^{(R)}$ is in a direction from right to left the *x-line* lies to the left of the *e-line* as shown in fig. 4.27(iv), *x-line (1)*. By a similar argument *x-line (2)*, parallel to the *c-line* is determined. The intersection of the two *x-lines* locates x, hence $_X\mathbf{a_B}^{(R)}$ is drawn, and the second *b-line* which fixes b. Given x and b, we can now use the image theorem to locate e and c, which, provided that the drawing is accurate, should be found to lie on the *e-* and *c-lines*.

Two other auxiliary points may be used instead of X. We could choose a point Y at the intersection of AE with AB, i.e., at A. The reader should verify for himself that in the velocity diagram, y lies at the intersection of the *e-* and *b-lines*. The other alternative is a point Z at the intersection of AB with CD. In the velocity diagram, z lies at the intersection of the *b-* and *c-lines*, that is to say at a. As Z is the instantaneous centre for BC this result is to be expected, fig. 4.28. Point e is now located by drawing ze perpendicular to ZE. The similarity between fig. 4.11(ii) and fig. 4.28(i) should be noted.

A general rule for locating auxiliary points for a tertiary link, such as that depicted in fig. 4.29, is that an auxiliary point lies at the intersection of the centre-lines of any convenient pair of the three members to which it is attached. In the velocity diagram, the associated point is then found to be located at the intersection of the corresponding pair of rays. It can be seen that we are not concerned with what happens at the far ends of the links PP', QQ', and RR'.

156 Dynamics of Mechanical Systems

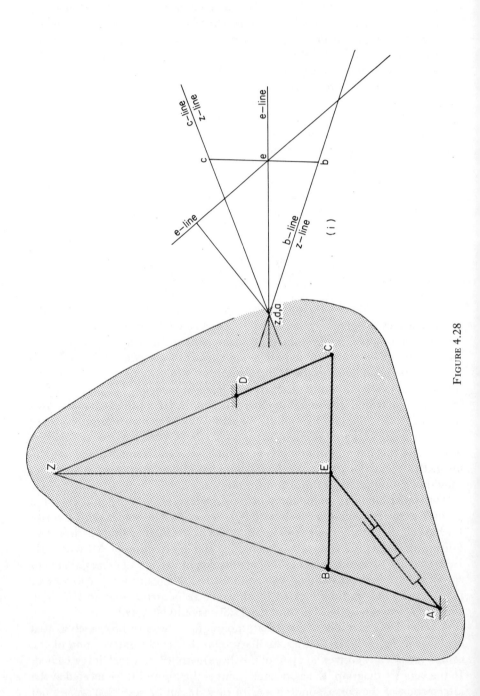

Figure 4.28

VELOCITIES AND ACCELERATIONS 157

If it happens that any pair, say Q′ and R′ are fixed hinges then the point of intersection Z of RR′ and QQ′ is the instantaneous centre for the link. This fact by itself in no way affects the procedure of the auxiliary point method. Nevertheless, if figs. 4.11(ii) and 4.28(i) are compared it will be seen that they are

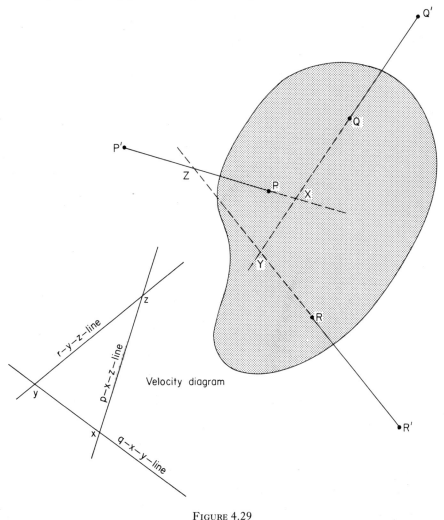

FIGURE 4.29

identical, so that the instantaneous centre method for velocities can be regarded as a special case of the auxiliary point method.

Equivalent mechanisms
Direct application of the methods so far developed in this chapter for constructing velocity and acceleration diagrams to the cam mechanism shown in fig. 4.30(a) is quite difficult. The problem of determining the velocity and

6*

acceleration of the follower can be very much simplified by an indirect approach in which the actual mechanism is replaced by another which, for small displacements from the given configuration, has the same transfer relationship. The selection of an appropriate equivalent mechanism proceeds as follows:

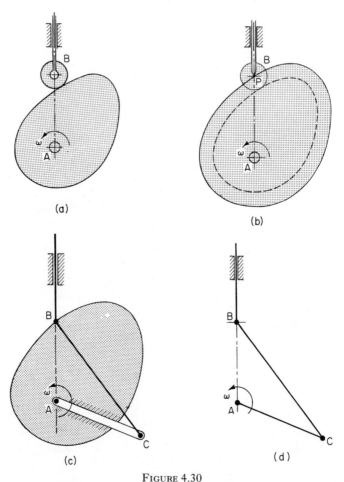

FIGURE 4.30

First, we observe that if we move the cam profile out by an amount equal to the radius of the roller, and replace the roller by a knife-edged follower we shall not alter the transfer relationship for the mechanism, and the follower velocity and acceleration for this mechanism will be the same as for the original one. This holds for all positions of the cam, so that as far as the overall transfer relationship is concerned the two systems are kinematically equivalent. The velocity and acceleration diagrams for system (b) are simpler than those for system (a). We can simplify the problem still more by a further substitution. If C is the centre of curvature of the cam profile at its point of contact P with the follower, we can replace the cam profile in the vicinity of P by a link CB which is hinged to the follower at B and to the cam at C, as in fig. 4.30(c). The final transformation to the mechanism in fig. 4.30(d) needs no elaboration. Provided that the radius of curvature of the profile is constant for some distance on either side of the point of contact P there can be little doubt that in the given configuration the velocity and acceleration of the two followers in fig. 4.30(b) and fig. 4.30(d) will respectively be the same. The velocity and acceleration diagrams for the cam mechanism and for the linkage mechanism are shown in fig. 4.31 (the velocity and acceleration diagrams for the cam mechanism are constructed side by side in fig. 4.24). There is nothing to choose by way of difficulty between the two velocity diagrams, but the acceleration diagram for the linkage mechanism is clearly easier to construct than the acceleration diagram for the cam itself; the latter involves a Coriolis component, the former does not.

It will be noted that the velocity and acceleration diagrams for the equivalent mechanisms are quite different from the velocity and acceleration diagrams for the original cam and follower. Nevertheless, we would expect the two sets of diagrams to give the same follower velocity and follower acceleration if, as we have presupposed, the radius of curvature of the cam profile is constant for some distance on either side of P. A formal proof that the two mechanisms give the same answers shows that it does not matter if the radius of curvature of the profile is varying continuously; see appendix 4.1 (page 171).

A different form of equivalent mechanism must be used when the radius of curvature of the profile of the cam is infinite. The locus of the centre of the follower relative to the cam is a straight line at a distance from its centre equal to the sum of the radii of the base circle and follower, fig. 4.32. In the equivalent mechanism this becomes the straight rod GQ which drives the follower through a sliding block and hinge. The velocity and acceleration diagrams for the equivalent mechanism are fairly straightforward, though in this case the Coriolis component is not avoided. However, it is simpler to proceed to the velocity and acceleration of the follower by direct analysis. From fig. 4.32,

$$x = a \sec \theta$$

The follower velocity is therefore given by

$$\frac{dx}{dt} = a \sec \theta \tan \theta \frac{d\theta}{dt}$$

$$= a\omega \frac{\sin \theta}{\cos^2 \theta} \tag{4.21}$$

and the follower acceleration is

$$\frac{d^2x}{dt^2} = a\omega^2 \frac{\cos^3 \theta + 2 \sin^2 \theta \cos \theta}{\cos^4 \theta}$$

$$= a\omega^2 \frac{1 + \sin^2 \theta}{\cos^3 \theta} \tag{4.22}$$

160 Dynamics of Mechanical Systems

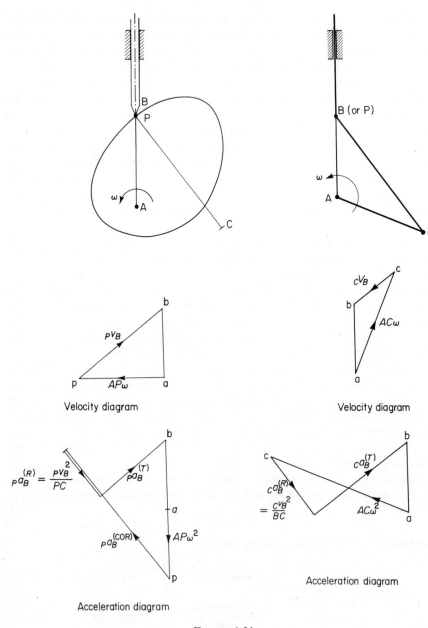

FIGURE 4.31

Another system for which consideration of the equivalent mechanism leads to simple analysis is the cam and flat follower, fig. 4.33. Using the same lettering as in the previous example, A is the axis of rotation of the cam, point P on the cam profile is in contact with point B on the follower, and C is the centre of

VELOCITIES AND ACCELERATIONS

curvature of the profile at P. Let us assume that the radius of curvature of the profile is constant for some distance on either side of P. In this case, the distance from C to the face of the follower is, for some time, constant. It follows that the velocity and acceleration of B must be the same as the components of the

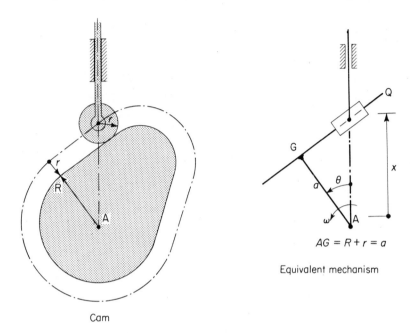

Cam

Equivalent mechanism

$AG = R + r = a$

FIGURE 4.32

velocity and acceleration of C along the axis of the follower. An equivalent mechanism which imparts the same velocity and acceleration to the follower is the *Scotch Yoke* mechanism, shown in fig. 4.33(b). It can be seen that if the cam speed ω is constant

$$x = AC \sin \theta$$

$$\frac{dx}{dt} = AC\omega \cos \theta = AN\omega \quad (4.23)$$

and

$$\frac{d^2x}{dt^2} = -AC\omega^2 \sin \theta = -CN\omega^2 \quad (4.24)$$

It can be shown that these expressions hold for the given instant even if the radius of curvature is changing continuously so that AC is not constant either in magnitude or in direction with respect to the cam. The effect of a sudden change in the radius of curvature, as happens when the profile is made up of circular arcs, can be seen by reference to fig. 4.33(c). The centre of curvature changes at P from C_1 for the portion SP of the cam profile to C_2 for the

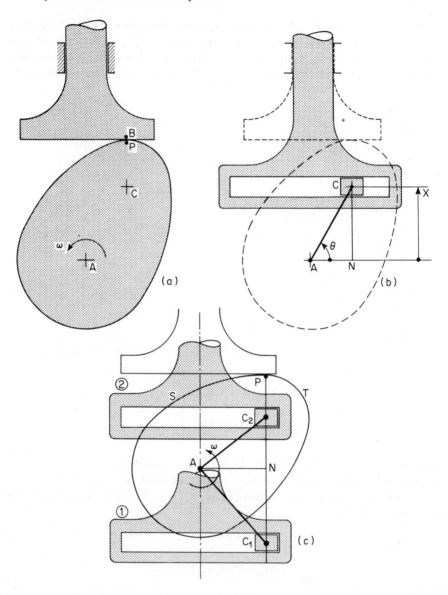

FIGURE 4.33

portion PT. Thus with the cam rotating in the anticlockwise direction there is a sudden change from mechanism 1 to mechanism 2. In this change there is no alteration in AN so that the follower velocity remains constant at $AN\omega$. The acceleration, however, changes from $NC_1\omega^2$ upwards to $NC_2\omega^2$ downwards.

VELOCITIES AND ACCELERATIONS

4.10 Relative accelerations (space mechanisms)
In three dimensions eqn. (4.20) becomes

$$\mathbf{a_B} = \mathbf{a_A} + \omega \times (\omega \times {}_A\mathbf{r_B}) + \dot{\omega} \times {}_A\mathbf{r_B} + {}_p\ddot{\mathbf{r}}_B + 2\omega \times {}_p\mathbf{v_B} \quad (4.25)$$

The difficulties of representing three-dimensional velocity relationships graphically has already been mentioned. These difficulties are obviously going to be magnified when accelerations are to be considered so that an analytical approach is even more likely to be preferred. As an example we will study the

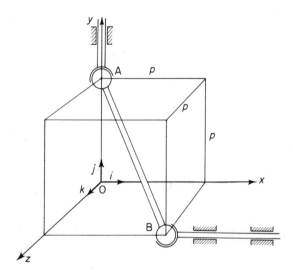

FIGURE 4.34

simple push-rod mechanism in fig. 4.34 (originally fig. 4.14) for which we have already deduced that $v_B = -v_A$, and that

$$\omega = \alpha_A\mathbf{r_B} + {}_A\mathbf{r_B} \times {}_A\mathbf{v_B}/AB^2 \quad (4.26)$$

$$= \alpha_A\mathbf{r_B} + v_A(\mathbf{i} - \mathbf{j} - 2\mathbf{k})/3p \quad (4.27)$$

(see page 138).

When A and B are points fixed to a rigid link eqn. (4.25) becomes

$$\mathbf{a_B} = \mathbf{a_A} + \omega \times (\omega \times {}_A\mathbf{r_B}) + \dot{\omega} \times {}_A\mathbf{r_B} \quad (4.28)$$

This equation does not immediately give an expression for $\mathbf{a_B}$ because $\dot{\omega}$ is unknown. However, as $\dot{\omega} \times {}_A\mathbf{r_B}$ is perpendicular to ${}_A\mathbf{r_B}$ we can eliminate $\dot{\omega}$ by forming the scalar product of all terms with ${}_A\mathbf{r_B}$, thus

$$(\mathbf{a_B} - \mathbf{a_A}) \cdot {}_A\mathbf{r_B} = [\omega \times (\omega \times {}_A\mathbf{r_B})] \cdot {}_A\mathbf{r_B} \quad (4.29)$$

Now if we let $\mathbf{a_B} = -a_B\mathbf{i}$ we have only to substitute for ω from eqn. (4.27) in (4.29) to obtain the value of a_B. The arbitrary rotation $\alpha_A\mathbf{r_B}$ of AB about its longitudinal axis obviously cannot affect the acceleration of B, and is found to disappear in the evaluation of the right-hand side of eqn. (4.28). Taking it for granted that this is going to happen we can replace ω in eqn. (4.28) by ω', its component perpendicular to ${}_A\mathbf{r}_B$ [i.e., the component ${}_A\mathbf{r}_B \times {}_A\mathbf{v}_B/AB^2$ in eqn. (4.26)]. It is then easy to see that

$$\omega' \times (\omega' \times {}_A\mathbf{r}_B) = -\omega'^2 \, {}_A\mathbf{r}_B$$

and hence that

$$(\mathbf{a}_B - \mathbf{a}_A) \cdot {}_A\mathbf{r}_B = -AB^2 \omega'^2 = -{}_A v_B^2 \tag{4.30}$$

This equation provides a more direct means of obtaining a_B than eqn. (4.29) and yields (for $a_A = 0$)

$$a_B = -2v_A^2/p$$

On rearranging the terms in eqn. (4.28) we have

$$\dot{\omega} \times {}_A\mathbf{r}_B = (\mathbf{a}_B - \mathbf{a}_A) - \omega' \times (\omega' \times {}_A\mathbf{r}_B)$$
$$= (\mathbf{a}_B - \mathbf{a}_A) + \omega'^2 \, {}_A\mathbf{r}_B \tag{4.31}$$

$\dot{\omega}$, like ω, has an arbitrary component about the longitudinal axis of AB. If we ignore this component we can readily determine $\dot{\omega}$ by taking the cross product of both sides of eqn. (4.31) with ${}_A\mathbf{r}_B$, for we then have

$$(\dot{\omega} \times {}_A\mathbf{r}_B) \times {}_A\mathbf{r}_B = -\dot{\omega} AB^2 = (\mathbf{a}_B - \mathbf{a}_A) \times {}_A\mathbf{r}_B$$

Hence
$$\dot{\omega} = -(\mathbf{a}_B - \mathbf{a}_A) \times {}_A\mathbf{r}_B/AB^2$$

In this example

$$\dot{\omega} = -\frac{2v_A^2}{3p^2}(\mathbf{j} + \mathbf{k})$$

4.11 Algebraic analysis of special mechanisms

It has been pointed out that relatively few mechanisms lend themselves to simple algebraic analysis, and that this is the reason why we have so far concentrated almost exclusively on graphical methods for determining velocity and acceleration relationships. We will conclude this chapter by considering three mechanisms which are of considerable practical importance and for which relatively simple algebraic analysis is feasible.

Quick-return mechanism

This mechanism is used as the basic drive for shaping machines and takes the form shown diagrammatically in fig. 4.35. Given a constant angular velocity of the driving crank BC, the mechanism enables the non-productive return stroke of the tool-head E to be made in appreciably less time than the working stroke; with a clockwise rotation of the crank, the working stroke is from left to right.

VELOCITIES AND ACCELERATIONS

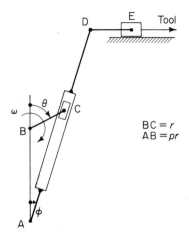

FIGURE 4.35

We can derive exact expressions for the angular velocity and acceleration of the swinging link AD as follows:

$$\tan \phi = \frac{r \sin \theta}{pr + r \cos \theta}$$

Differentiating with respect to time,

$$\sec^2 \phi \frac{d\phi}{dt} = \frac{(p + \cos \theta)\cos \theta + \sin^2 \theta}{(p + \cos \theta)^2} \cdot \frac{d\theta}{dt}$$

$$= \frac{1 + p \cos \theta}{(p + \cos \theta)^2} \cdot \frac{d\theta}{dt}$$

Now $\sec^2 \phi = 1 + \tan^2 \phi$, and $d\theta/dt = \omega$, hence the angular velocity of AD is,

$$\frac{d\phi}{dt} = \frac{1 + p \cos \theta}{(p + \cos \theta)^2} \times \frac{(p + \cos \theta)^2 \omega}{(p + \cos \theta)^2 + \sin^2 \theta}$$

$$= \frac{1 + p \cos \theta}{1 + 2p \cos \theta + p^2} \cdot \omega \qquad (4.32)$$

On differentiating again with respect to time we deduce the angular acceleration:

$$\frac{d^2\phi}{dt^2} = \frac{-(1 + 2p \cos \theta + p^2) \cdot p \sin \theta + (1 + p \cos \theta) \cdot 2p \sin \theta}{(1 + 2p \cos \theta + p^2)^2} \cdot \omega^2$$

$$= \frac{-p \sin \theta (p^2 - 1)}{(1 + 2p \cos \theta + p^2)^2} \cdot \omega^2 \qquad (4.33)$$

166 Dynamics of Mechanical Systems

These expressions for the angular velocity and acceleration of AD enable the horizontal component of the velocity and acceleration of D to be calculated and hence, provided that the mechanism is so proportioned that the obliquity of link DE is always small, the velocity and acceleration of the tool.

The slider-crank mechanism

This is perhaps the most important of all mechanisms and consists of a crank, a connecting-rod, and a piston moving inside a fixed cylinder. The purpose of the mechanism is commonly as a means of converting a reciprocating motion into a rotary motion, as for example in a reciprocating engine. But it is equally used as a means of converting rotary motion into reciprocating motion.

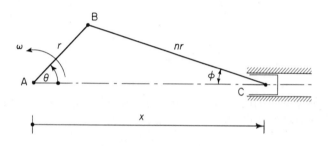

FIGURE 4.36

Although the slider-crank mechanism is an inversion of the quick return mechanism, for which it is possible to derive exact expressions for the velocity and acceleration of the output member, it is necessary to resort to approximate analysis to deduce usable expressions for the velocity and acceleration of the piston of the slider-crank mechanism in terms of the angular velocity of its crank. The approximation is based on the fact that in most applications the radius of the crank is appreciably shorter than the length of the connecting-rod.

In fig. 4.36,

$$x = r \cos \theta + nr \cos \phi$$

and

$$r \sin \theta = nr \sin \phi$$

Hence

$$x = r \cos \theta + nr \left(1 - \frac{1}{n^2} \sin^2 \theta \right)^{\frac{1}{2}}$$

On expanding the second term on the right-hand side of this equation by the binomial theorem we have

$$x = r \cos \theta + nr \left(1 - \frac{1}{2n^2} \sin^2 \theta - \frac{1}{8n^4} \sin^4 \theta - \frac{1}{16n^6} \sin^6 \theta - \cdots \right)$$

If $n > 3$ we can, with only small loss of accuracy, stop the series after the second term, so that on differentiating with respect to time

$$\frac{dx}{dt} = -r\omega \sin \theta - nr\omega \left(\frac{1}{n^2} \sin \theta \cos \theta + \cdots \right)$$

$$\approx -r\omega \left(\sin \theta + \frac{1}{2n} \sin 2\theta \right) \qquad (4.34)$$

where $\omega = d\theta/dt$ is the angular velocity of the crank. Assuming this to be constant, a second differentiation gives the acceleration

$$\frac{d^2x}{dt^2} \approx -r\omega^2 \left(\cos \theta + \frac{1}{n} \cos 2\theta \right) \qquad (4.35)$$

The first term on the right-hand side of this equation is referred to as the *primary acceleration* and it is independent of the ratio n of the length of the connecting-rod to the crank radius. The other term gives the *secondary acceleration*, which is inversely proportional to n. The effect of varying the value of n on the total acceleration is shown in fig. 4.37.

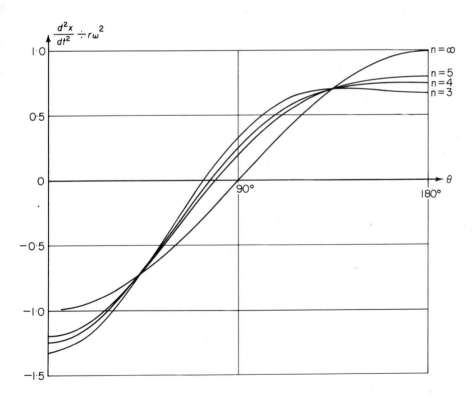

FIGURE 4.37

If more terms are considered in the binomial expansion we find that

$$\frac{d^2x}{dt^2} = -r\omega^2 \left\{ \cos\theta + \left(\frac{1}{n} + \frac{1}{4n^3} + \frac{15}{128n^5} + \cdots\right) \cos 2\theta \right.$$
$$- \left(\frac{1}{4n^3} + \frac{3}{16n^5} + \cdots\right) \cos 4\theta$$
$$\left. + \left(\frac{9}{128n^5} + \frac{45}{512n^7} + \cdots\right) \cos 6\theta + \cdots \right\} \quad (4.36)$$

The derivation of this equation is fairly complicated, and for this reason has been put in appendix 4.2. There are, however, various features of the equation itself which deserve attention. First, we note that apart from the first term, all the multiples of θ are even. Second, the coefficient of each trigonometric term, apart from the first, is an infinite series in powers of $1/n$, so that the approximations in eqn. (4.35) arise not only from the omission of the higher harmonics, but also from the neglect of the higher powers of $1/n$ in the coefficient of $\cos 2\theta$. It can be seen, however, that the error introduced is only about 3 per cent when $n = 3$, and decreases at a rate proportional to n^2. The final feature to notice is that amplitude of each of the higher harmonics is approximately $1/n^2$ the preceding harmonic.

Hooke's joint

Hooke's joint provides a means of connecting two shafts whose axes are non-parallel but intersecting, fig. 4.38. The cruciform member AA'OBB' is hinged to shafts 1 and 2 about axes AA' and BB' respectively. In order to determine the relationship between the input velocity ω_1 and the output velocity ω_2, we define the set of unit vectors shown in fig. 4.38(b). Vectors $\mathbf{i}$ and $\mathbf{j}$ are along the axes of the input and output shafts respectively. Vector $\mathbf{n}$ is the common normal to $\mathbf{i}$ and $\mathbf{j}$ and is the direction with which AA' is aligned when the mechanism is in its datum position. In the same position, BB' is aligned with vector $\mathbf{m}$ which is perpendicular to $\mathbf{j}$ in the plane of $\mathbf{i}$ and $\mathbf{j}$, and hence perpendicular to $\mathbf{n}$. If shaft 1 is rotated through an angle θ from the datum position we have

$$\overrightarrow{OA} \equiv \mathbf{r}_1 = \mathbf{n} \cos\theta + \mathbf{i} \times \mathbf{n} \sin\theta$$

If the corresponding rotation of shaft 2 is ϕ

$$\overrightarrow{OB} \equiv \mathbf{r}_2 = \mathbf{m} \cos\phi + \mathbf{j} \times \mathbf{m} \sin\phi$$

In all configurations, $\mathbf{r}_1$ and $\mathbf{r}_2$ are perpendicular to each other so that $\mathbf{r}_1 \cdot \mathbf{r}_2 = 0$, that is to say

$$(\mathbf{n} \cos\theta + \mathbf{i} \times \mathbf{n} \sin\theta) \cdot (\mathbf{m} \cos\phi + \mathbf{j} \times \mathbf{m} \sin\phi) = 0$$

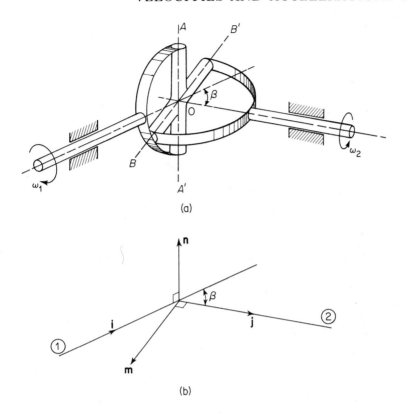

FIGURE 4.38

Now $\mathbf{j} \times \mathbf{m} = \mathbf{n}$, and $\mathbf{n} \times \mathbf{i}$ lies in the plane of $\mathbf{i}$ and $\mathbf{j}$ at an angle β to $\mathbf{m}$. Hence, $(\mathbf{i} \times \mathbf{n}) \cdot \mathbf{m} = -\cos \beta$. Substituting these relationships in the last equation, we have

$$\cos \theta \sin \phi - \cos \beta \sin \theta \cos \phi = 0$$

or
$$\tan \phi = \cos \beta \tan \theta$$

Differentiating with respect to θ, we find

$$\sec^2 \phi \, \frac{d\phi}{d\theta} = \cos \beta \sec^2 \theta$$

On substituting $(1 + \tan^2 \phi)$ for $\sec^2 \phi$ we have

$$\frac{d\phi}{d\theta} = \frac{\cos \beta}{\cos^2 \theta + \cos^2 \beta \sin^2 \theta}$$

$$= \frac{\cos \beta}{1 - \sin^2 \beta \sin^2 \theta}$$

If $d\theta/dt = \omega_1$ is constant, $d\phi/dt = \omega_2$ is given by

$$\omega_2 = \frac{\omega_1 \cos \beta}{1 - \sin^2 \beta \sin^2 \theta} \qquad (4.37)$$

We can see by inspection that $(\omega_2)_{min} = \omega_1 \cos \beta$ and that $(\omega_2)_{max} = \omega_1 \sec \beta$, corresponding to $\theta = 0$, π, 2π, etc. and $\theta = \pi/2$, $3\pi/2$, $5\pi/2$, etc. The angular acceleration of the output shaft is obtained by a second differentiation which yields

$$\frac{d\omega_2}{dt} = \frac{\omega_1^2 \sin^2 \beta \cos \beta \sin 2\theta}{(1 - \sin^2 \beta \sin^2 \theta)^2} \qquad (4.38)$$

Provided that β is small, the maximum acceleration occurs when $\theta = \pi/4$, $3\pi/4$, $5\pi/4$, etc. and has the approximate values $\pm(\omega_1 \sin \beta)^2$. Unless β is very small, the acceleration may give rise to troublesome inertia torques which can easily be avoided by using two Hooke's joints rather than one arranged back-to-back so that the variation in the velocity ratio of one is cancelled by the corresponding variation in the other. This arrangement is frequently used between the gearbox and the rear axle of an automobile transmission.

REFERENCES
1 BEYER, R. *Technische Raumkinomatik*, Springer-Verlag, 1963.

APPENDIX 4.1

Equivalent mechanisms

If the radius of curvature of the cam profile in fig. 4.39(a) is constant for some distance on either side of the point of contact P with the follower, the instantaneous motion of the follower must be the same as that of the slider in the equivalent mechanism given in fig. 4.39(b). It will now be shown that the two mechanisms give the same output velocity and acceleration even if the radius of curvature of the cam profile is varying continuously so that the given equivalent mechanism can only be instantaneously valid.

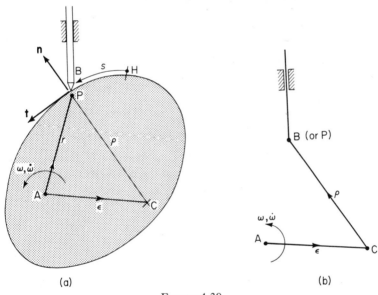

FIGURE 4.39

If s is the distance of the point of contact P round the profile from a convenient datum H, we have

$$\mathbf{v_B} = \mathbf{v_P} + {}_P\mathbf{v_B}$$
$$= \boldsymbol{\omega} \times \mathbf{r} + \dot{s}\mathbf{t}$$

But
$$\mathbf{r} = \boldsymbol{\varepsilon} + \boldsymbol{\rho}$$

and
$$\dot{s}\mathbf{t} = \boldsymbol{\omega}_{rel} \times \boldsymbol{\rho}$$

where ω_{rel} is the anticlockwise rate of rotation of the ray CP relative to the cam.

Hence
$$\mathbf{v}_B = \boldsymbol{\omega} \times (\boldsymbol{\varepsilon} + \boldsymbol{\rho}) + \boldsymbol{\omega}_{rel} \times \boldsymbol{\rho}$$
$$= \boldsymbol{\omega} \times \boldsymbol{\varepsilon} + (\boldsymbol{\omega} + \boldsymbol{\omega}_{rel}) \times \boldsymbol{\rho}$$

But $(\boldsymbol{\omega} + \boldsymbol{\omega}_{rel})$ is the absolute angular velocity of the ray CP, so that
$$\mathbf{v}_B = \boldsymbol{\omega} \times \boldsymbol{\varepsilon} + \boldsymbol{\omega}_{CP} \times \boldsymbol{\rho}$$

This equation applies equally to the equivalent linkage mechanism, fig. 4.39(b).

The acceleration of the cam follower is, from eqn. (4.20),
$$\mathbf{a}_B = \mathbf{a}_P + {}_P\ddot{\mathbf{r}}_B + {}_P\mathbf{a}_B^{(Cor)}$$

To make the situation completely general, let us assume that the cam has an angular acceleration $\dot{\boldsymbol{\omega}}$ so that, on substituting for the various terms on the right-hand side of this expression for $\mathbf{a}_B$, we have

$$\mathbf{a}_B = (-\omega^2 \mathbf{r} + \dot{\boldsymbol{\omega}} \times \mathbf{r}) + \left(\ddot{s}\mathbf{t} - \frac{\dot{s}^2}{\rho}\mathbf{n}\right) + 2\boldsymbol{\omega} \times (\dot{s}\mathbf{t})$$
$$= \left[-\omega^2(\boldsymbol{\varepsilon} + \boldsymbol{\rho}) + \dot{\boldsymbol{\omega}} \times (\boldsymbol{\varepsilon} + \boldsymbol{\rho})\right] + (\dot{\boldsymbol{\omega}}_{rel} \times \boldsymbol{\rho} - \omega_{rel}^2 \boldsymbol{\rho}) + \left[2\boldsymbol{\omega} \times (\boldsymbol{\omega}_{rel} \times \boldsymbol{\rho})\right]$$
$$= -\omega^2 \boldsymbol{\varepsilon} + \dot{\boldsymbol{\omega}} \times \boldsymbol{\varepsilon} + (\dot{\boldsymbol{\omega}} + \dot{\boldsymbol{\omega}}_{rel}) \times \boldsymbol{\rho} - (\omega^2 + 2\omega\,\omega_{rel} + \omega_{rel}^2)\boldsymbol{\rho}$$

Now, again $(\boldsymbol{\omega} + \boldsymbol{\omega}_{rel}) = \boldsymbol{\omega}_{CP}$, and $(\dot{\boldsymbol{\omega}} + \dot{\boldsymbol{\omega}}_{rel}) = \dot{\boldsymbol{\omega}}_{CP}$, hence
$$\mathbf{a}_B = -\omega^2 \boldsymbol{\varepsilon} + \dot{\boldsymbol{\omega}} \times \boldsymbol{\varepsilon} + \dot{\boldsymbol{\omega}}_{CP} \times \boldsymbol{\rho} - \omega_{CP}^2 \boldsymbol{\rho}$$

This equation holds for the equivalent linkage mechanism at the given instant, and is quite independent of any variation in the equivalent mechanism due to variation in the curvature of the cam profile.

APPENDIX 4.2

Velocity and acceleration relationships for the slider-crank mechanism

From fig. 4.36,

$$x = r\cos\theta + nr\left(1 - \frac{1}{n^2}\sin^2\theta\right)^{\frac{1}{2}}$$

Since x is clearly a periodic function of θ we would expect it to be possible to express x in terms of θ as a Fourier series. It is clear from the above expression that x does not alter with a reversal of the sign of the angle θ, i.e., that x is an even function of θ, so that the relevant Fourier series cannot contain sine terms and must be of the form

$$x = a_o + a_1\cos\theta + a_2\cos 2\theta + a_3\cos 3\theta + \cdots$$

As we are interested primarily in the velocity and acceleration of the piston, the term a_o is of little interest since it is eliminated when x is differentiated. The general term of the series is $a_k\cos k\theta$, and to determine a_k we must evaluate

$$a_k = \frac{1}{\pi}\int_0^{2\pi} x\cos k\theta\, d\theta$$

Since the integral

$$\int_0^{2\pi}\cos k\theta\left(1 - \frac{1}{n^2}\sin^2\theta\right)^{\frac{1}{2}} d\theta$$

cannot be evaluated in terms of simple functions, e.g., $\sin\theta$, $\cos\theta$, e^θ, etc., we will expand the bracketed term by means of the binomial theorem and then integrate term by term. Hence

$$a_k = \frac{1}{\pi}\int_0^{2\pi}\cos k\theta \left\{ r\cos\theta + nr\left[1 - \frac{1}{2}\cdot\frac{1}{n^2}\sin^2\theta \right.\right.$$
$$- \frac{1}{2!}\cdot\frac{1}{2}\cdot\frac{1}{2}\left(\frac{1}{n^2}\sin^2\theta\right)^2$$
$$\left.\left. - \frac{1}{3!}\cdot\frac{1}{2}\cdot\frac{1}{2}\cdot\frac{3}{2}\left(\frac{1}{n^2}\sin^2\theta\right)^3 - \cdots \right]\right\} d\theta$$

Now

$$\frac{1}{\pi}\int_0^{2\pi} r\cos k\theta \cos\theta\, d\theta = \begin{cases} r & \text{for } k = 1 \\ 0 & \text{for } k \neq 1 \end{cases}$$

and

$$\frac{1}{\pi}\int_0^{2\pi} nr\cos k\theta\, d\theta = 0$$

so that we can now concentrate on the term $1/(2n^2)\sin^2\theta$ and those which follow it.

The general term for this series is

$$b_m \sin^{2m}\theta$$

where
$$b_m = \frac{\frac{1}{2}(\frac{1}{2} - 1)(\frac{1}{2} - 2)\cdots(\frac{1}{2} - m + 1)}{m! \times n^{2m}}$$

Let us consider

$$I_m = \int_0^{2\pi} \cos k\theta \sin^{2m}\theta\, d\theta$$

Integrating by parts twice over we obtain

$$I_m = \frac{m^2}{k^2} I_m - \frac{2m(2m - 1)}{k^2} I_{m-1}$$

and so deduce the reduction formula

$$I_m = \frac{2m(2m - 1)}{[(2m)^2 - k^2]} I_{m-1}$$

where
$$I_{m-1} = \int_0^{2\pi} \cos k\theta \sin^{2(m-1)}\theta\, d\theta$$

If $2m < k$, or if k is odd, there is no difficulty in applying the reduction formula m times over to obtain

$$I_m = \frac{2m(2m - 1)(2m - 2)\ldots(2m - \overline{2m - 1})}{[(2m)^2 - k^2][(2m - 2)^2 - k^2]\ldots[(2m - \overline{2m - 2})^2 - k^2]} I_{2m-2m}$$

$$= \frac{(2m)!}{[(2m)^2 - k^2]\ldots[4^2 - k^2][2^2 - k^2]} \int_0^{2\pi} \cos k\theta\, d\theta$$

$$= 0$$

It follows that, apart from the term $r\cos\theta$ already dealt with as a special case, the Fourier series contains only even values of k. Furthermore, the contribution of the term $b_m \sin^{2m}\theta$ to the Fourier series is limited to those terms for which k is either less than or equal to $2m$. This means, for example, that the coefficient of $\cos 6\theta$ cannot involve terms $1/n^2$ and $1/n^4$, but only $1/n^6$, $1/n^8$, etc.

VELOCITIES AND ACCELERATIONS

If

$$2m = k$$

we have

$$I_{k/2} = \int_0^{2\pi} \cos k\theta \sin^k \theta \, d\theta$$

Making use of the result in the preceding paragraph, we can show that

$$I_{k/2} = -\frac{1}{2^2} I_{k/2-1}$$

and hence that

$$I_{k/2} = (-1)^{k/2} \frac{1}{2^k} \int_0^{2\pi} 1 \cdot d\theta = (-1)^{k/2} \frac{\pi}{2^{k-1}}$$

If $2m > k$, we can use the reduction formula to determine I_m by starting with $I_{k/2}$ and building up. If, for example, we wish to determine the coefficient of $\cos 4\theta$, we have from the equation for $I_{k/2}$

$$I_2 = (-1)^2 \frac{\pi}{2^3} = \frac{\pi}{8}$$

Then from the relationship between I_m and I_{m-1}

$$I_3 = \frac{6 \cdot 5}{(6^2 - 4^2)} I_2 = \frac{3\pi}{16}$$

$$I_4 = \frac{8 \cdot 7}{(8^2 - 4^2)} I_3 = \frac{35\pi}{256}$$

and so on. Values of I_m/π for various values of k are given in the table below.

m	$k = 2$	$k = 4$	$k = 6$	$k = 8$	b_m
1	$-\dfrac{1}{2}$	0	0	0	$-\dfrac{1}{2n^2}$
2	$-\dfrac{1}{2}$	$\dfrac{1}{8}$	0	0	$-\dfrac{1}{8n^2}$
3	$-\dfrac{15}{32}$	$\dfrac{3}{16}$	$-\dfrac{1}{32}$	0	$-\dfrac{1}{16n^6}$
4	$-\dfrac{7}{16}$	$\dfrac{7}{32}$	$-\dfrac{1}{16}$	$\dfrac{1}{128}$	$-\dfrac{5}{128n^8}$

To obtain a_k ($k \neq 0$ or 1) in the series for the piston displacement

$$x = a_0 + a_1 \cos \theta + a_2 \cos 2\theta + a_4 \cos 4\theta + \cdots$$

we form the series

$$a_k = \frac{nr}{\pi} \sum b_m I_m$$

The corresponding coefficient in the series for piston velocity is $b_k = -k\omega a_k$, and for piston acceleration is $c_k = -k^2\omega^2 a_k$. Values of $-c_k/r\omega^2$ are given below.

	$k = 2$	$k = 4$	$k = 6$	$k = 8$
$\dfrac{1}{n} \times$	1	0	0	0
$\dfrac{1}{n^3} \times$	$\dfrac{1}{4}$	$-\dfrac{1}{4}$	0	0
$\dfrac{1}{n^5} \times$	$\dfrac{15}{128}$	$-\dfrac{3}{16}$	$\dfrac{9}{128}$	0
$\dfrac{1}{n^7} \times$	$\dfrac{35}{512}$	$-\dfrac{35}{258}$	$\dfrac{45}{512}$	$-\dfrac{5}{512}$

5

FORCE RELATIONSHIPS IN MECHANISMS
II Inertia forces

5.1 Introduction
The sole purpose of the acceleration analysis in the last chapter was to enable us to calculate the inertia forces which we brought into play by the motion of the members of a mechanism. The inertia forces manifest themselves in various ways; as stresses within the members, as driving forces which are required to maintain the motion, and as forces on the bearings and the structure on which the mechanism is supported. Evaluation of the stresses and bearing loads is an essential step in the selection of optimum sizes of the members of the mechanism, even though ultimately other considerations, such as cost of manufacture and assembly, and ease of maintenance, may be of over-riding importance.

In addition to calculating inertia effects, we must consider ways in which they can be mitigated. If it is possible to so arrange the motion of two members of a mechanism that their inertia forces cancel each other out, the mechanism will still suffer internal inertia stresses but no resultant force is transmitted to the structure on which the mechanism is mounted. If all inertia forces are in equilibrium with each other the mechanism is said to be *balanced*.

We will assume, in this chapter, that the members of the mechanism are sufficiently stiff for the deformations, which always accompany the inertia stresses, to be neglected.

Before we consider actual mechanisms it is necessary to make some observations on the basic mechanics of rigid bodies.

5.2 Rigid body mechanics
In mechanics, as in other things, the law of diminishing returns makes itself felt. Simple concepts can often take one a long way in dealing with practical problems, as the concepts become more general, and hence more powerful, they demand a much greater intellectual effort on the part of the student, but at the same time tend to find practical application only in rather specialized fields. We have an illustration of this in the last chapter where, as the reader will no doubt readily concede, the theory of three-dimensional mechanisms is seen to be much more difficult than the theory of the plane mechanisms which are far more commonly used in practice. With these thoughts in mind, there

178 Dynamics of Mechanical Systems

seems to be little point in giving a general revision of three-dimensional rigid body mechanics. We shall assume that the reader knows most, if not all, of the basic mechanics which we require and for the most part will be willing to accept without rigorous proof what appears to him to go beyond elementary theory.

5.3 D'Alembert's principle

According to Newton's second law of motion, the acceleration a of a particle of mass m under the action of a single force P is given by the equation

$$P = ma \qquad (5.1)$$

We shall find it more convenient to express this equation in the form

$$P - ma = 0 \qquad (5.2)$$

and to refer to the quantity $-ma$ as the inertia force, or *D'Alembert force*, of the particle. Equation (5.2) is then expressed verbally by saying that the vector sum of the applied force F and the inertia force $-ma$ is zero. Thus, the equation of motion for the particle, eqn. (5.1), becomes an equilibrium equation, eqn. (5.2). This idea expressed more generally is known as *D'Alembert's principle*: a mechanical system must be in equilibrium under the action of its applied and D'Alembert forces.

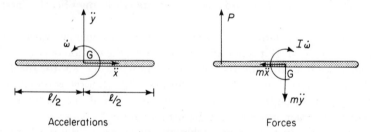

FIGURE 5.1

In the case of a rigid body in plane motion, whose centre of gravity has an acceleration a and whose angular acceleration is $\dot{\omega}$, the D'Alembert effects are a force $-ma$ and a couple $-I\dot{\omega}$, where I is the moment of inertia of the body about the axis of rotation through its centre of gravity. An example is given in fig. 5.1 of a uniform rod of length l and mass m lying on a smooth horizontal plane, and subjected to a horizontal force P at one end perpendicular to its axis. With an isolated member, the directions of the coordinate axes which are used to describe the motion are arbitrary; we have chosen x to be along the axis of the rod as a matter of convenience. The rod must be in equilibrium under the action of the forces shown, so that resolving forces, and taking moments about G, we have

$$-m\ddot{x} = 0, \quad P - m\ddot{y} = 0, \quad Pl/2 + \tfrac{1}{12}ml^2\dot{\omega} = 0$$

Hence $\qquad \ddot{x} = 0, \quad \ddot{y} = P/m, \quad \dot{\omega} = -6P/ml$

It will be noted that we have drawn two diagrams, one showing the accelerations, and the other showing the forces. This is not essential, but helps to avoid the confusion which can arise when everything is put on a single diagram.

There are two points which should be emphasized. First, it is strongly recommended that the reader should acquire the habit of showing all the applied and inertia forces in the force diagram, even though some may be of little direct interest in the problem being considered. Thus we have thought it wise to show the D'Alembert force $-m\ddot{x}$ in fig. 5.1, even though in this example it is obviously zero; we are less likely to forget it on some other occasion when it is not zero. Second, as a general rule in this book, the moment of inertia will be defined only in relation to an axis through the centre of gravity. When one is dealing with the rotation of a single body about a fixed axis it is very often convenient to refer to the moment of inertia about the fixed axis of rotation, but this concept so often causes difficulty it is better to avoid it. In fig. 5.2, a rectangular lamina rotates in a horizontal plane about a fixed hinge at O. The lamina is subjected to the force P and the hinge reactions R_1 and R_2. It is tempting to take moments about O and simply write

$$Pa - I_0\ddot{\theta} = 0$$

where I_O is the moment of inertia about an axis through O. This temptation should, however, be resisted. In this particular problem the equation is quite correct, but the danger of applying the same reasoning wrongly to a similar problem in which the hinge O may move (perhaps constrained by a spring) is so real that it is much safer to consider the inertia torque $I_G\ddot{\theta}$, plus an inertia force $ma\ddot{\theta}$, acting at G, as shown in fig. 5.2. We now have, taking moments about O,

$$Pa - I_g\ddot{\theta} - (ma\ddot{\theta}) \times a = 0$$

or
$$Pa - (I_g + ma^2)\ddot{\theta} = 0$$

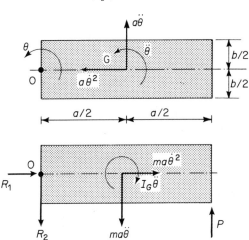

Figure 5.2

180 Dynamics of Mechanical Systems

and the equation is precisely the same as that obtained by the line of reasoning which we wish to discourage. The use of I_O breaks down completely if the hinge reactions R_1 and R_2 are required. To obtain these forces there is no alternative to the presentation in fig. 5.2, where we have

$$R_1 + ma\dot{\theta}^2 = 0 \quad \text{and} \quad R_2 + ma\ddot{\theta} - P = 0$$

It is pointless to engage in argument about the actual existence of inertia forces. D'Alembert equations can be written with the tongue in the cheek or otherwise, as it pleases, it makes no difference to the calculations. There is only one reason for using D'Alembert's principle: it simplifies the analysis of difficult problems. An important practical consequence of this simplification is that the chances of error are reduced.

5.4 Rotary balancing

The inertia force due to a concentrated mass m which is attached eccentrically to a rotating shaft is $-mr\omega^2$, where r is the distance of the mass from the axis of rotation and ω the angular velocity of the shaft. To illustrate just what this statement means in terms of a common engineering situation we will evaluate the force for one of the low-pressure blades of a 500-MW steam turbine. If the mean diameter of the blading is 2·5 m, and the mass of a single blade is 9 kg, the anchorage force for one blade at a shaft speed of 3000 rev/min is

$$9 \times 1\cdot25 \times \left(\frac{3000}{60} \times 2\pi\right)^2 = 1\cdot11 \times 10^6 \text{ N}\dagger$$

Due to the axial symmetry of a turbine rotor the inertia forces due to the blades balance each other and are not transmitted to the bearings which support the rotor. We shall now see how such a state of balance can be achieved for a shaft which carries a number of eccentric masses which are not symmetrically arranged.

Figure 5.3(a) shows a rotor which consists of a number of masses connected by light radial arms to a shaft. There is no restriction on the size of the masses as seen in the end view, but it will be assumed that all the centres of gravity lie in a single transverse plane, and that the thickness of the masses in the longitudinal direction (i.e., parallel to the shaft axis) is small compared with the radii r_1, r_2, r_3, etc. from the shaft axis to the centres of gravity of the masses. As the shaft rotates with constant angular velocity ω, the force $\mathbf{f}_i$ which is generated in the arm which connects a typical mass m_i to the shaft is given by the equation

$$\mathbf{f}_i - \omega^2 m_i \mathbf{r}_i = 0$$

We can write a similar equation for each of the masses. Adding all the equations we have

$$\sum \mathbf{f}_i - \omega^2 \sum m_i \mathbf{r}_i = 0 \tag{5.3}$$

† 10^4 N $\approx$ 1 tonf.

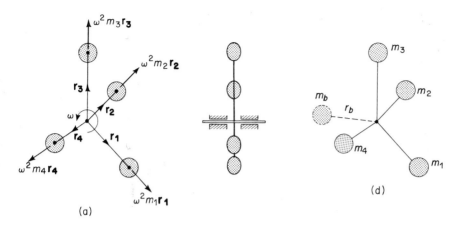

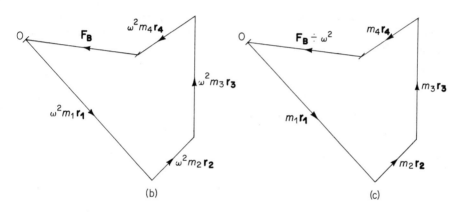

FIGURE 5.3

Now $-\sum \mathbf{f}_i$ is the resultant form applied to the shaft by the arms which carry the concentrated masses. The shaft itself is in equilibrium under the action of the arm forces and the bearing forces $\mathbf{F}_B$, so that

$$\mathbf{F}_B - \sum \mathbf{f}_i = 0 \tag{5.4}$$

Adding eqns. (5.3) and (5.4) we have

$$\mathbf{F}_B - \omega^2 \sum m_i \mathbf{r}_i = 0 \tag{5.5}$$

as the equilibrium equation for the whole system. It should be noted that this equation involves only the bearing forces and the inertia forces, the internal reactions f_i having balanced each other out. Whatever system we consider, this cancellation of the internal reactions always happens.

To evaluate eqn. (5.5) we can either express the $\mathbf{r}_i$ in terms of a convenient pair of unit vectors, and hence evaluate $\mathbf{F}_B$ numerically, or we can draw a vector

182 Dynamics of Mechanical Systems

polygon to scale as in fig. 5.3(b). In practice, it is easier and more usual to draw this diagram with all the vectors divided by ω^2, as in fig. 5.3(c). Note that the direction for $\mathbf{F_B}$ is such as to close the polygon and so represents the resultant force applied to the shaft by the bearings. The force applied to the bearings is equal and opposite.

The force on the bearing can be eliminated by the addition of an extra mass m_b, at the appropriate radius and angular position, as in fig. 5.3(d), where $\mathbf{r_b}$ is parallel to $\mathbf{F_B}$ and the mass and radius are so chosen that $-\omega^2 m_b \mathbf{r_b} = \mathbf{F_B}$. We then have

$$-\omega^2 m_b \mathbf{r_b} - \omega^2 \sum m_i \mathbf{r_i} = 0$$

or
$$m_b \mathbf{r_b} + \sum m_i \mathbf{r_i} = 0 \qquad (5.6)$$

With the addition of this extra mass the shaft is said to be balanced. Equation (5.6) shows that the balancing mass which is required does not depend on ω, so that if the rotor is balanced for one speed, it is balanced for all speeds.

The physical significance of eqn. (5.6) is that for a balanced rotor the centre of gravity of all the masses, including the balancing mass, lies on the axis of the shaft. This means that if the shaft is mounted horizontally in frictionless bearings it will rest in equilibrium in any angular position. For this reason, a rotor for which eqn. (5.6) is satisfied is said to be *statically balanced*.

A statically balanced rotor is also balanced in respect of angular acceleration. In fig. 5.4 an angular acceleration of $\dot{\omega}$ gives rise to the inertia forces and inertia couples as shown. If the resultant bearing force is $\mathbf{F'_B}$ we have

$$\mathbf{F'_B} - m_b \dot{\omega} \times \mathbf{r_b} - \sum m_i \dot{\omega} \times \mathbf{r_i} = 0$$

or
$$\mathbf{F'_B} - \dot{\omega} \times (m_b \mathbf{r_b} + \sum m_i \mathbf{r_i}) = 0$$

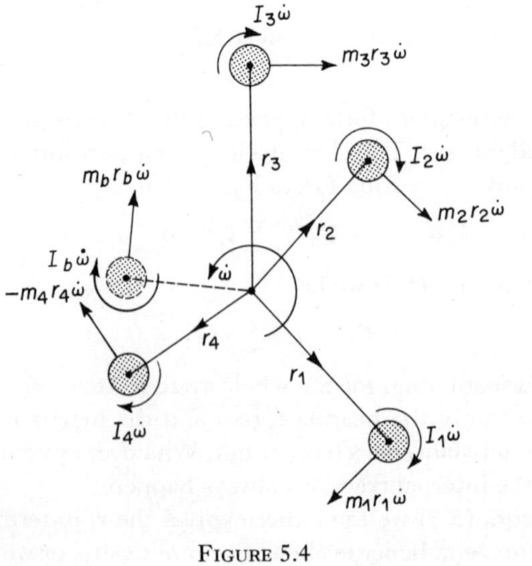

FIGURE 5.4

FORCE RELATIONSHIPS IN MECHANISMS II

But, from eqn. (5.6)
$$m_b \mathbf{r}_b + \sum m_i \mathbf{r}_i = 0$$
hence
$$\mathbf{F}'_B = 0$$

Equations (5.5) and (5.6) can be expressed somewhat differently when it is recognized that
$$\sum m_i \mathbf{r}_i = M \mathbf{r}_g$$
where $M = \sum m_i$ is the total mass of the rotor and $\mathbf{r}_g$ defines the position of the centre of gravity relative to the axis of rotation. We then have

$$\mathbf{F}_B - \omega^2 M \mathbf{r}_g = 0 \qquad (5.5\text{(a)})$$

for an unbalanced rotor,

and
$$m_b \mathbf{r}_b + M \mathbf{r}_g = 0 \qquad (5.6\text{(a)})$$

for one which is balanced.

In order to avoid subjecting bearings to unnecessary forces, and vibration of the supporting structure, rotors such as the armatures of electrical motors, blowers of centrifugal pumps and blowers, fans, motor-car wheels, etc., are required to run in a balanced condition. Manufacturing tolerances generally result in an actual rotor being out of balance to some extent. If the speed of operation is modest this inherent lack of balance may be acceptable, but for high-speed operation it is necessary to check and correct the state of balance. The crudest check is simply to support the rotor in its bearings and see if it is heavy on one side, generally friction in the bearings will prevent this test from revealing all but a gross lack of balance. For small rotors, balancing machines are available which measure directly the amount of imbalance and its angular position relative to an arbitrarily chosen radial datum. A less direct method which can, if need be, be carried out with the rotor *in situ* and is applicable to all sizes of rotor, is as follows.

The rotor is supported in bearings on a flexible mounting and run at a speed well above the resonance speed, i.e., the speed at which the amplitude of vibration is a maximum. The amplitude of the resulting vibration a_0 is then measured (sections 6.17 and 9.7). This procedure is then repeated at the same speed with a small mass attached eccentrically to the rotor at a convenient point; let the resulting amplitude of vibration be a_1. The small mass is then removed and attached to the diametrically opposite point at the same radius; at the appropriate speed the amplitude of vibration is now a_2.

The original out-of-balance $M\mathbf{r}_g$ can be represented by vector $\overrightarrow{OA}$ in fig. 5.5. On adding the additional mass m at radius r the resulting out-of-balance becomes $\overrightarrow{OB}$, where $\overrightarrow{AB} = m\mathbf{r}$. The effect of moving the mass to the diametrically opposite point is to reverse the vector $m\mathbf{r}$, giving $\overrightarrow{OC}$ as resultant. It is reasonable to assume that the amplitude of vibration measured in each of the three tests is proportional to the resultant out-of-balance.† We have therefore, in fig. 5.5(a),

$$OA = ka_0, \quad OB = ka_1, \quad OC = ka_2$$

where k is a suitable scale factor. To draw the diagram to scale as it is presented in fig. 5.5(a) would require a trial and error method. A neater solution is

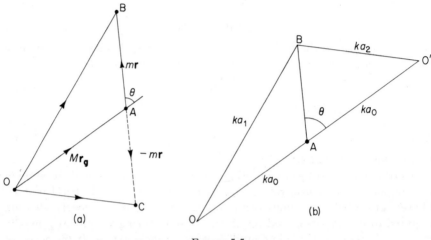

FIGURE 5.5

obtained by pivoting triangle OAC about A to bring C into coincidence with B, as shown in fig. 5.5(b). This diagram can be drawn to scale without difficulty. Direct measurement gives the angle θ between $\mathbf{r}_g$ and $\mathbf{r}$. To obtain the magnitude of $M\mathbf{r}_g$ we have

$$\frac{|M\mathbf{r}_g|}{|m\mathbf{r}|} = \frac{OA}{AB}$$

Since m and r are known, $M r_g$ can be calculated. In drawing triangle OBO′, fig. 5.5(b), a choice must be made between drawing the diagram as shown or its mirror image. Both diagrams give the same value for $M r_g$ but different signs to the value of θ. The correct sign is determined by experiment. This ambiguity can be removed by using three equally spaced positions for the mass m instead of two (see exercise 5.8). It is entirely a matter of taste which method one uses.

Equation (5.6) is sufficient to establish the equilibrium state of a rotor only if all the masses lie in the same plane perpendicular to the axis of the shaft. In fig. 5.6, the two small masses are coplanar with the axis of the shaft but are not

† It is shown in section 9.4 that the rotor tends to rotate about its centre of gravity at high speed so that

$$a_0 = |\mathbf{r}_g|, \quad a_1 = \left|\frac{M\mathbf{r}_g + m\mathbf{r}}{M + m}\right|, \quad \text{and} \quad a_2 = \left|\frac{M\mathbf{r}_g - m\mathbf{r}}{M + m}\right|.$$

Provided $m \ll M$ the error in the given method is small.

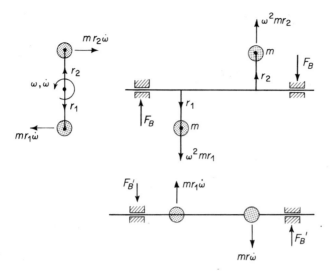

FIGURE 5.6

in the same plane perpendicular to the axis of the shaft. If the two masses are equal, and if the arms are of the same length, $\sum mr = 0$ and the shaft is statically balanced. Nevertheless, it is clear that the inertia forces which arise from the angular velocity ω are going to cause bearing reactions $\mathbf{F_B}$, and that an angular acceleration $\dot{\omega}$ results in bearing reactions $\mathbf{F'_B}$.

There are two methods of investigating the state of balance for long rotors, i.e., rotors for which the distribution of mass along the shaft must be considered. If we wish to calculate the bearing forces, one method is to determine the forces at the two bearings by each of the eccentric masses individually. The resultant force on a bearing is the vector sum of the forces due to the individual masses. In fig. 5.7, the inertia force $m\mathbf{r}\omega^2$ due to the single concentrated mass is in equilibrium with the bearing reactions

$$\mathbf{F_A} = -m\mathbf{r}\omega^2 b/l, \quad \text{and} \quad \mathbf{F_B} = -m\mathbf{r}\omega^2 a/l$$

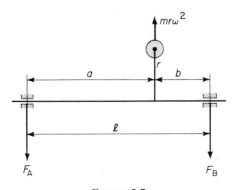

FIGURE 5.7

186 Dynamics of Mechanical Systems

In fig. 5.8, a shaft carries three concentrated masses on arms set at 120 degrees to each other. Using the result already given for a single concentrated mass, the vector diagrams are drawn to determine the resultant bearing reactions. Since

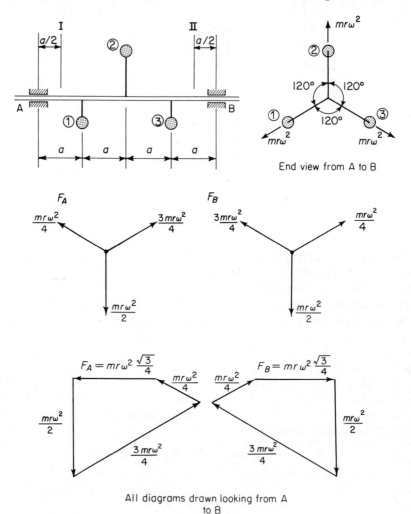

FIGURE 5.8

the shaft is statically balanced (the resultant inertia force is zero) we would expect these two reactions to constitute a couple, as indeed they do, the magnitude of the couple being $\sqrt{3}\, mr\omega^2/a$. To balance this shaft, two balancing masses are needed and which may be situated in any two conveniently located transverse planes. If the chosen planes are I and II, as indicated in fig. 5.8, we repeat the construction already given but compute the bearing reactions as if the bearings were located in planes I and II. It is left, as an exercise to the

FORCE RELATIONSHIPS IN MECHANISMS II 187

reader, to show that the equivalent bearing reactions are both of magnitude $mr\omega^2/\sqrt{3}$. Balancing masses, for which $m_b r_b = mr/\sqrt{3}$, can now be added in planes I and II to reduce the overall inertia effects to zero. As well as being statically balanced the rotor is now said to be *dynamically balanced*.

An alternative method of determining the bearing reactions at A and B (or of determining the balance masses required in I and II) is to compute the moment about A, say, of all the inertia forces and equate this with the moment due to the bearing reaction at B. The reaction at A can then be determined either by forming the vector sum of all the forces on the rotor, or by taking moments about B. The safest thing to do is to determine both $\mathbf{F_A}$ and $\mathbf{F_B}$ by taking moments, and then to use the fact that the vector sum of all the forces should be zero as a check on the calculations.

The calculation itself can be done numerically or graphically. If numerical calculation is preferred, the vertical and horizontal components of $\mathbf{F_B}$ are determined by taking moments about respective horizontal and vertical axes through axes through A. The calculation in terms of the unit vectors $\mathbf{i}$, $\mathbf{j}$, and $\mathbf{k}$ in fig. 5.9, can be set out as given in table 5.1.

Table 5.1

		Moments about A		Moments about B	
(1)	(2)	(3)	(4)	(5)	(6)
Mass	Inertia force $\mathbf{f} \div mr\omega^2$	Moment arm $\mathbf{e} \div a$	Moment $\mathbf{e} \times \mathbf{f} \div mr\omega^2 a$	Moment arm $\mathbf{e} \div a$	Moment $\mathbf{e} \times \mathbf{f} \div mr\omega^2 a$
1	$-\dfrac{\mathbf{j}}{2} - \mathbf{k}\dfrac{3}{2}$	$\mathbf{i}$	$\mathbf{i} \times \left(-\dfrac{\mathbf{j}}{2} - \mathbf{k}\dfrac{3}{2}\right)$	$-3\mathbf{i}$	$\mathbf{i} \times \left(\mathbf{j}\dfrac{3}{2} + \mathbf{k}\dfrac{3}{2}\right)$
2	$\mathbf{j}$	$2\mathbf{i}$	$\mathbf{i} \times (2\mathbf{j})$	$-2\mathbf{i}$	$\mathbf{i} \times (-2\mathbf{j})$
3	$-\dfrac{\mathbf{j}}{2} + \mathbf{k}\dfrac{3}{2}$	$3\mathbf{i}$	$\mathbf{i} \times \left(-\mathbf{j}\dfrac{3}{2} + \mathbf{k}\dfrac{3\sqrt{3}}{2}\right)$	$-\mathbf{i}$	$\mathbf{i} \times \left(\dfrac{\mathbf{j}}{2} - \mathbf{k}\dfrac{\sqrt{3}}{2}\right)$
$\sum$	0		$\mathbf{i} \times (\mathbf{k}\sqrt{3})$ $= 4a\mathbf{i} \times \mathbf{F_B} \div mr\omega^2 a$		$\mathbf{i} \times (\mathbf{k}\sqrt{3})$ $= -4a\mathbf{i} \times \mathbf{F_A} \div mr\omega^2 a$

Column (2) lists the inertia forces for the individual masses with the common factor $mr\omega^2$ omitted, column (3) gives the distances along the shaft measured from A, and column (4) gives the individual moments about A. The sum of column (4) equals the moment about A of $\mathbf{F_B}$ since

$$\overrightarrow{AB} \times \mathbf{F_B} - \sum \mathbf{e} \times \mathbf{f} = 0$$

188 Dynamics of Mechanical Systems

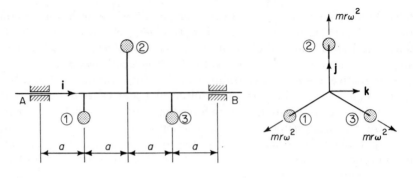

FIGURE 5.9

In our example

$$\sum \mathbf{e} \times \mathbf{f} = \mathbf{i} \times (\mathbf{k}\sqrt{3})mr\omega^2 a$$

and

$$\overrightarrow{AB} \times \mathbf{F_B} = 4a\mathbf{i} \times \mathbf{F_B} = \mathbf{i} \times (4\mathbf{F_B})$$

It is not normally permissible to say that if

$$\mathbf{a} \times \mathbf{b} = \mathbf{a} \times \mathbf{c}$$

then $\mathbf{b} = \mathbf{c}$, because in forming the vector products $\mathbf{a} \times \mathbf{b}$ and $\mathbf{a} \times \mathbf{c}$ we eliminate any components of $\mathbf{b}$ and $\mathbf{c}$ in the $\mathbf{a}$ direction. If, however, we know that $\mathbf{b}$ and $\mathbf{c}$ have no components in the $\mathbf{a}$ direction then the conclusion that $\mathbf{b} = \mathbf{c}$ is valid. Applying this reasoning to the problem in hand we conclude that

$$4a\mathbf{F_B} = \sqrt{3}\, mr\omega^2 a \mathbf{k}$$

or

$$\mathbf{F_B} = \frac{\sqrt{3}}{4} mr\omega^2 \mathbf{k}$$

Similarly, we obtain

$$\mathbf{F_A} = -\frac{\sqrt{3}}{4} mr\omega^2 \mathbf{k}$$

It should be noted that in column (5) of the table the moment arms are negative. This is because the relevant distances from B are in the sense opposite to that for which $\mathbf{i}$ is taken as positive.

As a final check, we note that

$$\mathbf{F_A} + \mathbf{F_B} - \sum mr\omega^2 = 0$$

where $\sum mr\omega^2$ is the sum of the elements in column (2).

In the above example, equal masses are set at 120 degrees to each other and are evenly spaced along the shaft. With a less regular distribution of the masses the calculation is rather less neat, and a graphical solution may be preferred.

FORCE RELATIONSHIPS IN MECHANISMS II

Each of the bracketed terms in column (4) of table 5.1 consists of an inertia force vector ($m\mathbf{r}\omega^2$) multiplied by the distance from A, about which moments are being taken, to the plane of motion of the appropriate mass. The sum of the modified inertia forces can be evaluated by drawing a vector polygon, as

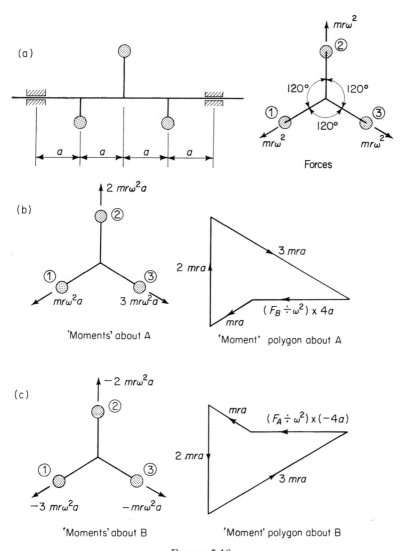

FIGURE 5.10

shown in fig. 5.10, where the product of an inertia force and its moment arm is referred to as an inertia moment. It will be noted that in drawing the moment polygon all the moments have been divided by the common factor ω^2. The title 'moment polygon' is not strictly correct for the given diagram because the effect of forming the vector product with the unit vector **i** in table 5.1 has been

190 Dynamics of Mechanical Systems

ignored. The effect of this operation is simply to rotate all the vectors of the moment polygon through 90 degrees about the axis of the shaft. Nothing is gained by drawing the moment polygon with this rotation, indeed the chances of confusion and error are increased.

In column (5) of table 5.1 the moment arms are entered as negative, on the ground that the distances are measured in the sense opposite to that taken as positive for the unit vector **i**. In column (6) the negative sign has been contained inside the brackets, that is to say it is incorporated in the inertia moments. To allow for this, the inertia moments about B, fig. 5.10(c), carry a negative sign. A simple rule which gives the signs correctly is: an inertia moment is positive if in the side view of the shaft the relevant mass is to the right of the point about which moments are being taken, and negative if it is to the left. The application of this rule to a shaft with overhung masses is illustrated in fig. 5.11.

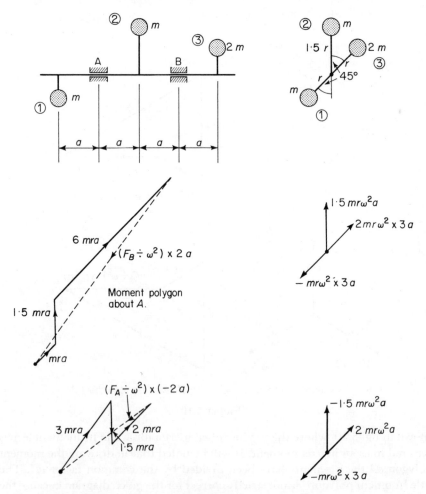

FIGURE 5.11

In practice, rotors are rarely of the form depicted in the diagrams which have been used to illustrate the theory of balancing. When plane rotors were considered it was pointed out that, provided the axial dimension of an eccentric mass is small compared with the radial dimensions of the rotor, it is correct to calculate the inertia force as if all the mass were concentrated at the centre of gravity of the mass, and that the actual shape or extent of the mass does not matter. With long rotors we have made the restriction that the eccentric masses shall be concentrated. Since practical rotors usually have their mass well distributed, we must examine the consequences of this.

Following on our treatment of plane rotors, the theory so far developed for long rotors clearly still applies if the concentrated masses are spread out in planes perpendicular to the axis of rotation. In particular, it is valid for a rotor which consists of a number of eccentrically mounted discs; provided that the plane of each disc is perpendicular to the axis of rotation, as for example in

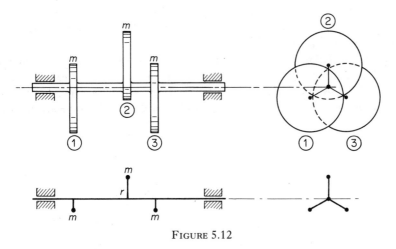

FIGURE 5.12

fig. 5.12, which shows two rotors which are equivalent to each other in respect of their overall state of balance. Expanding this idea, any rotor can be visualized as being composed of a number of slices arranged perpendicular to the axis of rotation, each disc can then be treated as a concentrated mass, and the state of balance investigated. The masses of individual discs may be coalesced if their centres of gravity lie on a line which is parallel to the axis of rotation. Figure 5.13 shows a conical rotor which is rotated about an axis parallel to, and distance ε from, its polar axis. The inertia forces due to the individual slices into which the rotor is divided are all parallel and their sum $\sum \delta m \varepsilon \omega^2 = \omega^2 \varepsilon \sum \delta m$ is clearly equal to $M \varepsilon \omega^2$ acting through the centre of gravity of the rotor. When the centres of gravity of the elementary slices do not all lie on a line parallel to the axis of rotation, the inertia effects, in general, are equivalent to a moment about the centre of gravity in addition to the force $M \varepsilon \omega^2$. The calculation of this moment for rotors which it is not reasonable to treat as consisting of a

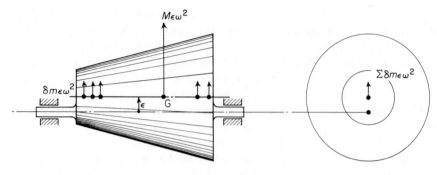

FIGURE 5.13

number of concentrated masses is complicated. Fortunately, it is rarely necessary to do such calculations. In practice, most rotors have axial symmetry and such unbalance as exists occurs as a result of manufacturing tolerances, and the problem of determining what corrections are needed is solved by experiment rather than calculation.

In order to balance a long rotor, two conveniently sited planes must be selected where balance masses can be added or alternatively excess material removed. The rotor is then supported in bearings which are carried by a cradle, fig. 5.14, which rests on springs. The motion of the cradle and rotor is further

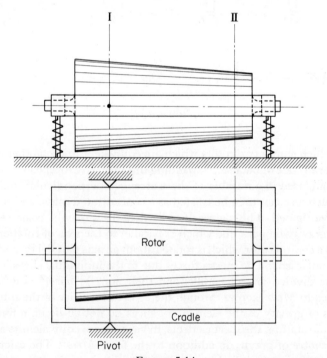

FIGURE 5.14

constrained by transverse pivots whose axis can be located at any point along the centre line of the rotor. The pivot axis can, in particular, be located in one of the two planes I and II in which balancing masses are to be added. If we consider the resultant out-of-balance to be equivalent to two eccentric masses, one in each of the balancing planes, it can be seen that when the rotor is turned the inertia force due to the equivalent mass in plane I, fig. 5.14, will be reacted entirely at the transverse pivot. The equivalent mass in plane II, on the other hand, will cause the rotor to oscillate about the pivot. The equivalent out-of-balance mass in plane II is now determined by adding a trial mass in two or

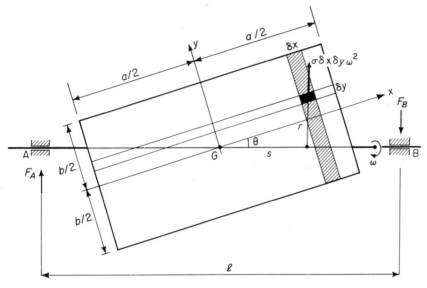

FIGURE 5.15

three positions in that plane, noting the amplitude of the resulting vibrations for each position when the rotor is spun at high speed, and following the graphical procedure explained earlier in relation to the balancing of a plane rotor. When a satisfactory degree of balance has been achieved in respect of plane II, either by adding (or more usually removing) metal, the procedure is repeated with the pivot located in plane II. As a final check, the rotor is then rotated with the transverse pivot removed completely.

To conclude the subject of rotary balancing, we will calculate the bearing forces induced when a rectangular plate which is rotated about a fixed axis which passes through its centre of gravity G, but is inclined at an angle θ (fig. 5.15) to its centre-line Gx.

As the mass centre of the lamina lies on the axis of rotation there is no resultant inertia force, and the bearing reactions are done solely to an unbalanced couple. To evaluate this couple we calculate first the inertia force due to the small shaded rectangular element shown in fig. 5.15 and integrate over the area of the lamina to obtain the overall effect of all such forces.

The mass of the element is $\sigma\,\delta x\,\delta y$ where σ is the mass per unit area and δx and δy are increments of length in the directions of the axes Gx and Gy. The inertia force due to the element is therefore $\sigma m r \omega^2\,\delta x\,\delta y$, where r is the distance of the element from the axis of rotation. If s is the distance along this axis, the moment about G is

$$\delta Q = \sigma r s \omega^2\,\delta x\,\delta y$$

It can readily be shown that r and s expressed in terms of the coordinates x and y of the element are

$$r = x\sin\theta + y\cos\theta$$
$$s = x\cos\theta - y\sin\theta$$

so that, substituting for r and s,

$$\delta Q = \sigma\omega^2[(x^2 - y^2)\cos\theta\sin\theta + xy(\cos^2\theta - \sin^2\theta)]\,\delta x\,\delta y$$

On integrating with respect to y, between the limits $y = \pm b/2$, we obtain

$$\delta Q = 2\sigma\omega^2\left\{\left[x^2\frac{b}{2} - \frac{1}{3}\left(\frac{b}{2}\right)^3\right]\cos\theta\sin\theta\right\}\delta x$$

as the moment due to the lightly shaded strip. Integration with respect to a, between the limits $x = \pm a/2$, gives

$$Q = \frac{1}{12}\sigma\omega^2(a^3 b - ab^3)\cos\theta\sin\theta$$

$$= \frac{1}{24}M\omega^2(a^2 - b^2)\sin 2\theta$$

The bearing reactions are in the directions indicated with

$$F_A = F_B = Q/l$$

This method of evaluating bearing reactions from 'first principles' has the great merit of being straightforward, and does not rely on any difficult concepts. It can, however, be tedious to apply to complicated rotors, and except in very simple cases becomes much too unwieldy to apply when the axis of rotation of the rotor changes its direction with time.

5.5 Gyroscopic effects — circular rotor

A change in the direction of the spin-axis of a rotor brings into play inertia effects which are generally referred to as gyroscopic effects. A gyroscope consists, essentially, of a flywheel which is rotated at high speed about the axis of its shaft, and is supported so that the direction of the axis can change. Navigational gyroscopes are based on the principle that unless a torque is applied to the rotor about an axis perpendicular to its spin axis the direction of the spin axis will not change. In case this seems merely to state what is obvious

FORCE RELATIONSHIPS IN MECHANISMS II 195

we must amplify this statement a little to state that even if the spin axis changes its direction at a constant rate, that is, the axis itself has a constant angular velocity, a torque must be applied. Furthermore, the axis of the torque is perpendicular to the axis about which the rotation occurs. This somewhat mysterious behaviour is readily illustrated by a toy gyroscope, fig. 5.16. The

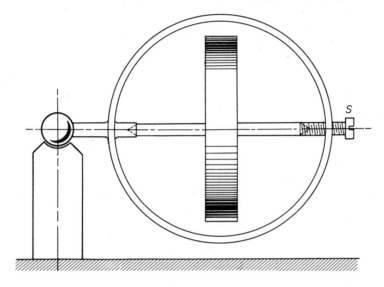

FIGURE 5.16

gyroscope is supported on a pedestal by a spherical joint, and with the rotor spinning at high speed the whole gyroscope will rotate slowly about the pedestal with the rotor axis describing a horizontal plane, apparently in defiance of gravity. The rotation of this spin axis about the vertical axis is called *precession*. The addition of an extra weight hung on the adjusting screw S leads apparently to an instantaneous increase in the rate of precession without any vertical movement.

To explain this behaviour we will consider a thin rotor which is supported so that it is free to rotate about any axis in its own plane. This can be achieved by the gimbal type of mounting shown in fig. 5.17(a), where the bearings BB allow rotation about a horizontal diameter of the rotor, and CC allow rotation about a vertical diameter. The essential difference between the system in fig. 5.17 and the one shown in fig. 5.16 is that the centre of gravity of the rotor is fixed in space instead of moving. The addition of a weight at A_1 causes equal and opposite forces to be applied to the rotor through the bearings A_1 and A_2, and the consequent motion is a steady rotation about the axis CC. We now have the situation defined in fig. 5.17(b), where Q is the torque applied to the rotor, ω_s is its spin velocity, and ω_p is the velocity of precession.

At any given instant the resultant angular velocity of the rotor is the vector sum Ω of ω_s and ω_p, fig. 5.18. The inertia forces which act on the elementary masses which comprise the rotor are then as shown. The inertia force for a particular elementary mass for distance r from the instantaneous axis of rotation is $\delta m\, r\Omega^2$. This force can be split into components $\delta m\, y\Omega^2$ and $\delta m\, x\Omega^2$. The overall effect of the $\delta m\, x\Omega^2$ components acting on all the particles of the disc is zero; the $\delta m\, y\Omega^2$ components reinforce each other to give a couple about an axis through the centre O of the rotor, perpendicular to the spin and pre-

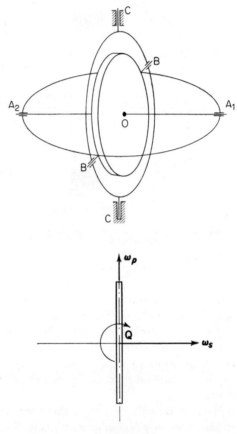

FIGURE 5.17

cession axes, as indicated in fig. 5.18. To some extent, the mystery of the gyroscopic couple is explained: it is due in part to the out-of-balance effect of the rotor turning about an oblique axis. This is not the whole story because the instantaneous axis of rotation is continually changing its direction as the rotor precesses. The rate of change of the resultant velocity vector Ω is an angular acceleration which implies the existence of a second set of inertia forces and a second contribution to the resulting gyroscopic couple. The two components of the gyroscopic couple $-\mathbf{Q}$, which counterbalances the applied couple $\mathbf{Q}$ in fig. 5.17, are calculated separately as follows:

FORCE RELATIONSHIPS IN MECHANISMS II

(a) Out-of-balance couple

The moment of the inertia force due to the element δm about the transverse axis BB is

$$\delta Q_1 = -\delta m\, y\Omega^2 \times y \tan\theta$$

where θ is the angle which the instantaneous axis of rotation makes with the spin axis, fig. 5.18, the positive direction for torques being as defined by the direction of Q in fig. 5.17.

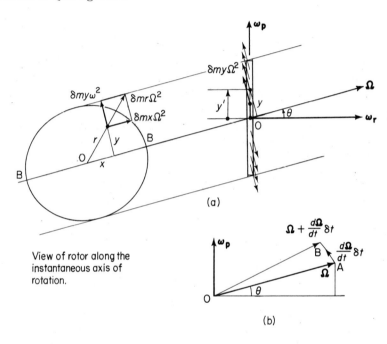

View of rotor along the instantaneous axis of rotation.

(a)

(b)

FIGURE 5.18

Substituting

$$y = y' \cos\theta$$

where y' is the distance of the element from BB, measured in the plane of the rotor, we have

$$\delta Q_1 = -\delta m\, y'^2 \Omega^2 \sin\theta \cos\theta$$

Summing for the whole disc, and putting $\Omega \cos\theta = \omega_s$, $\Omega \sin\theta = \omega_p$, we have

$$Q_1 = \omega_s \omega_p \sum \delta m\, y'^2$$
$$= -\omega_s \omega_p I_{BB}$$

where I_{BB} is the moment of inertia of the disc about a diameter.

(b) Angular acceleration couple

The disc is subjected to an angular acceleration because the resultant angular velocity vector Ω changes its direction with time. Effectively, the vector Ω rotates about the axis of ω_p with an angular velocity ω_p.

The rate of change of Ω is therefore

$$\frac{d\Omega}{dt} = \omega_p \times \Omega$$

which is a vector of magnitude

$$\omega_p \Omega \sin(90° - \theta) = \omega_p \Omega \cos\theta$$

We can deduce from the rule for vector products that the direction of this vector is along BB, fig. 5.18(a), and into the plane of the paper in the edge view of the disc. The same conclusion is reached from a consideration of fig. 5.18(b).

At time t the axis of rotation is OA. After a time interval δt the axis has moved to OB. The change in the velocity vector Ω is $(d\Omega/dt)\delta t$ which is normal to the plane of the paper and of magnitude $\Omega \cos\theta \, \omega_p \delta t$. The angular acceleration is therefore $\omega_p \Omega \cos\theta$ in a clockwise direction in relation to the edge view of the disc in fig. 5.18. The resulting inertia torque is

$$Q_2 = -\omega_p \Omega I_{BB} \cos\theta$$
$$= -\omega_p \omega_s I_{BB}$$

Q_1 and Q_2 reinforce each other to give a total inertia torque

$$Q = -2I_{BB}\omega_p\omega_s \qquad (5.7)$$

As the rotor is a thin circular disc, the perpendicular axis theorem tells us that the moments of inertia I_{BB} about a diameter is one half the polar moment of inertia J, so that

$$Q = -J\omega_p\omega_s \qquad (5.8)$$

It should perhaps be emphasized that the gyroscopic torque which we have deduced is the inertia (or D'Alembert) torque. The torque which must be applied to the rotor to produce the precession is equal and opposite to the inertia torque. Too much significance should not be attached to the fact that the torque which we have calculated carries a negative sign, this sign arises from the way in which the situation is defined in fig. 5.18. If the spin and precession axes are interchanged the gyroscopic torque is positive. A firm rule which gives the correct sign for the gyroscopic inertia torque in any situation is

$$Q = -J\omega_p \times \omega_s \qquad (5.9)$$

Unfortunately, confusion can arise because it is very easy to forget whether the equation refers to the inertia torque or the applied torque. It is also easy to reverse the sign inadvertently by putting ω_p and ω_s in the wrong order. A method of determining the sign of the torque which avoids these difficulties will be developed below, where an alternative derivation of the equation for the torque is given.

In deducing eqn. (5.8) it has been assumed that the spin axis and the axis of precession intersect at the centre of the disc. It can be shown, by a similar analysis, that the same equation holds when the precession takes place about an axis which is not in the plane of the disc, as for example happens in the case of the toy gyroscope depicted in fig. 5.16. This fact is important because a long cylindrical rotor spinning about its longitudinal axis can be considered to be composed of a number of these discs. By adding the gyroscopic couples due to all the discs we again arrive at eqn. (5.8), but with J now referring to the polar moment of inertia of a long rotor.

5.6 Angular momentum

So far, our method of determining the forces which must be applied to a rotor which has specified motion has been to consider the rotor to be composed of elementary particles, to calculate the accelerations of these particles and hence to find the inertia force associated with each particle, and finally to sum all these forces to obtain their resultant force and couple. This method has the merit that it is easy to appreciate how the forces arise. Its main disadvantage is that it becomes awkward when applied to rotors which are not very simple in form.

A much neater method of calculating inertia effects uses the momentum equations allied with vector algebra. We start by considering a single particle of mass m subjected to a force $\mathbf{F}$. The equation of the motion of the particle is

$$\mathbf{F} = m \frac{d\mathbf{v}}{dt}$$

Provided that the mass of the particle is constant, this equation can be written in the form

$$\mathbf{F} = \frac{d}{dt}(m\mathbf{v}) \tag{5.10}$$

that is to say, the force on the particle equals the rate of change of its momentum, which is the usual form in which Newton's second law is stated. Momentum is a vector quantity since it is simply the product of a vector quantity $\mathbf{v}$ and a scalar quantity m. It follows that momentum can be resolved and compounded according to the parallelogram law. The moment of $\mathbf{F}$ about a fixed point O is $\boldsymbol{\rho} \times \mathbf{F}$, where $\boldsymbol{\rho}$ is the position vector of the particle relative to O, and this must equal the moment of the vector $d/dt(m\mathbf{v})$ about O so that we have

$$\boldsymbol{\rho} \times \mathbf{F} = \boldsymbol{\rho} \times \frac{d}{dt}(m\mathbf{v})$$

This equation can be written

$$\boldsymbol{\rho} \times \mathbf{F} = \frac{d}{dt}(\boldsymbol{\rho} \times m\mathbf{v})\dagger \qquad (5.11)$$

$\boldsymbol{\rho} \times m\mathbf{v}$ is the moment of momentum of the particle about O, so that the moment $\mathbf{Q}$ about a fixed point O of the resultant force $\mathbf{F}$ acting on a particle is the rate of change of the moment of momentum about O. We shall use the symbol $\mathbf{h}$ to denote moment of momentum, and write eqn. (5.11) in the shortened form

$$\mathbf{Q} = \frac{d\mathbf{h}}{dt} \qquad (5.12)$$

In a system, which consists of several particles, an equation in the form of (5.12) can be stated for each particle. These equations can then be added to give

$$\mathbf{Q} = \sum \frac{d\mathbf{h}}{dt} \qquad (5.13)$$

An individual particle is subjected to a force of interaction with the other particles of the system, possibly as well as an externally applied force. The forces of interaction for the whole system must be in equilibrium, so need not be included in eqn. (5.13), and $\mathbf{Q}$ can be interpreted as referring only to externally applied moments. The only problem is, how do we calculate the moment of momentum $\mathbf{h}$ for a rigid body?

As it would not be appropriate here to embark on a full presentation of the theory of rigid body mechanics, we shall proceed by a plausible argument rather than a strict logic.

Let us consider first a lamina rotating with angular velocity ω about an axis in its plane, fig. 5.19, and through its centre of gravity. By assuming that the lamina is statically balanced we shall avoid excessive algebra. A set of Cartesian axes $Gxyz$ are drawn through G, Gx is along the axis of rotation, Gy is in the plane of the lamina and perpendicular to Gx, and Gz is perpendicular to the lamina. The momentum of a small element δm situated at the point (x, y) is $\delta m\, y$ and is in the z direction. The moment of this momentum about G has components $(\delta m\, y\omega).y$ about Ox, $-(\delta m\, y\omega).x$ about Oy, and zero about Oz. In terms of the unit vectors $\mathbf{i}$, $\mathbf{j}$, and $\mathbf{k}$ we have

$$\delta \mathbf{h}_G = \delta m\, y^2 \omega \mathbf{i} - \delta m\, xy\omega \mathbf{j}$$

Summing for the whole lamina

$$\mathbf{h}_G = \omega\left(\sum \delta m\, y^2\right)\mathbf{i} - \omega\left(\sum \delta m\, xy\right)\mathbf{j}$$

---

† Because $\dfrac{d}{dt}(\boldsymbol{\rho} \times m\mathbf{v}) = \dfrac{d\boldsymbol{\rho}}{dt} \times m\mathbf{v} + \boldsymbol{\rho}\dfrac{d}{dt}(m\mathbf{v})$, $\dfrac{d\boldsymbol{\rho}}{dt} = \mathbf{v}$, and $\mathbf{v} \times \mathbf{v} = 0$.

FORCE RELATIONSHIPS IN MECHANISMS II

$\sum \delta m \, y^2 = I_{xx}$ is the *moment of inertia* of the lamina about the axis Gx. $\sum \delta m \, xy = I_{xy}$ is the *cross product of inertia* with respect to the axes Gxy. From eqn. (5.13),

$$\mathbf{Q}_G = \frac{d\mathbf{h}_G}{dt} = (I_{xx}\mathbf{i} - I_{xy}\mathbf{j})\frac{d\omega}{dt} + \omega\left(I_{xx}\frac{d\mathbf{i}}{dt} - I_{xy}\frac{d\mathbf{j}}{dt}\right)$$

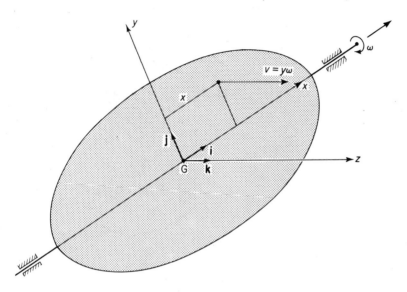

FIGURE 5.19

Now the unit vector $\mathbf{i}$ is in a fixed direction along the axis of rotation Gx so that $d\mathbf{i}/dt = 0$. The vector $\mathbf{j}$ rotates with the lamina with an angular velocity ω and so is constantly changing, with $d\mathbf{j}/dt = \omega\mathbf{k}$, so we have finally

$$\mathbf{Q}_G = \frac{d\mathbf{h}_G}{dt} = (I_{xx}\mathbf{i} - I_{xy}\mathbf{j})\frac{d\omega}{dt} - \omega^2 I_{xy}\mathbf{k} \qquad (5.14)$$

The first term on the right-hand side of this equation is a driving torque about the axis of rotation, and is simply the product of the moment of inertia about that axis and the angular acceleration. The other terms represent couples due to bearing reactions. It will be noted that even if the angular velocity of the lamina is constant there is still a bearing reaction couple unless the product of inertia I_{xy} is zero. When this happens, the rotor is dynamically balanced.

The product of inertia for a lamina is zero if either of the axes Gx or Gy is an axis of symmetry, fig. 5.20. For each element situated with coordinates (x, y) there is an identical one with coordinates $(x, -y)$. In these circumstances, $\sum \delta m \, xy$ must be zero. An immediate consequence of this is that if a lamina is dynamically balanced by virtue of symmetry about its axis of rotation, it must also be dynamically balanced for rotation about the perpendicular axis through its centre of gravity. If $I_{xy} = 0$, the pair of axes Gx and Gy are known as the *principal axes* at G, and the moments of inertia about these axes are *principal moments of inertia*. It can be shown that, if $I_{yy} \neq I_{xx}$, there is no other pair of orthogonal axes through G for which $I_{xy} = 0$. It can also be shown that there must be one pair of principal axes at G, though in the absence of symmetry the task of determining them is rather tedious.

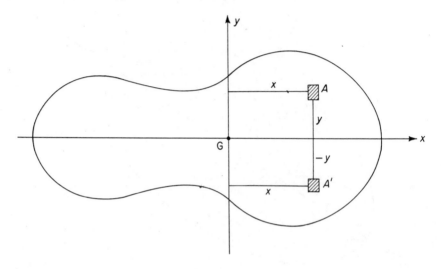

FIGURE 5.20

In the rather special case where two principal moments of inertia are equal, it is found that the principal axes are not unique, but on the contrary that $I_{xy} = 0$ for every pair of orthogonal axes through G, apart from the obvious case of a circular lamina, the simplest example of this is a square lamina.

Although it is useful as a demonstration that as a result of rotation about the **i** axis torques are induced about the **j** and **k** axes, eqn. (5.14) is not a very convenient one to apply in practice because of the labour involved in evaluating I_{xx} and I_{xy}. Provided that the lamina is sufficiently symmetrical for the principal axes to be determined by inspection, irrespective of whether or not one of them coincides with the axis of rotation, there is an easier method of determining $\mathbf{Q_G}$. We will take as an example a rectangular lamina rotating about an axis which makes an angle θ with its centre line, fig. 5.21. The out-of-balance couple for such a lamina was determined on page 194 by summing the inertia forces on its elements.

Axes I and II are axes of symmetry and hence principal axes; the principal moments of inertia are $I_\mathrm{I} = mb^2/12$ and $I_\mathrm{II} = ma^2/12$.

The angular velocity vector ω can be resolved into components $\omega \cos \theta$ along axis I, and $\omega \sin \theta$ along axis II. Since moment of momentum is a vector quantity it too can be resolved into components, or conversely determined by adding components. The component along axis I is $h_\mathrm{I} = I_\mathrm{I}\omega \cos \theta$, and the component along axis II is $h_{II} = I_{II}\omega \sin \theta$. We will not combine these two components at this stage but note that both change their direction as a result of the rotation of the lamina. It is to be noted also that both change if ω changes

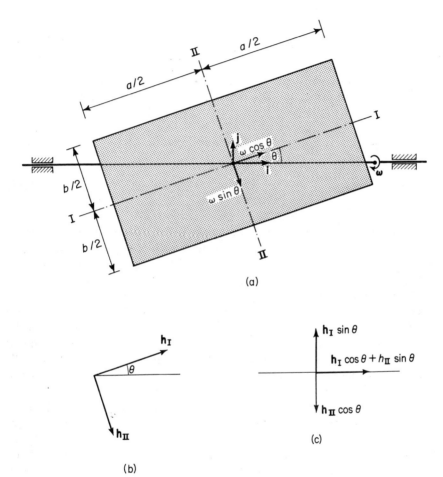

FIGURE 5.21

in magnitude. The changes in these components can best be appreciated by resolving them along, and at right-angles to, the axis of rotation. The resultant moment of momentum along the axis of rotation is

$$h_I \cos \theta + h_{II} \sin \theta$$
$$= (I_I \cos^2 \theta + I_{II} \sin^2 \theta)\omega$$

and so
$$\mathbf{h_i} = (I_I \cos^2 \theta + I_{II} \sin^2 \theta)\omega \mathbf{i}$$

This component is constant.

The resultant moment of momentum perpendicular to the axis of rotation in the j direction is

$$h_\mathrm{I} \sin\theta - h_\mathrm{II} \cos\theta$$

and so
$$\mathbf{h_j} = (I_\mathrm{I} - I_\mathrm{II}) \cos\theta \sin\theta \omega \mathbf{j}$$

This component is constant in magnitude but, due to the rotation of the lamina, changes in direction. The rate of change is in the $\mathbf{k}$ direction and is

$$\frac{d\mathbf{h}}{dt} = (I_\mathrm{I} - I_\mathrm{II}) \cos\theta \sin\theta \omega^2 \mathbf{k}$$

$$= -\frac{1}{24} M\omega^2(a^2 - b^2) \sin^2\theta \mathbf{k} \tag{5.15}$$

The negative sign shows that an anticlockwise torque must be applied about the $\mathbf{k}$-axis. In fig. 5.21, the positive direction for $\mathbf{k}$ is out of the plane of the paper towards the reader so that in this view the torque to be applied to the rotor by the bearings is clockwise.

This reasoning can be extended to apply to the rotation of a lamina about an axis not in its plane. It is evident that there can be no product of inertia with respect to an axis perpendicular to the plane of a lamina and an axis in the plane of the lamina, so that the axis perpendicular to the lamina through its centre of gravity is a principal axis. The perpendicular axis theorem tells us that the moment of inertia I_III about this axis equals the sum of the other two principal moments of inertia. There are now three components of the moment of momentum which must be considered: if I_I and I_II are the principal moments of inertia about axes in the plane of the lamina, the three components are $I_\mathrm{I}\omega_\mathrm{I}$, $I_\mathrm{II}\omega_\mathrm{II}$, and $I_\mathrm{III}\omega_\mathrm{III}$, where ω_I, ω_II, and ω_III are the components of the angular velocity vector along the three principal axes. It can be shown that any solid body has three mutually perpendicular principal axes at its centre of gravity, so that in this case also the components of the moment of momentum are $I_\mathrm{I}\omega_\mathrm{I}$, $I_\mathrm{II}\omega_\mathrm{II}$, and $I_\mathrm{III}\omega_\mathrm{III}$. The computational difficulties in calculating these components, and their rates of change, may be considerable.

Although the use of momentum theory has so far been presented in relation to rotation of a body about a fixed axis, there is no essential difference when the axis of rotation is itself in motion.

Figure 5.22 shows an aircraft propeller which is spinning with a constant angular velocity ω relative to the aeroplane. The aeroplane itself is turning in a horizontal circle with a constant angular velocity Ω about the vertical axis Oz. The principal axes of the propeller are Ox, II–II, and III–III, and the corresponding moments of inertia I_1, I_2, and I_3. The components of angular velocity about these axes are ω, $\Omega \sin\theta$, and $\Omega \cos\theta$ respectively, and the

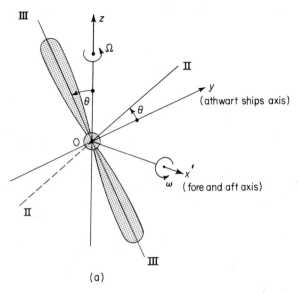

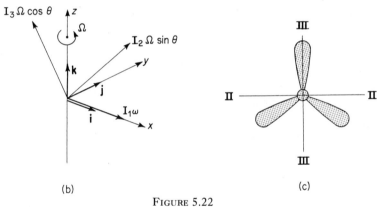

FIGURE 5.22

corresponding moments of momentum are as shown in fig. 5.22(b). In terms of the unit vectors **i**, **j**, and **k**, the moment of momentum is

$$\mathbf{h} = I_1\omega\mathbf{i} + (I_2\Omega \sin\theta \cos\theta - I_3\Omega \cos\theta \sin\theta)\mathbf{j}$$
$$+ (I_2\Omega \sin^2\theta + I_3\Omega \cos^2\theta)\mathbf{k}$$
$$= I_1\omega\mathbf{i} + \tfrac{1}{2}(I_2 - I_3)\Omega \sin 2\theta \cdot \mathbf{j} + (I_2 \sin^2\theta + I_3 \cos^2\theta)\Omega\mathbf{k}$$

This moment of momentum changes with time due to the changing value of θ, and the changing direction of the vectors **i** and **j**.

$$\mathbf{Q} = \frac{d\mathbf{h}}{dt} = I_1\omega\frac{d\mathbf{i}}{dt} + \tfrac{1}{2}(I_2 - I_3)\Omega \sin 2\theta \frac{d\mathbf{j}}{dt} - (I_2 \sin^2\theta + I_3 \cos^2\theta)\Omega\frac{d\mathbf{k}}{dt}$$
$$+ (I_2 - I_3)\Omega \frac{d\theta}{dt} \cos 2\theta \cdot \mathbf{j} + 2(I_2 - I_3)\Omega \frac{d\theta}{dt} \sin\theta \cos\theta \cdot \mathbf{k}$$

206 Dynamics of Mechanical Systems

Now
$$\frac{d\mathbf{i}}{dt} = \Omega\mathbf{j}, \quad \frac{d\mathbf{j}}{dt} = -\Omega\mathbf{i}, \quad \frac{d\mathbf{k}}{dt} = 0, \quad \text{and} \quad \frac{d\theta}{dt} = \omega$$

so that

$$\mathbf{Q} = \frac{d\mathbf{h}}{dt} = -\tfrac{1}{2}(I_2 - I_3)\Omega^2 \sin 2\theta \cdot \mathbf{i} + \left[I_1 + (I_2 - I_3)\cos 2\theta\right]\Omega\omega\mathbf{j}$$
$$+ (I_2 - I_3)\Omega\omega \sin 2\theta \cdot \mathbf{k} \qquad (5.16)$$

In normal circumstances, the angular velocity Ω of the aircraft is very small compared with the spin velocity ω of the propeller, so that the $\mathbf{i}$ component of $\mathbf{Q}$ is usually negligible compared with the other two. We are left, therefore,

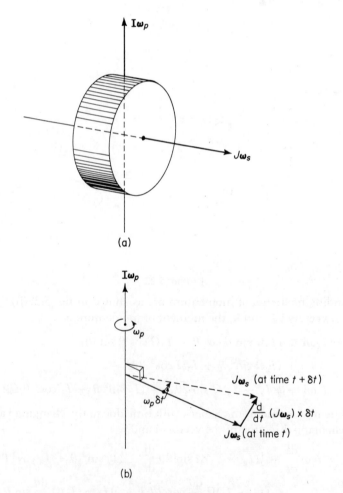

FIGURE 5.23

with a steady torque $I_1 \Omega \omega$ about Oy, and harmonically varying torques of magnitude $(I_2 - I_3)\Omega \omega$ about Oy and Oz. It will be noted that the frequency of torque variation is twice the spin velocity of the propeller. In the case of the two-bladed propeller indicated in fig. 5.22, $I_3 \ll I_2$. With a three-bladed propeller, as in fig. 5.22(c), $I_3 = I_2$, and only the steady torque $I_1 \Omega \omega$ exists.

This result will be recognized as being identical in form with eqn. (5.8) which gives the gyroscopic torque for a circular rotor. The use of momentum to derive this result is much simplified if account is taken of axial symmetry right from the start of the analysis, rather than introducing it as a special case at the end. In fig. 5.23 the rotor has a constant spin velocity ω_s about its polar axis and precesses with constant angular velocity ω_p about the vertical axis. At the given instant the moment of momentum has the two components $J\omega_s$ along the spin axis and $I\omega_p$ about the vertical axis. Because of the axial symmetry the moment of momentum about the vertical axis does not change as the rotor turns. The moment of momentum about the spin axis is constant in magnitude but changes its direction as a result of the precession, the magnitude of its rate of change being $(J\omega_s)\omega_p$. This is demonstrated in fig. 5.23(b) where the vectors are depicted at time t and time $t + \delta t$. In the interval δt the vector $J\omega_s$ rotates about the vertical axis through an angle $\omega_p \delta t$. The change in $J\omega_s$ is a vector of magnitude $J\omega_s \omega_p \delta t$ perpendicular to $J\omega_s$ and in the horizontal plane (i.e., perpendicular to ω_p). The rate of change of $J\omega_s$, and hence the magnitude of the torque required to cause the precession is $J\omega_s\omega_p$. The simplest way of determining the correct direction of this torque is by means of a sketch, such as that in fig. 5.23(b).

So far, we have concentrated on the steady precession of a gyroscope due to a steady torque. The behaviour in response to a varying torque will be considered in a later chapter.

5.7 Balancing of reciprocating masses

In rotary balancing we have the problem of eliminating an out-of-balance force which is constant in magnitude, but varying in direction. In balancing the inertia forces induced by a reciprocating motion, such as that of a piston of an automobile engine, we have to deal with a force which is constant in direction but varying in magnitude.

It was shown in chapter 4 that the acceleration a of the piston of a simple slider-crank mechanism is given by the equation

$$a = -r\omega^2(\cos\theta\, A_2 \cos 2\theta + A_4 \cos 4\theta + \cdots)$$

where r is the crank radius, ω the crank speed, and θ is the angle which the crank makes with the line of stroke. Because the speed is constant we can replace θ by ωt and then

$$a = -r\omega^2(\cos \omega t + A_2 \cos 2\omega t + A_4 \cos 4\omega t + \cdots) \quad (5.17)$$

208 Dynamics of Mechanical Systems

The inertia force due to the mass of the piston is

$$F = mr\omega^2(\cos \omega t + A_2 \cos 2\omega t + A_4 \cos 4\omega t + \cdots) \qquad (5.18)$$

The terms of this series which vary harmonically at frequencies equal to crank speed, twice crank speed, four times crank speed, etc., are referred to as

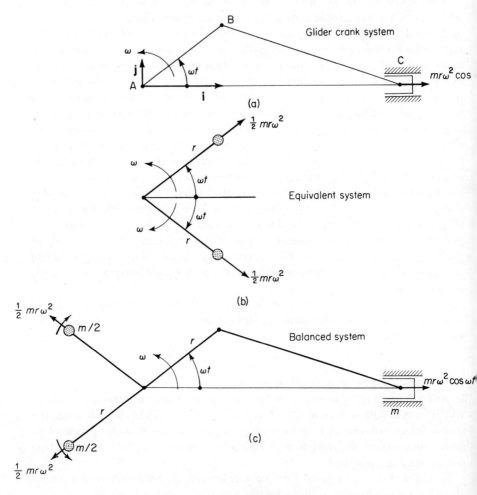

FIGURE 5.24

first order, second order, fourth order, etc., forces. It is convenient to consider the component forces separately, taking first the first order or primary force $mr\omega^2 \cos \omega t$, as shown in fig. 5.24. In terms of the unit vectors $\mathbf{i}$ and $\mathbf{j}$ the primary inertia force is

$$\mathbf{F}_1 = \mathbf{i} mr\omega^2 \cos \omega t \qquad (5.19)$$

FORCE RELATIONSHIPS IN MECHANISMS II

Adopting the somewhat artificial, but nevertheless frequently very useful, trick of introducing a pair of forces in equilibrium (in this case $\pm \mathbf{j}\tfrac{1}{2}mr\omega^2 \sin \omega t$) we can put

$$\mathbf{F}_1 = \tfrac{1}{2}m\omega^2(\mathbf{i} \cos \omega t + \mathbf{j} \sin \omega t) + \tfrac{1}{2}m\omega^2(\mathbf{i} \cos \omega t - \mathbf{j} \sin \omega t)$$
$$= \tfrac{1}{2}m\mathbf{r}\omega^2 + \tfrac{1}{2}m\mathbf{r}'\omega^2 \qquad (5.20)$$

where $\qquad \mathbf{r} = r(\mathbf{i} \cos \omega t + \mathbf{j} \sin \omega t)$

and $\qquad \mathbf{r}' = r(\mathbf{i} \cos \omega t - \mathbf{j} \sin \omega t)$

It follows that two masses of $m/2$, both at the same radius r and rotating in opposite directions at the crank speed ω, generate a resultant inertia force equal to the primary force due to a piston of mass m. The primary force can therefore be completely balanced by a pair of contrarotating masses as indicated in fig. 5.24(c). It is not essential for the masses to be placed at crank radius r; a pair of masses $\tfrac{1}{2}m_b$ at the same radius r_b will serve as well, provided $m_b r_b = mr$.

To balance the secondary component of the inertia force a pair of masses rotating at twice the crank speed are needed. The balancing masses in this case are $\tfrac{1}{2}A_2 m/4$, where A_2 is the coefficient of $\cos 2\omega t$ in the Fourier series for the piston acceleration. In general, the nth order inertia force requires a pair of masses $\tfrac{1}{2}A_n m/n^2$ rotating at speed $n\omega$, though in practice it is unlikely that any components of order higher than 2 will be of concern. The use of a pair of masses to achieve primary balance may depend on whether a contrarotating shaft can readily be provided. In the case of the secondary component, the use of two extra shafts must be contemplated.

When a system has a number of reciprocating masses the inertia effects may be self-balancing, and it is the extent to which this may happen that we will now consider.

In an in-line engine the pistons all move along parallel axes, which lie in a plane containing the axis of the common crankshaft. If, in an end-view of the crankshaft, the cranks are seen to be at angles α_2, α_3, etc. to one crank which is taken as a datum, fig. 5.25, the resultant primary inertia force due to the piston masses is

$$F_1 = m_1 r_1 \omega^2 \cos \omega t + m_2 r_2 \omega^2 \cos(\omega t + \alpha_2)$$
$$+ m_3 r_3 \omega^2 \cos(\omega t + \alpha_3) + \cdots \qquad (5.21)$$

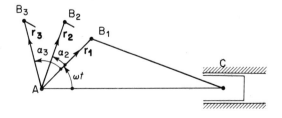

FIGURE 5.25

Following on eqn. (5.20), we can write this as

$$\mathbf{F}_1 = \tfrac{1}{2}m_1\mathbf{r}_1\omega^2 + \tfrac{1}{2}m_2\mathbf{r}_2\omega^2 + \tfrac{1}{2}m_3\mathbf{r}_3\omega^2 + \cdots$$
$$+ \tfrac{1}{2}m_1\mathbf{r}'_1\omega^2 + \tfrac{1}{2}m_2\mathbf{r}'_2\omega^2 + \tfrac{1}{2}m_3\mathbf{r}'_3\omega^2 + \cdots$$
$$= \tfrac{1}{2}\omega^2 \sum_{i=1}^{n} m_i\mathbf{r}_i + \tfrac{1}{2}\omega^2 \sum_{i=1}^{n} m_i\mathbf{r}'_i \qquad (5.22)$$

where n is the number of pistons.

The system is balanced if $\mathbf{F} = 0$. Since $\mathbf{r}_i$ and $\mathbf{r}'_i$ are symmetrically placed about AC,

$$\omega^2 \sum_{i=1}^{n} m_i\mathbf{r}_i \quad \text{and} \quad \omega^2 \sum_{i=1}^{n} m_i\mathbf{r}'_i$$

must both vanish when $\mathbf{F} = 0$, so that the condition for the resultant inertia force to be zero is simply that

$$\omega^2 \sum_{i=1}^{n} m_i\mathbf{r}_i$$

should be zero.

But $m_i\mathbf{r}_i\omega^2$ is the inertia force due to a mass m_i rotating with angular velocity ω at radius $\mathbf{r}_i$. Calculation of the state of primary balance of the reciprocating parts of an in-line engine is therefore the same as the calculation for the state of balance of an equivalent rotor, where the rotor masses and radii are equal to the piston masses and crank radii, with the same relative crank positions. If we restrict our consideration to engines in which all the piston masses are equal, and the cranks have a regular angular spacing with respect to each other the resultant primary out-of-balance force is zero if there are two or more pistons. With a four-cylinder two-stroke petrol engine, for example, the four cranks are at 90 degrees to each other so that the primary force polygon is a square, fig. 5.26. The order in which the cylinders fire does not affect the inertia forces, but we will assume that the firing order is 1–2–3–4.

To determine the primary out-of-balance moment we again adopt the procedure developed for rotating masses. Since the primary forces have no resultant they must constitute a couple, so that it does not matter which point along the axis of the crankshaft we choose to take moments about. This is illustrated in fig. 5.26, where two moment polygons are drawn, one for moments about the junction of crank 1 with the crankshaft, and the other for moments about the mid-point of the crankshaft. In both cases, the resultant moment is $mr\omega^2 a\sqrt{8}$, where a is the distance between adjacent cranks. In order to understand just what is meant by this moment, we must remember that although fig. 5.26 shows a simple rotor with masses m, strictly speaking we need two contrarotating rotors with masses $m/2$ to simulate the primary forces due to the pistons. Inertia forces perpendicular to the line of stroke thereby cancel each other leaving a resultant along the line of stroke. The maximum component which the moment vector has along the line of stroke is in this case $mr\omega^2 a\sqrt{8}$, and this happens when θ, which is the angle which crank 1 makes with the line of stroke, is 135 or 315 degrees. The reader may well be at a loss to understand how a system of forces in a particular direction, i.e., along the line of stroke, can give rise to a moment vector in the same direction. It must be

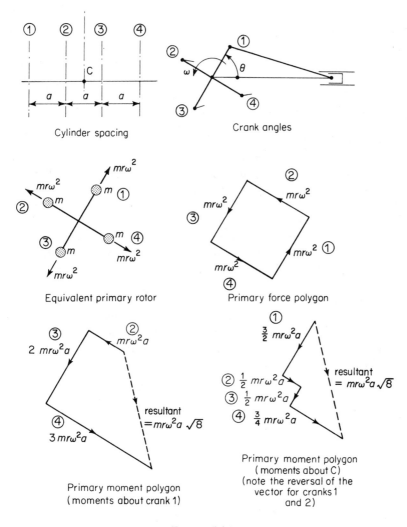

FIGURE 5.26

remembered, however, that to simplify the construction of the moment vector polygon we have omitted a 90-degree rotation of the moment vector (page 189). This means that, although the maximum primary couple occurs when the resultant moment vector coincides with the line of stroke, the axis of this moment is perpendicular to the line of stroke.

So far, we have dealt only with the primary force, that is the inertia force which is due to the first term in the series which expresses the acceleration of a piston. The resultant secondary force is

$$F_2 = m_1 r_1 \omega^2 A_2 \cos 2\omega t + m_2 r_2 \omega^2 A_2 \cos 2(\omega t + \alpha_2)$$
$$+ m_3 r_3 \omega^2 A_2 \cos 2(\omega t + \alpha_3) + \cdots \quad (5.23)$$

Comparing eqns. (5.21) and (5.23) for the primary and secondary forces respectively we see that the only difference is that in eqn. (5.23) a common factor A_2 has been introduced, and all the angles have been multiplied by two. Starting from eqn. (5.21) we showed that the primary forces balance if an equivalent rotor is balanced. By analogy, we deduce that the secondary forces are balanced if a second equivalent rotor is balanced. As the secondary forces occur with twice the frequency of the primary forces the equivalent secondary rotor turns at twice the speed of the engine crank. The crank radii and the size of the masses may be assumed to be chosen so as to have the appropriate inertia force $A_2 mr\omega^2$, the actual values being of no interest. To allow for the doubling of all the angles which occurs in eqn. (5.23), the angles between the cranks of the equivalent secondary rotor must be twice the corresponding angles between the engine cranks.

The equivalent secondary rotor and associated inertia forces for the four-cylinder engine, which we are studying by way of example, are shown in fig. 5.27. Clearly, the resultant secondary force is zero. The secondary out-of-balance couple has its maximum value of $2A_2 mr\omega^2 a$ whenever the resultant of the secondary couple polygon is parallel to the line of stroke, that is to say, when the cranks are aligned with the line of stroke. Now when crank 1 of the engine is at an angle θ to the line of stroke, crank 1 of the equivalent secondary rotor must be at an angle 2θ to the line of stroke; *vide* the doubling of angles in eqn. (5.23). Thus the secondary couple is a maximum when

$$2\theta = 0°, 180°, 360°, 540°, 720°, \text{etc.}$$

or
$$\theta = 0°, 90°, 180°, 270°, 360°, \text{etc.}$$

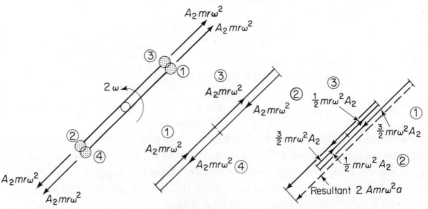

FIGURE 5.27

FORCE RELATIONSHIPS IN MECHANISMS II

The argument which we have outlined in relation to second order inertia effects can be extended to deal with inertia effects of any order. Thus to deal with fourth order effects we multiply all the crank angles by four. This brings all the cranks into phase so that the resultant inertia force is $4A_4 mr\omega^2$. The line of action of this force is parallel to the cylinder centre-lines through the centre point of the crankshaft. The coefficient A_4 is 0·01 when the ratio of the lengths of the connecting-rod and crank is 3:1 and 0·003 when the ratio is 4·5:1, so that the fourth and high order inertia effects are unlikely to be of much practical importance.

Changing the firing order has no effect on the resultant inertia forces. The couples may, however, be affected. In fig. 5.28, the primary and secondary couple polygons are drawn for the firing order 1–3–2–4. The primary couple

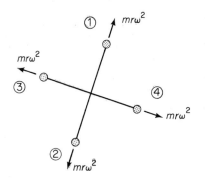

Equivalent primary rotor

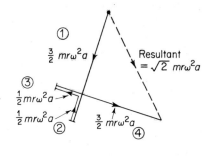

Primary couple polygon

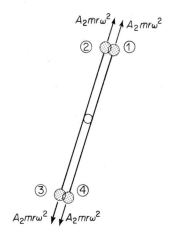

Equivalent secondary rotor

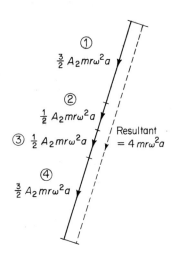

Secondary couple polygon

FIGURE 5.28

214 Dynamics of Mechanical Systems

is one-half the primary couple for the 1–2–3–4 arrangement, but the secondary couple is twice as large. Since the secondary effects are approximately $1/n$ the primary effects in magnitude, where n is the connecting-rod length divided by the crank radius, the 1–3–2–4 arrangement is clearly preferable, and it is the one normally used in practice.

With a four-stroke four-cylinder in-line engine it is usual for the two outer cranks to be in phase with each other and 180 degrees out of phase with the two inner cranks. The crankshaft is thus symmetrical about its mid-point and consequently moments balance about the mid-point for all orders of inertia forces. The resultant primary force is zero. The reader should verify that for all higher order forces the resultant is $4A_n mr\omega^2$.

In a two-cylinder Vee-engine, fig. 5.29, the centre-lines of the two cylinders intersect on the axis of the crankshaft, and the two connecting-rods hinge on a common crank-pin. The inertia effects due to the reciprocating parts of this engine are best investigated algebraically. The components V_1 and H_1 of the resultant primary force are

$$V_1 = mr\omega^2\left[\cos(\theta - \alpha) + \cos(\theta + \alpha)\right]\cos\alpha$$
$$= 2mr\omega^2 \cos^2\alpha \cos\theta \qquad (5.24)$$

and
$$H_1 = mr\omega^2\left[\cos(\theta - \alpha) - \cos(\theta + \alpha)\right]\sin\alpha$$
$$= 2mr\omega^2 \sin^2\alpha \sin\theta \qquad (5.25)$$

If the angle between the cylinder centre-lines is 90 degrees we have

$$V_1 = mr\omega^2 \cos\theta \qquad (5.24(a))$$

and
$$H_1 = mr\omega^2 \sin\theta \qquad (5.25(a))$$

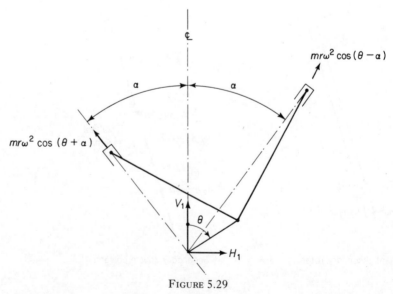

FIGURE 5.29

so that the primary inertia effect is equivalent to that of a mass m situated at the crank-pin. For such an engine, primary balance is readily achieved by adding a mass m_b at radius r_b opposite to the crank-pin, such that $m_b r_b = mr$.

In general, the primary effects are equivalent to a mass m rotating with the crank at radius r, and a second mass $m \cos 2\alpha$ rotating in the opposite direction to the crank also at radius r. Second and higher order forces can be treated similarly.

5.8 Inertia effects due to plane motion of a rigid body

The inertia effects due to plane motion of a rigid body are equivalent to a force $-M\mathbf{a}$ through the centre of gravity and a couple $-I_G \alpha$, where α is the angular acceleration of the body. In order to determine the degree of overall balance for a particular mechanism it is necessary to evaluate the inertia force and couple for each of the members of the mechanism for a series of positions throughout the range of movement, and to determine the line of action and the magnitude of the resultant force for each position. For the large body of mechanisms which operate cyclically, the resultant inertia effects vary periodically with time and can be expressed in the form of a Fourier series, and the main components can be balanced by masses rotating eccentrically at the appropriate speeds.

In view of the emphasis which has already been given in this chapter to the reciprocating engine it is reasonable to take the connecting-rod of such an engine as an example of a body in general plane motion. Procedure along the lines just indicated to determine the overall effects of the inertia of the connecting-rods in a multi-cylinder engine can hardly be faced with enthusiasm. The calculations are bound to be long and tedious, and it is natural to pause to consider whether this labour can be shortened. A little thought suggests that it might be avoided altogether; suppose that we can consider the distributed mass of a connecting-rod to be replaced by two concentrated masses, one at each end. We could then lump the mass at the crank-end in with the rotating mass of the crankshaft, and the mass at the piston-end with the mass of the piston. There would then be no need to consider the connecting-rods separately and the state of balance of the engine could be assessed by the methods already described for dealing with rotating and reciprocating masses respectively. For this line of reasoning to be valid the original connecting-rod and its replacement must be *dynamically equivalent*. This means (fig. 5.30) that the two bodies must have

1. the same total mass, i.e., $m_b + m_c = m$ \hfill (5.26)
2. the same centre of gravity $m_b b = m_c c$ \hfill (5.27)
3. the same moment of inertia $m_b b^2 + m_c c^2 = mk^2$ \hfill (5.28)

where k is the radius of gyration of the actual connecting-rod about its centre of gravity.

Unfortunately, we have three equations to satisfy but only variables to

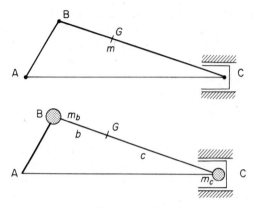

FIGURE 5.30

adjust, namely the two masses m_b and m_c. The lengths b and c are determined by the position of the centre of gravity of the connecting-rod in relation to the bearings. We can select m_b and m_c to satisfy the first two equations, and if this is done the proper allowance is thereby made for the acceleration of the centre of gravity. To do this we require

$$m_b = \frac{c}{b+c} m \quad \text{and} \quad m_c = \frac{b}{b+c} m$$

When these values for m_b and m_c are substituted in the third equation, we find

$$bc = k^2$$

If this relationship happens to be satisfied the two masses m_b and m_c are dynamically equivalent to the original connecting-rod. Due to the bearings, the mass of a connecting-rod tends to be concentrated at the two ends, so that we would not expect the product bc to be very different from k^2. If, however, there is a difference, it means that our 'equivalent' connecting-rod will have an inertia couple $-m(k^2 - bc)\alpha$ about an axis parallel to the crankshaft axis unaccounted for. This discrepancy may well be negligible, but it is interesting to investigate the extent to which we can expect the overall error to be self cancelling in a multi-cylinder in-line engine.

Referring to fig. 5.31, we have

$$\sin \phi = \frac{r}{nr} \sin \theta$$

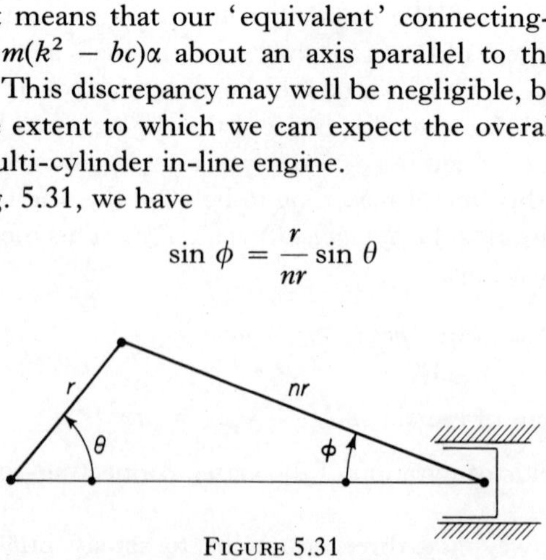

FIGURE 5.31

On differentiating with respect to time this becomes

$$\cos\phi \frac{d\phi}{dt} = \frac{1}{n}\cos\theta \frac{d\theta}{dt}$$

so that

$$\frac{d\phi}{dt} = \frac{\omega}{n}\cos\theta(1 - \sin^2\phi)^{-\frac{1}{2}}$$

On substituting for $\sin\phi$, expanding the bracketed term by use of the binomial theorem, and using appropriate trigonometrical relationships, this equation becomes

$$\frac{d\phi}{dt} = \frac{\omega}{n}\cos\theta + \frac{1}{8n^2}(\cos\theta - \cos 3\theta) + \cdots$$

$$= \frac{\omega}{n}\left(1 + \frac{1}{8n^2} + \cdots\right)\cos\theta - \left(\frac{1}{8n^2} + \cdots\right)\cos 3\theta - \cdots$$

Assuming ω to be constant, we can differentiate this expression with respect to time to obtain

$$\alpha = \frac{d^2\phi}{dt^2} = -\frac{\omega^2}{n}\left(1 + \frac{1}{8n^2} + \cdots\right)\sin\theta - \left(\frac{3}{8n^2} + \cdots\right)\sin 3\theta - \cdots$$

$$= -\frac{\omega^2}{n}(B_1\sin\theta - B_3\sin 3\theta - B_5\sin 5\theta - \cdots) \quad (5.29)$$

For two or more cylinders the primary couples due to the first term in this series must balance, in exactly the same way as the primary forces due to the piston masses balance. The third order couples will be very small (of the order of 3 per cent of the primary couple), they will in any case be in balance for any two-cylinder or four-cylinder engine, and for a six-cylinder two-stroke engine. For a six-cylinder four-stroke engine, however, the third order couples do not balance.

5.9 Transmission of inertia forces

Balance of the resultant inertia forces and couples for a particular piece of machinery does not mean that the inertia effects have been entirely eliminated. The inertia force due to the acceleration of an individual piston of a multi-cylinder engine still induces a side thrust on the cylinder wall, reactions at the connecting-rod and crankshaft bearings, and a torque in the crankshaft.

The problem of evaluating, say, the bearing reactions at the piston and crank due to the inertia of the connecting-rod does not involve us in any principles beyond those used in chapter 3. Inertia forces can be treated in exactly the same way as other forces, and we may appeal either to the methods of simple statics or virtual work to determine unknown reactions. There are, however, one or two points of detail which merit attention. To illustrate these points we will examine the crank and rocker mechanism shown in fig. 5.32. The velocity and acceleration diagram for the mechanism in the given configuration are drawn for a constant crank speed ω_i. The accelerations **a** of the centres of gravity of the links and their angular accelerations α can be deduced from the acceleration diagram and hence the D'Alembert force and torque for each member are determined. We will now consider how the driving torque Q_i needed to sustain this motion, and the consequent bearing reactions $\mathbf{R_A}$ and $\mathbf{R_D}$, are determined.

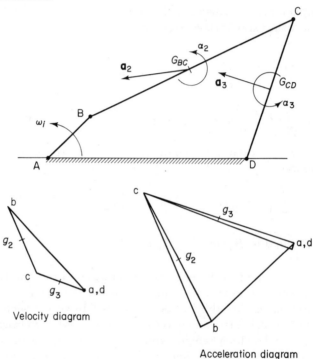

FIGURE 5.32

The method of virtual work is the most convenient method to use when one particular reaction is to be determined, and the rest are of no concern. The most likely reaction to merit individual attention is the driving torque Q_i. To determine this torque we can allow a small (virtual) rotation $\delta\theta_i$ of the crank AB and hence formulate the work equation

$$Q_i\,\delta\theta_i - \sum m\mathbf{a}.\boldsymbol{\delta} - \sum I\alpha\,\delta\phi = 0$$

where $\boldsymbol{\delta}$ is the virtual displacement of a centre of gravity and $\delta\phi$ is the virtual rotation of a member. If the virtual displacements are allowed to take place in a time δt, we have, on allowing both the displacements and δt tend to zero,

$$Q_i\cdot\frac{d\theta_i}{dt} - \sum m\mathbf{a}.\frac{d\boldsymbol{\delta}}{dt} - \sum I\alpha\,\frac{d\phi}{dt} = 0$$

and instead of virtual displacements we have virtual velocities. Now the virtual velocity $d\theta_i/dt$ can be of arbitrary amplitude so that we may, if we wish, set it equal to the actual angular velocity ω_i of the input crank. Furthermore, although we have referred to virtual displacements and velocities, the displacements with which we are dealing are the displacements which actually occur due to the rotation of the crank. It follows that if we allow $d\theta_i/dt = \omega_i$, the $d\boldsymbol{\delta}/dt$ and $d\phi/dt$ will be the velocities and angular velocities to be obtained from the velocity diagram. So we have

$$Q_i\omega_i - \sum m\,\mathbf{a}\cdot\mathbf{v} - \sum I\alpha\omega = 0 \qquad (5.30)$$

The scalar products $m\,\mathbf{a}\cdot\mathbf{v}$ are evaluated by measuring on the velocity diagram the component of $\mathbf{v}$ in the direction of $\mathbf{a}$. Care must be exercised to take proper account of signs.

The calculation of the bearing reactions and driving torque by the application of simple statics could proceed as follows:

A relationship between the horizontal and vertical components of the reaction at D can be obtained by taking moments about hinge C of the forces which act on CD, and equating the total to zero. A second relationship can be obtained by taking moments about hinge B of the forces which act on BC and CD, and equating the total to zero. The two equations thus obtained enable the reaction at D to be determined. Q_i then follows by taking moments about A, and finally $\mathbf{R_A}$ is evaluated by drawing a polygon of forces.

It is easier, however, to consider the inertia effects for each of the links separately, fig. 5.33. The inertia force $-m_1\mathbf{a}_1$ due to AB causes a reaction $\mathbf{R}_{A1}$ at A, and that is all. In the case of the connecting-rod BC, the first step is to combine the inertia force $m_2\mathbf{a}_2$ acting through the centre of gravity and the inertia couple $I_2\dot{\omega}_2$ into a single force $\mathbf{F}_2$, which is equal in magnitude and direction to $m_2\mathbf{a}_2$, but offset from it by an amount $p_2 = I_2\dot{\omega}_2/m_2 a_2$. BC is in equilibrium under the action of $\mathbf{F}_2$ and the hinge reactions at B and C. As the hinge reactions at C and D are the only forces which are at present considered to act on CD, these forces must both act along the centre line of CD. The situation with regard to crank AB is different in that it is subjected to a torque Q_{i2} at A as well as the reactions at its ends. This means that the reaction at hinge B is not, in general, directed along AB. Consideration of the equilibrium of link BC tells us that the reaction at B must be such that it passes through H_2, the point at which the force $\mathbf{F}_2$ intersects DC. A triangle of forces can now be drawn for BC and so the reactions at B and C are determined. The reaction at A is parallel to BH_2 and equal in magnitude to the reaction at B. The driving torque $Q_{i2} = R_{A2}\,d_2$, where d_2 is the distance by which A is offset from BH_2. A similar argument enables the hinge reactions due to the inertia of CD to be determined. The resultant reactions at A and D are determined by drawing two vector polygons. The resultant torque at A is $Q_i = Q_{i2} + Q_{i3}$.

220 Dynamics of Mechanical Systems

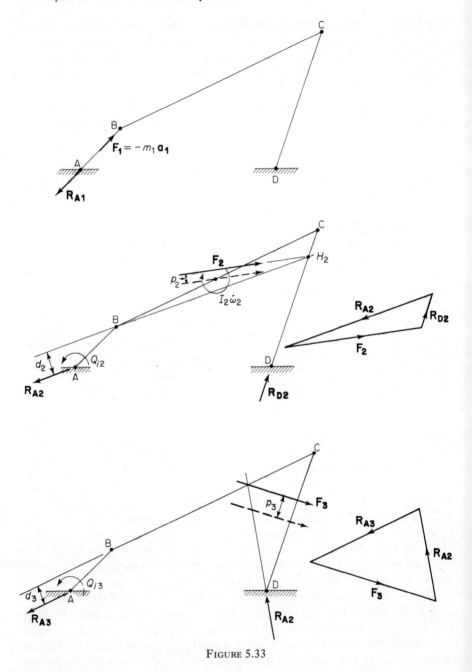

FIGURE 5.33

5.10 Dynamic response of mechanisms

So far, we have been concerned with calculating the forces which are applied to a mechanism, whether as driving forces or passive reactions, when it moves in a specified way. The converse problem of determining the motion of a mechanism under the action of a given set of forces will now be considered. We have shown that for the crank and rocker mechanism in fig. 5.33, the driving torque Q_i is given by

$$Q_i \omega_i - \sum m \mathbf{a} \cdot \mathbf{v} - \sum I \alpha \omega = 0 \qquad (5.30(a))$$

Now
$$\mathbf{a} = \frac{d\mathbf{v}}{dt} \quad \text{and} \quad \alpha = \frac{d\omega}{dt}$$

so that we can write

$$Q_i \omega_i - \sum m \left(\frac{d\mathbf{v}}{dt}\right) \cdot \mathbf{v} - \sum I \left(\frac{d\omega}{dt}\right) \omega = 0$$

or
$$Q_i \omega_i - \frac{d}{dt} \sum m \mathbf{v} \cdot \mathbf{v} - \frac{d}{dt} \sum \tfrac{1}{2} I \omega^2 = 0$$

and finally, on evaluating all the scalar products,

$$Q_i \omega_i = \frac{d}{dt} \sum \tfrac{1}{2} m v^2 + \frac{d}{dt} \sum \tfrac{1}{2} I \omega^2 \qquad (5.31)$$

$\sum \tfrac{1}{2} m v^2$ is the kinetic energy due to the linear motion of the centres of gravity of the various members, and $\sum \tfrac{1}{2} I \omega^2$ is the kinetic energy due to their rotations. Equation (5.31) tells us that the rate at which work is done by the driving torque θ_i equals the rate of change of kinetic energy of the mechanism. For a mechanism with a single degree of freedom, all the velocities and angular velocities are proportional to each other, and in particular all are proportional to the input angular velocity ω_i. We can therefore write eqn. (5.31) in the form

$$Q_i \omega_i = \frac{d}{dt} (\tfrac{1}{2} I^* \omega_i^2) \qquad (5.32)$$

where I^* is the *effective inertia* of the mechanism with respective to the input velocity ω_i. The effective inertia varies with the configuration, and to evaluate it for any particular configuration we draw the velocity diagram for an arbitrary input velocity ω_i, calculate the total kinetic energy and divide by $\tfrac{1}{2}\omega_i^2$. The way in which I^* varies with the position of the input crank for a particular mechanism is shown in fig. 5.34: the reader is invited to verify the calculation for a few of the points on this curve.

If the ω_i on the left-hand side of eqn. (5.31) is written as $d\theta_i/dt$ we can immediately integrate the equation to give

$$\int Q_i \, d\theta_i = \text{change in } (\tfrac{1}{2} I^* \omega_i^2)$$

that is to say: the work done by the input torque equals the change in kinetic energy. As an example, let us assume that the mechanism in fig. 5.34 starts

222 Dynamics of Mechanical Systems

from rest with $\theta_i = 0$ and with Q_i constant. When $\theta_i = 180$ degrees we see from the graph that $I^* = 0.9 \times 10^{-3}$ kg m² so that

$$Q_i \pi = \tfrac{1}{2}\omega_i^2 \times 0.9 \times 10^{-3}$$

hence $\qquad \omega_i = 83\sqrt{Q_i}$ rad/s

The above method of calculation is possible provided that Q_i is either constant, or a known function of θ_i. In the latter case, the work done by Q_i must be evaluated by analytical or graphical integration, but otherwise there

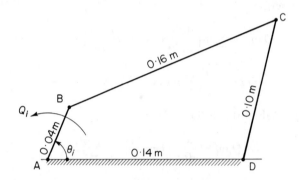

Uniform links with mass 1 kg/m.

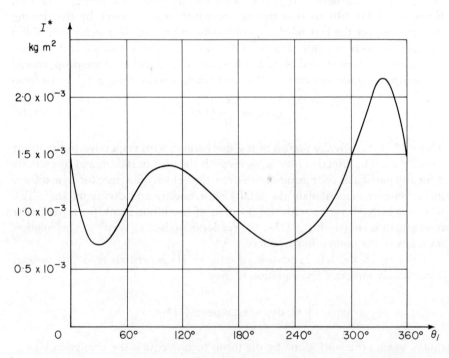

Figure 5.34

FORCE RELATIONSHIPS IN MECHANISMS II 223

is no difference from the simple example just considered. By repeating the calculation for a series of intermediate points, the variation of ω_i with θ_i can be determined. Figure 5.35 is drawn for $Q_i = 1$ N m. The instantaneous acceleration at any stage, e.g., $\theta_i = 30$ degrees, can be obtained from the graph, for

$$\frac{d^2\theta_i}{dt^2} = \frac{d}{dt}\left(\frac{d\theta_i}{dt}\right)$$

$$= \frac{d\theta_i}{dt}\frac{d}{d\theta_i}\left(\frac{d\theta_i}{dt}\right)$$

$$= \omega_i \frac{d\omega_i}{d\theta_i} \qquad (5.33)$$

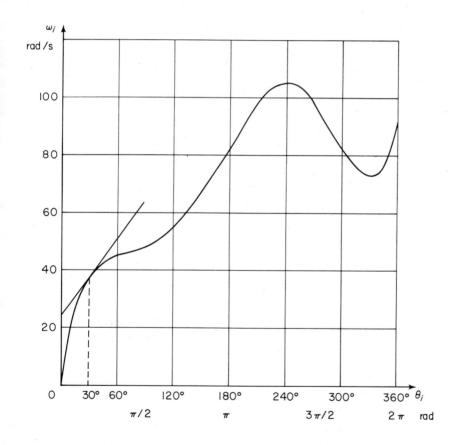

FIGURE 5.35

$d\omega_i/d\theta_i$ is determined by drawing a tangent to the curve at $\theta_i = 30$ degrees and measuring its slope, it is in fact 25 s^{-1}. Hence $d\omega_i/dt = 950$ rad/s^2.

224 Dynamics of Mechanical Systems

The time taken to move from one configuration to another can also be obtained. We note first that

$$dt = \frac{d\theta_i}{d\theta_i/dt} = \frac{d\theta_i}{\omega_i} \quad (5.34)$$

The time taken for a small angular movement is therefore given approximately, by dividing the displacement by the mean velocity for the interval. This can be done graphically by measuring the angle ϕ, fig. 5.36. For a large change of θ_i,

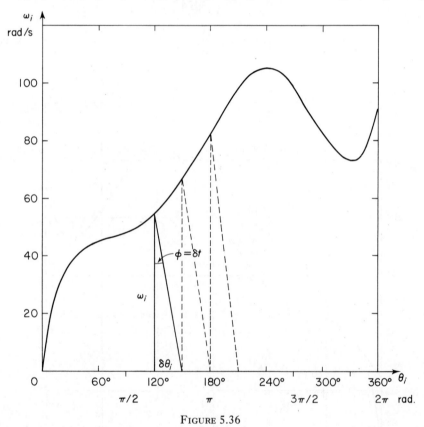

FIGURE 5.36

it is necessary to subdivide the movement up into a number of smaller intervals, and to sum the increments of time.

Inspection of fig. 5.35 reveals that it is not practicable to evaluate $\omega_i \, d\omega_i/d\theta_i$ at the origin, so that some alternative method must be used to find the initial acceleration of a mechanism which starts from rest. Equation (5.32) is again used, but it is treated somewhat differently:

$$Q_i \omega_i = \frac{d}{dt}(\tfrac{1}{2}I^*\omega_i^2)$$

On expanding the right-hand side we have

$$Q_i \omega_i = I^* \omega_i \frac{d\omega_i}{dt} + \tfrac{1}{2}\omega_i^2 \frac{dI^*}{dt}$$

or
$$Q_i = I^* \frac{d\omega_i}{dt} + \tfrac{1}{2}\omega_i \frac{dI^*}{dt}$$

$$= I^* \frac{d\omega_i}{dt} + \tfrac{1}{2}\omega_i^2 \frac{dI^*}{d\theta_i} \tag{5.35}$$

If the mechanism starts from rest, $\omega_i = 0$, so that

$$\left(\frac{d\omega_i}{dt}\right)_{t=0} = \frac{Q_i}{I^*} \tag{5.36}$$

If Q_i is known as a function of time, rather than as a function of θ_i, it is not practicable to proceed to a direct integration of eqn. (5.32); again eqn. (5.35) provides a feasible method of solution. First we rewrite it in the form

$$I^* \frac{d\omega_i}{dt} = Q_i - \tfrac{1}{2}\omega_i^2 \frac{dI^*}{d\theta_i}$$

We must start from an initial state in which θ_i and ω_i are known so that the right-hand side of this equation, and hence $d\omega_i/dt$, are calculable. We can therefore predict what the value of ω_i and θ_i will be after the lapse of a finite time interval. Since Q_i is known as a function of time the new value of Q_i is known and so again the right-hand side of the equation can be evaluated, a new value of $d\omega_i/dt$ obtained, and the cycle is then repeated. In this way, the subsequent motion of the system can be predicted step by step. The calculation is tedious and subject to accumulative errors, but can easily be programmed for solution by digital computer. This technique can of course be used if Q_i is constant, but generally direct integration of eqn. (5.32) is preferable.

Equation (5.35) also enables the response of the system to an impulse to be deduced. Let us suppose that Q_i is constant for a time interval Δt. We can integrate eqn. (5.35) over this time interval thus:

$$\int_0^{\Delta t} Q_i \, dt = \int_0^{\Delta t} I^* \frac{d\omega_i}{dt} dt + \int_0^{\Delta t} \tfrac{1}{2}\omega_i^2 \frac{dI^*}{d\theta_i} dt$$

$$= \int_1^2 I^* \, d\omega_i + \int_1^2 \tfrac{1}{2}\omega_i \, dI^*$$

If we now allow $\Delta t \to 0$, while keeping $\int_0^{\Delta t} Q_i \, dt$ finite, our input becomes an impulsive torque. Since we cannot contemplate an instantaneous change in the

226 Dynamics of Mechanical Systems

configuration of the mechanism I^* must remain constant over the period Δt, so that

$$\int_1^2 I^* \, d\omega_i \to I^*(\omega_{i2} - \omega_{i1}) \quad \text{as } \Delta t \to 0$$

and

$$\int_1^2 \tfrac{1}{2}\omega_i \, dI^* \to 0 \quad \text{as } \Delta t \to 0$$

In the limit we have

$$\left(\int_0^{\Delta t} Q_i \, dt\right)_{\Delta t \to 0} = I^*(\omega_{i2} - \omega_{i1}) \tag{5.37}$$

5.11 Linear systems

As the previous section shows, there is, in general, no easy solution to determining the motion of a mechanism which results from a disturbing force. Whether the equation of motion is expressed in the form of eqn. (5.32) or its extended form eqn. (5.35), we have to use numerical and semi-graphical techniques to solve it: a simple analytical solution exists only in special cases. One particular case which is of some importance and for which a simple solution does exist is that of a mechanism for which I^* is constant and independent of the configuration. When this happens, eqn. (5.35) reduces to

$$Q_i = I^* \frac{d\omega_i}{dt}$$

If Q_i is constant or a specified function of time this is a linear differential equation which can be integrated directly to give ω_i and θ_i as functions of time.

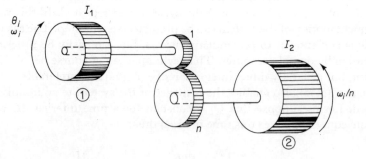

FIGURE 5.37

An example of a linear system is given in fig. 5.37, which shows two rotors connected through gearing, e.g., a turbine driving an electric generator. The kinetic energy in terms of the velocity ω_i of rotor 1 is

$$\tfrac{1}{2}I_1\omega_i^2 + \tfrac{1}{2}I_2(\omega_i/n)^2$$

The effective inertia at rotor 1 is therefore

$$I_1 + I_2/n^2$$

Inertia effects in linear systems are considered in greater detail in the next chapter which deals specifically with linear systems.

5.12 Flywheel design

The purpose of a flywheel is to limit the speed fluctuations of a shaft, most commonly the crankshaft of an engine, which result from variations of driving torque, load torque, and effective inertia. As in the previous paragraph we have to determine the way in which a mechanism responds to stated forces; in the case of an engine, these forces arise from the case pressure within the engine cylinder and the load torque against which the engine drives. The problem is, however, not quite as before in that we are not concerned here with the way that the motion develops from a fully specified initial state, but to determine the way in which the speed varies periodically about a mean.

As an example, we will determine the size of flywheel necessary to keep the speed fluctuation of a single-cylinder gas engine under normal running conditions within 5 per cent of the mean speed. The driving torque on the work shaft due to gas pressure can be calculated from an indicator diagram as in fig. 5.38(a), which also shows how the stroke of the indicator piston is related to the crank angle. The driving torque for a particular configuration is readily determined by multiplying the force on the piston by the distance AM, fig. 5.38(b). The justification for this statement is that the resultant of the gas force P and reaction R between the cylinder wall and the piston must be along the axis CB of piston rod and through M, where the force can again be resolved into its components R and P: the moment of these components about A is $P \cdot \text{AM}$. The crank torque Q_G due to gas pressure is plotted in fig. 5.39(a).

The crank torque due to the inertia of the connecting-rod and piston could be determined precisely by the methods outlined in section 5.10, but since the variation in crank torque due to inertia effects might well be expected to be considerably less than the variation due to gas pressure we shall use an approximate method. First, we shall assume that we can replace the actual connecting-rod by an equivalent connecting-rod consisting of two concentrated masses

$$m_c = 30 \times \frac{0.34}{0.87} = 11.7 \text{ kg at the piston}$$

and

$$m_b = 30 \times \frac{0.53}{0.87} = 18.3 \text{ kg at the big end}$$

The mass of the piston itself is 20·9 kg so that the total equivalent mass at the piston is $m_p = 32.6$ kg.

The total kinetic energy V of the engine is

$$V = \tfrac{1}{2} m_p v_p^2 + \tfrac{1}{2} I \omega^2$$

228 Dynamics of Mechanical Systems

Stroke	0·38 m
Bore	0·18 m
Length of connecting rod	0·87 m
Mass of connecting rod	30 kg
Dist. of c.g. from big end	0·34 m
Mass of piston	20·9 kg
Mean speed	300 rev/min.

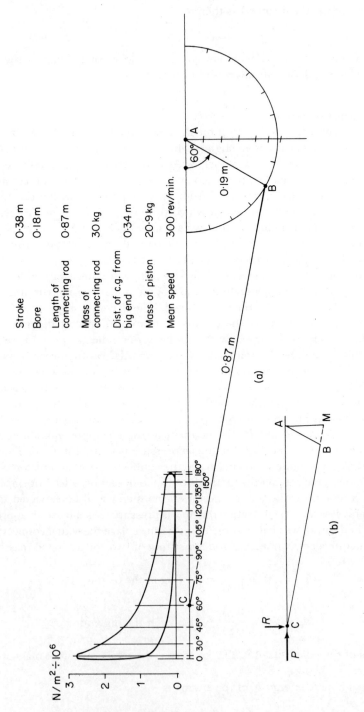

FIGURE 5.38

where v_p is the piston velocity, r is the crank radius, ω is the crank speed, and I is the moment of inertia of the flywheel and crankshaft assembly.

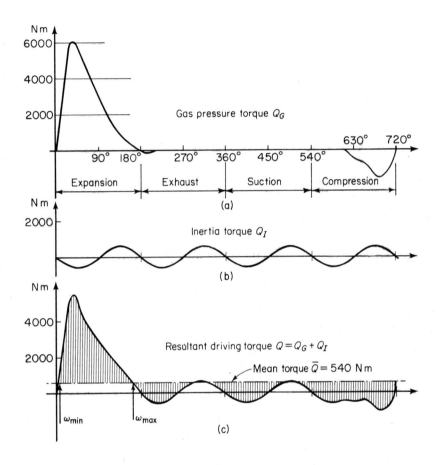

FIGURE 5.39

From eqn. (4.34), the velocity of the piston is given by

$$v_p = -r\omega\left(\sin\theta + \frac{1}{2n}\sin 2\theta + \cdots\right)$$

so that

$$v_p^2 = r^2\omega^2\left(\sin^2\theta + \frac{1}{n}\sin 2\theta \sin\theta + \frac{1}{4n^2}\sin^2 2\theta + \cdots\right)$$

$$= r\omega^2\left[\tfrac{1}{2}(1 - \cos 2\theta) + \frac{1}{2n}(\cos\theta - \cos 3\theta) + \cdots\right]$$

Now n, the ratio of the length of the connecting-rod to crank radius, is in this case 4·6, so that it is reasonable to approximate further and simply write

$$v_p^2 = \tfrac{1}{2}r\omega^2(1 - \cos 2\theta)$$

We now have for the total kinetic energy

$$V = \frac{\omega^2}{2}\left[\tfrac{1}{2}m_p r^2(1 - \cos 2\theta) + I\right]$$

The work equation for the system is

$$\begin{Bmatrix}\text{power input due to}\\ \text{gas-pressure}\end{Bmatrix} = \begin{Bmatrix}\text{rate of change of}\\ \text{kinetic energy}\end{Bmatrix} + \begin{Bmatrix}\text{power}\\ \text{output}\end{Bmatrix}$$

If the output torque is Q_o, this equation is

$$Q_G\omega = \frac{dV}{dt} + Q_o\omega$$

$$= \omega\frac{d\omega}{dt}\left[\tfrac{1}{2}m_p r^2(1 - \cos 2\theta) + I\right]$$

$$+ \frac{\omega^2}{2}\left[m_p r^2 \sin 2\theta \frac{d\theta}{dt}\right] + Q_o\omega$$

Now $d\theta/dt = \omega$, so that

$$Q_G - \tfrac{1}{2}m_p r^2 \omega^2 \sin 2\theta = \frac{d\omega}{dt}\left[\tfrac{1}{2}m_p r^2(1 - \cos 2\theta) + I\right] + Q_o$$

Assuming that $m_p r^2 \ll I$, we have

$$Q_G - \tfrac{1}{2}m_p r^2 \omega^2 \sin 2\theta = Q_o + I\frac{d\omega}{dt} \qquad (5.38)$$

The first term on the left-hand side of this equation is the torque due to gas pressure. The second represents the torque required to overcome the inertia of the reciprocating parts. With a change of sign we shall refer to this component as the inertia torque Q_I, thus

$$Q_I = -\tfrac{1}{2}m_p r^2 \omega^2 \sin 2\theta$$

The sum of these two components will be referred to as the driving torque Q,

$$Q = Q_G + Q_I$$

and is the torque available at the crankshaft to overcome the load torque Q_o and to accelerate the flywheel and rotating parts attached directly to the crankshaft.

If the flywheel is sufficiently large to make the speed fluctuations small we

FORCE RELATIONSHIPS IN MECHANISMS II

can evaluate the inertia torque Q_I without much error by assuming ω to be constant. For the given system

$$Q_I = -580 \sin 2\theta \text{ N m}$$

The net driving torque can now be computed throughout the cycle and is shown in fig. 5.39(c). The difference between the driving torque and the output torque is absorbed in acceleration of the flywheel

$$Q - Q_o = I\frac{d\omega}{dt} \tag{5.39}$$

If the mean speed is constant there can be no net increase in the kinetic energy of the flywheel over a complete cycle, so that the average value of $Q - Q_o$ must be zero. The mean driving torque is 540 Nm. Assuming that the output torque is constant at this value, the difference between the driving and output torques throughout the cycle is represented by the shaded area in fig. 5.39(c).

It can be seen that for

$$5° < \theta < 162°$$

the driving torque is greater than the output torque so that the crankshaft and flywheel accelerate during this interval. Except for two short periods, during the rest of the cycle the driving torque is less than the output torque so that there is a net drain of energy from this system and the flywheel decelerates. It follows that when $\theta = 5$ degrees the speed is a minimum, and that it is a maximum when $\theta = 162$ degrees.

Equation (5.39) can be written

$$Q - Q_o = I\frac{d\theta}{dt} \cdot \frac{d\omega}{d\theta}$$

$$= I\omega \frac{d\omega}{d\theta}$$

Integrating with respect to θ over the interval $\theta = 5$ degrees to $\theta = 162$ degrees

$$\int_{\theta=5°}^{\theta=162°} (Q - Q_o)\, d\theta = \tfrac{1}{2}I(\omega_{max}^2 - \omega_{min}^2)$$

$$= I\left(\frac{\omega_{max} + \omega_{min}}{2}\right)(\omega_{max} - \omega_{min})$$

Because the speed variation is small we can set $\tfrac{1}{2}(\omega_{max} + \omega_{min}) = \bar{\omega}$, the mean speed, so that

$$\frac{\omega_{max} - \omega_{min}}{\bar{\omega}} = \frac{1}{I\bar{\omega}^2}\int_{\theta=5°}^{\theta=162°}(Q - Q_o)\, d\theta \tag{5.40}$$

We have decided that the speed shall be kept within 5 per cent of the mean speed so that $(\omega_{max} - \omega_{min})/\bar{\omega} = 0{\cdot}1$; $\bar{\omega} = 300$ rev/min $= 31{\cdot}4$ rad/s; from fig. 5.39(c)

$$\int_{\theta=5°}^{\theta=162°} (Q - Q_o)\, d\theta = 13\,700 \text{ J}$$

Inserting these figures into eqn. (5.40) we find that the necessary moment of inertia for the flywheel and crankshaft assembly is $I = 139$ kg m².

Having determined the size of flywheel in this manner it is possible to calculate the way in which the velocity actually varies throughout a cycle. Bearing in mind the variations in the indicator diagram which occur in practice, the uncertainty as to the amount of friction, and the unlikelihood that the load will be precisely constant at the value which we have calculated, a more refined calculation is hardly justified.

5.13 Inertia stresses

The classical example of a failure due to excessive inertia stresses is the disintegration of a flywheel. The analysis of the inertia stresses induced in a rotating flywheel is fairly complicated and will not be considered here. A similar, equally important, but somewhat simpler example of a mechanical

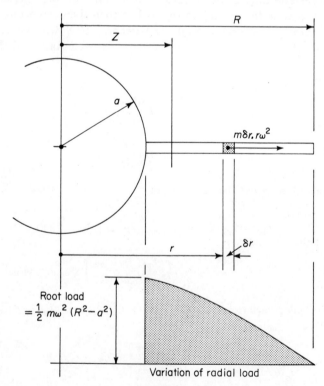

FIGURE 5.40

component in which inertia stresses are a limiting factor in the design is provided by the rotor blade of a gas turbine. We will consider two systems of stress: the direct radial stress due to the radial acceleration of the blade elements, and the bending stresses due to the gas pressures on the blade and angular acceleration of the rotor. In order to avoid considerable complication we will assume that the cross-section of the blade does not vary with radius, and that the blade is untwisted. Neither of these assumptions hold for a practical blade and in practice the calculations must make allowance for this.

(a) Radial loading
Assuming that the mass per unit length of blade is m the inertia on an element $m\,\delta r$ at radius r due to its radial acceleration is $(m\,\delta r)r\omega^2$, where ω is the angular velocity of the rotor. At radius z the force is

$$\int_{r=z}^{r=R} mr\omega^2\,dr = \tfrac{1}{2}m\omega^2(R^2 - z^2)$$

If the blade material has density ρ, the stress is

$$\sigma = \tfrac{1}{2}\rho\omega^2(R^2 - z^2)$$

This tensile stress causes the blade to extend radially. The elastic extension is unlikely to be troublesome, but if the blade is operating at a high temperature the creep strain which occurs under sustained loading is a matter of considerable importance. At the present time it is the creep properties of the blade material which sets an upper limit on the temperature at which turbines operate.

The force at the blade root is

$$\tfrac{1}{2}m\omega^2(R^2 - a^2)$$

This force can also be evaluated by considering the overall equilibrium of the blade:
The resultant D'Alembert force acting at the centre of gravity of the blade is

$$Mr_g\omega^2 = m(R - a)\cdot\left(\frac{R + a}{2}\right)\omega^2$$

$$= \tfrac{1}{2}m\omega^2(R^2 - a^2)$$

The terminal forces, as determined by the internal distribution of stress, should always be checked by comparison with the forces obtained for a consideration of overall equilibrium.

(b) Transverse loading with rotor acceleration
Angular acceleration of the rotor induces transverse inertia forces on the elements of a blade. These forces clearly cause bending and shear stresses, but before we can calculate these stresses we must know how the acceleration is produced. We will consider two possibilities: first, the acceleration is produced

234 Dynamics of Mechanical Systems

by a driving torque applied at the hub. For simplicity, we will assume that the aerodynamics forces on the blade are negligible. This situation is that of an axial compressor just starting up.

The shear force on a section at radius z is

$$S = \int_z^R m r \alpha \, dr = \tfrac{1}{2} m \alpha (R^2 - z^2)$$

the bending moment is

$$M = \int_z^R m r \alpha (r - z) \, dr$$
$$= m \alpha (\tfrac{1}{3} R^3 - \tfrac{1}{2} R^2 z + \tfrac{1}{6} z^3)$$
$$= \tfrac{1}{6} m \alpha (R - z)^2 (2R + z)$$

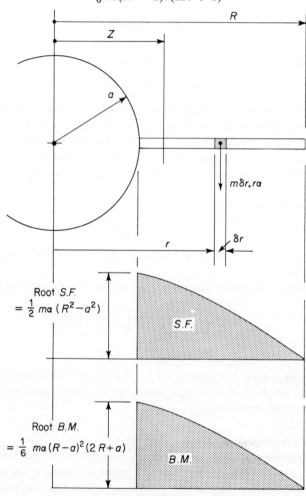

FIGURE 5.41

FORCE RELATIONSHIPS IN MECHANISMS II

The bending moment and shear force both increase steadily from the tip to the root of the blade. Verification of the root values by considering the resultant D'Alembert effects is left as an exercise.

The second example which we shall take is when the angular acceleration α of the rotor is produced by a uniformly distributed transverse loading p on the blade, with no resisting torque applied to the rotor shaft. To simplify the algebra we will assume that the hub is of negligible radius.

The first thing which we must determine is the relationship between the gas loading p on the blade and the angular acceleration α. Taking moments about the hub, fig. 5.42(a), we have

$$\tfrac{1}{2}pR^2 = \tfrac{1}{12}mR^3\alpha + \tfrac{1}{2}m^2\alpha \times R/2$$
$$= \tfrac{1}{3}mR^3\alpha$$

so that
$$p = \tfrac{2}{3}mR\alpha$$

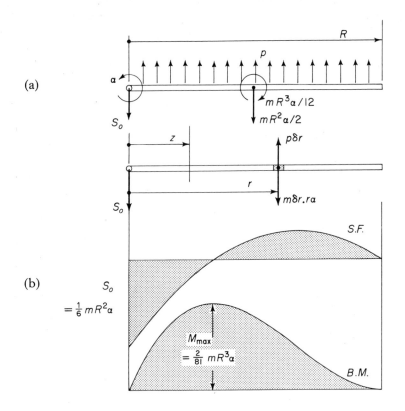

FIGURE 5.42

The shear force on a section at radius z is

$$S = \int_z^R (mr\alpha - p)\, dr$$

$$= \int_z^R m\alpha(r - \tfrac{2}{3}R)\, dr$$

$$= \frac{m\alpha}{6}(R - z)(3z - R)$$

The bending moment at the same section is

$$M = \int_z^R m\alpha(r - \tfrac{2}{3}R)(r - z)\, dr = \frac{m\alpha}{6} z(R - z)^2$$

The distribution of shear force and bending moment along the blade is shown in fig. 5.42(b). It should be noted that the shear force S_o at $z = 0$ equals the reaction at the hub obtained by resolving forces in fig. 5.42(a), and that S is zero at the blade tip. The bending moment is zero at the hub, and at the tip. The maximum bending moment occurs at $z = R/3$, where the shear force is zero.

6
LINEAR SYSTEMS
I First order systems

6.1 Introduction
So far, our study of mechanical systems has been confined to mechanisms which are assumed to be composed of rigid members. There are two important consequences of this assumption: first, whatever the load transmitted by the system, the output displacement is uniquely determined by the input displacement: secondly, time does not enter into the relationship between the input displacement and the response, the response is always in step with the input. Of course, no real mechanism is made up of entirely rigid members, and so in practice neither of the two foregoing consequences is fully realized. All that can be said is that there is a large body of mechanisms for which the flexibility of the members is so small that any consequences which it may have are unimportant.

For the rest of this book we shall be concerned with mechanical systems for which the flexibility does have an important effect on the behaviour of the system. In order to keep our subject within reasonable bounds we shall have to restrict severely the type of system which we shall consider, and the type of input for which we shall attempt to predict the response. We shall concentrate, almost exclusively, on linear systems.

6.2 Linear systems defined
A linear system is defined as a system to which the principle of superposition applies. Broadly, this means that the effect of doubling the magnitude of the input or disturbance applied to a system is to double the magnitude of the response; by doubling the torque applied to a rotor, we double the acceleration and the angle turned through in a given time. Note, however, that this does not mean that we halve the time taken to rotate through a given angle (fig. 6.1); in this context we shall be primarily concerned with what has happened in a given time interval, and only incidently with the time taken to achieve a given response.

This concept of the principle of superposition is directly analogous to that used in the theory of structures where a linear structure is one for which displacement is proportional to load, and usually a good enough test for linearity is to see if the ordinates of separate response curves (fig. 6.1), are at all proportional to the magnitudes of the respective inputs. With the introduction of the time element, which is not relevant in the structural context of the principle of superposition, a rather more searching test is required to determine whether this principle applies, than is provided by the simple test of proportionality.

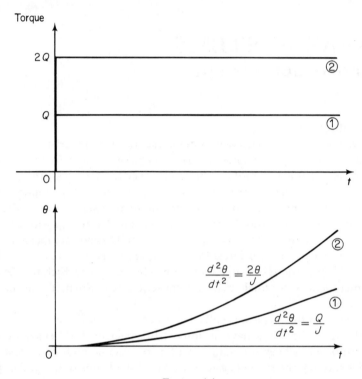

FIGURE 6.1

Formally, the principle of superposition requires that if one input $x_i(t)$ gives rise to a response $x_o(t)$, and a separate input $y_i(t)$ gives rise to a response $y_o(t)$, then the combined input $Ax_i(t) + By_i(t)$ gives rise to a resultant response $Ay_o(t) + By_o(t)$. We can readily verify that the principle of superposition does apply in this sense in the case of the response of a system which consists simply of a rotor of moment of inertia J to a torque Q.

The rotation due to a torque Q applied at time $t = 0$ is

$$\theta = \frac{1}{2}\frac{Q}{J}t^2$$

The rotation due to a torque Q applied at time $t = \tau$ is

$$\theta = 0 \qquad \text{if } t \leqslant \tau$$

and

$$\theta = \frac{1}{2}\frac{Q}{J}(t - \tau)^2 \quad \text{if } t > \tau$$

LINEAR SYSTEMS I First order systems

If the principle of superposition holds, the combined response due to Q at $t = 0$ and an additional Q at $t = \tau$ is

$$\theta = \frac{1}{2}\frac{Q}{J}t^2 \qquad \text{for } t \leq \tau$$

and

$$\theta = \frac{1}{2}\frac{Q}{J}t^2 + \frac{1}{2}\frac{Q}{J}(t-\tau)^2 \qquad \text{for } t > \tau$$

as in fig. 6.2.

A calculation of the angle of rotation for $t > \tau$ which does not invoke the principle of superposition proceeds as follows:

At $t = \tau$
$$\theta_\tau = \frac{1}{2}\frac{Q}{J}\tau^2$$

and
$$\omega_\tau = \frac{Q}{J}\tau$$

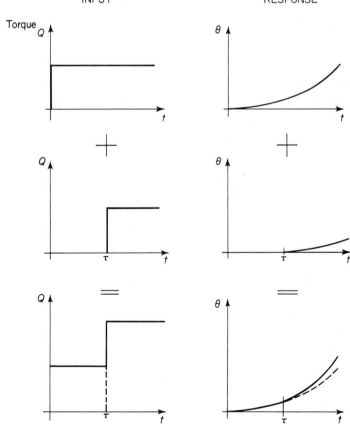

FIGURE 6.2

For $t > \tau$ the angular acceleration is $2Q/J$ and

$$\theta = \theta_\tau + \omega_\tau t' + \frac{1}{2}\left(\frac{2Q}{J}\right)t'^2$$

where $\qquad t' = t - \tau$

Hence $\qquad \theta = \frac{1}{2}\frac{Q}{J}t^2 + \frac{Q}{J}(t-\tau) + \frac{Q}{J}(t-\tau)^2$

$$= \frac{1}{2}\frac{Q}{J}t^2 + \frac{1}{2}\frac{Q}{J}(t-\tau)^2$$

as before.

The two inputs at $t = 0$ and $t = \tau$, respectively, are of the same 'shape', but clearly we could apply the principle of superposition in this way whatever the separate inputs might be.

In general, the principle of superposition applies if the differential equation which governs the response of the system is of the form

$$(a_n D^n + a_{n-1} D^{n-1} + \cdots + a_1 D + a_0)x_o$$
$$= (b_m D^m + b_{m-1} D^{m-1} + \cdots + b_0)x_i \qquad (6.1)$$

where D denotes differentiation with respect to time, and the coefficients a and b are either constant or functions of time, and are independent of x_i and x_o. We shall confine our attention to cases where a and b are constants.

The differential equation of the system which we have just taken as an example,

$$\frac{d^2\theta}{dt^2} = \frac{Q}{J}$$

clearly is of the above general form. To prove that the principle of superposition holds for a system whose governing differential equation is of the stated form we have merely to note that if the coefficients a and b are constants we can add the two equations

$$(a_n D^n + \cdots + a_0)x_o = (b_m D^m + \cdots + b_0)x_i$$
$$(a_n D^n + \cdots + a_0)y_o = (b_m D^m + \cdots + b_0)y_i$$

in any proportions that we wish to give

$$(a_n D^n + \cdots + a_0)(Ax_o + By_o) = (b_m D^m + \cdots + b_0)(Ax_i + By_i)$$

Hence an input $(Ax_i + By_i)$ gives rise to a response $(Ax_o + By_o)$.

6.3 Transfer relationships

The expression *transfer relationship* is a convenient way of referring, in general terms, to the relationship between the input and response of a system. The

LINEAR SYSTEMS I First order systems

transfer relationship for a particular system operating in a certain way is defined by the differential equation for the system. A system can be classified according to the form taken by its differential equation.

6.4 Proportional elements

A proportional element, or system, is one for which the transfer relationship takes the form

$$x_o = k x_i \qquad (6.2)$$

Some examples of simple proportional elements are to be seen in fig. 6.3, the gear pair calls for no special comment; the spring has been included to

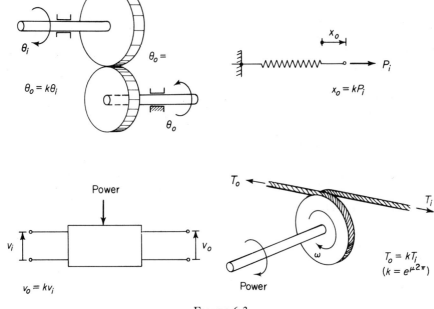

FIGURE 6.3

show that the input and the response used need not be physically the same type of variable; the electronic amplifier differs from the first two examples in that they are *passive* devices with no power supply, other than that provided by the input, whereas it is an *active* device in the sense that there is an amplification of power between the input and response; the capstan or torque amplifier is included to make the point that power amplification is not the prerogative of electrical systems. It is perhaps not always appreciated just how much power amplification can be achieved by purely mechanical means. If we take the coefficient of friction μ to be 0·2, then k for a single winding on the capstan is $e^{2\pi \cdot 0 \cdot 2} = 3·5$. For 6 turns k is $3·5^6 \approx 2000$.

The main feature which distinguishes proportional elements from higher order elements or systems is that time is not a relevant variable. The response always matches the input exactly, whereas with other types of system the response varies with time.

It is important to realize that in labelling an element as a proportional element we are idealizing it, and that in real life ideal proportional elements are very rare, if indeed they exist at all. A simple push-rod, fig. 6.4, may be taken

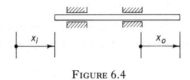

FIGURE 6.4

as an example to illustrate this point. If the rod is sensibly rigid it seems to be self-evident that the displacements of the two ends will at all times be related by the equation

$$x_o = x_i$$

that is to say, the push-rod is a proportional device for which $k = 1$. Now to move the rod at all within a finite period of time it must be accelerated, and because no real push-rod is massless, a force must be applied to move it. The stresses thereby induced in the rod cause it to be strained, and so in general the two ends of the rod do not have precisely the same displacement. Provided that the difference between x_o and x_i is very small compared with the total movement of the rod we can usually ignore it. If the difference is not negligible then our simple picture of a rigid member will not do.

The problem of making valid idealizations, or the 'right approximations', is extremely important; it is also often very difficult, and we shall return to it again later.

6.5 Integrating elements

In fig. 6.5 we have a hydraulic ram to which oil is admitted through a calibrated control valve. Adjustment of the flow can be thought of as being achieved by means of a lever which moves over a graduated scale. When this scale reading is x_i the flow-rate is

$$q = kx_i$$

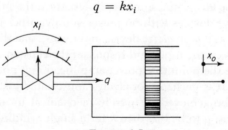

FIGURE 6.5

The speed of the ram is

$$\frac{dx_o}{dt} = \frac{q}{A}$$

where A is the area of the piston,

so that

$$\frac{dx_o}{dt} = \frac{kx_i}{A} = \frac{x_i}{T}$$

where T is a constant which has the dimensions of time. On integrating the equation we have

$$x_o = \int \frac{x_i}{T} dt \qquad (6.3)$$

so that x_o is the time integral of x_i.

Provided that it is the relationship between x_o and x_i which is important, the control valve and ram can be regarded as an integrating device. But, if we are interested in using the valve as a means of controlling the ram velocity, rather than its displacement, it is a proportional device. We cannot label a piece of hardware seen in isolation as being this or that type of element, it can only be classified in the contexts of particular applications.

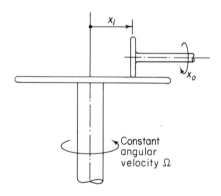

FIGURE 6.6

The use of the word integrating naturally conjures up the idea that its prime purpose is as a calculating device. In fact, this is not the case, in automatic control systems it is frequently necessary to include an integrating element in order to achieve a certain type of performance from the system, and although the integrating element does indeed integrate the signal presented to it there is no direct interest in the fact that it is behaving as a calculating device. Nevertheless, integrating devices have been used specifically for the purpose of solving differential equations, and it is of some interest to note how this is done, fig. 6.6.

244 Dynamics of Mechanical Systems

A turntable rotates at constant speed Ω and drives a small wheel by friction. The angular velocity Dx_o of the small wheel depends on the distance x_i of its point of contact from the centre of the turntable, so we have

$$\frac{dx_o}{dt} = \Omega \frac{x_i}{r}$$

or

$$x_o = \int \frac{\Omega x_i}{r} dt$$

where r is the radius of the small wheel.

6.6 Simple exponential lag
The next type of element to be mentioned is that for which the governing differential equation is of the form

$$(1 + TD)x_o = x_i \tag{6.4}$$

This type is referred to either as a *first order* system, or as a *simple exponential lag*. The coefficient T has the dimensions of time and is called the *time constant* of the system. Examples of first order systems are given in fig. 6.7.

1 Rotor
The input is a torque Q_i applied to the shaft of a rotor whose inertia is resisted by a friction torque which is proportional to the speed ω_o, which we take as the response. The input torque must therefore equal the sum of the friction and inertia torque so that

$$Q_i = \lambda \omega_o + J D\omega_o$$

or

$$Q_i/\lambda = (1 + TD)\omega_o$$

where

$$T = J/\lambda$$

2 Electrical network
Assuming that no current is drawn from the output terminals of this network we have

$$i = C Dv_o$$

and

$$v_i = v_o + Ri$$

Hence

$$v_i = (1 + RC D)v_o$$

3 Thermometer
It is a matter of common experience that a thermometer which is suddenly plunged into, say, a bath of hot liquid takes some time to settle down to give a steady reading. This is because the temperature which the thermometer indicates is really the temperature of its own liquid, and it takes time for the

thermometer liquid to absorb heat from its surroundings. Provided that the temperature difference is not too large, the rate at which heat is conducted to the thermometer liquid is $k(\theta_i - \theta_o)$, where θ_o is the temperature of the

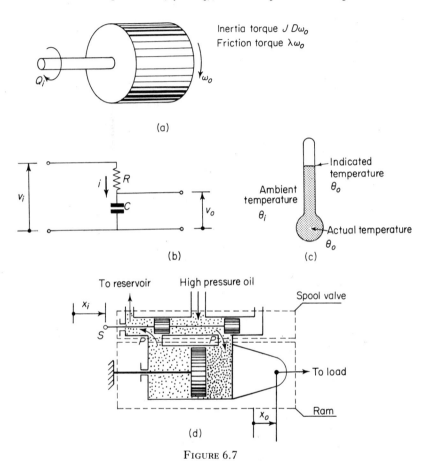

FIGURE 6.7

thermometer liquid and θ_i is the temperature which is to be measured. The rate at which the temperature of the thermometer liquid, and hence its volume, increases is $C\,D\theta_o$, where C is the thermal capacity of the thermometer, so that

$$k(\theta_i - \theta_o) = C\,D\theta_o$$

or
$$\theta_i = (1 + T\mathrm{D})\theta_o$$

where
$$T = c/k$$

4 Hydraulic relay

The hydraulic relay represents a considerable advance in physical complexity on the previous examples. It has two main components, the spool-valve and the ram. The input signal is the displacement x_i which moves the double piston of the spool-valve and so uncovers two ports P in the wall of the spool-valve cylinder. This admits high-pressure oil to one side of the ram piston and allows the escape of oil at a lower pressure from the other side of the ram piston. Since the ram piston is anchored, it is the ram cylinder which moves as a result of the flow of oil. The effect of this is to close the ports and so reduce the supply of oil to the ram. Motion ceases when the displacement of the ram cylinder is $x_o = x_i$, and the ports are fully covered.

Apart from fairly small hydrodynamic and friction effects the resultant force on the piston of the spool-valve is zero so that the input power is very small. The pressure drop across the ram piston depends upon the piston velocity, but for very low speeds is only slightly less than the supply pressure which typically might be 15×10^6 N/m².

Provided that the load against which the ram is operating is constant the flow-rate through the ports depends only on the port opening $(x_i - x_o)$. With suitably shaped ports the flow is proportional to $(x_i - x_o)$, so that if A is the area of the ram we have

$$q = k(x_i - x_o) = A\,\mathrm{D}x_o$$

and
$$x_i = (1 + T\,\mathrm{D})x_o$$

where
$$T = A/k$$

In fact, valve ports are not usually designed to give strict proportionality between the flow and the opening, but this is a complication which we shall leave on one side.

6.7 Goodness of theory

In formulating the mathematical equations for a real system, it is necessary to make certain idealizations and approximations in order to obtain an equation which can be solved without too much difficulty. In principle, we can obtain a more accurate picture of the behaviour of a system by increasing the accuracy with which we define the properties of its elements, but in practice it can soon happen that the equations become intractable. The question then naturally arises: how do we know whether the equations which we have derived in any particular case provide an accurate description of the behaviour of the system?

The only certain way of testing the goodness of a theory is to make a direct comparison between the behaviour predicted by the equations and the behaviour observed in an experiment. It may be profitable to develop a more refined theory which takes account of an acknowledged idealization in a simplified representation of a system, and to compare the predictions of the two theories. Such a comparison enables an assessment to be made as to whether the refinement is worth the extra effort. A natural reaction is to decide that, having developed a better theory it might as well be used. If the particular element concerned is to be used in isolation there is no objection to using the most accurate mathematical representation that can be developed. Very often,

LINEAR SYSTEMS I First order systems 247

however, the element will be used in conjunction with others and any additional complications in the equations for individual elements adds to the complexity of the final equations for the whole system. In these circumstances, there is everything to be said for simplifying the equations for the individual elements as much as possible.

6.8 Standard inputs

Before we can make a comparison between the behaviour predicted by theory with that observed in an experiment we must specify the conditions under which the experiment is to be conducted; we must specify the input to the system. Taking as an example a first order system, for which the governing equation is

$$(1 + TD)x_o = x_i$$

this means that we must specify the input x_i as a function of time.

In principle, x_i can be any function of time which we care to choose. In practice, the following considerations narrow the choice to one or two possibilities:

1 We must choose x_i so that the mathematical solution of the differential equation is not too difficult.
2 It is necessary to make x_i reasonably convenient from the experimental standpoint.
3 It is desirable also to choose a form of input which is a fair idealization of the type of input to which the element might be subjected in practice.

In order to fix ideas a little let us think for a moment about the types of input to which a thermometer might be expected to respond:

(i) The thermometer records the steady air temperature and is suddenly plunged into a bath of hot liquid. This type of disturbance is idealized as a *step* input, where the change in x_i is considered to be instantaneous. Mathematically, we define it as

$$x_i = 0 \qquad \text{for } t < 0$$
$$x_i = \text{constant} \quad \text{for } t > 0$$

The thing which concerns us here is the time taken for the thermometer to settle down to a steady reading.

(ii) The thermometer is immersed in a bath of liquid whose temperature is rising steadily. This is known as a *ramp* input, mathematically

$$x_i = 0 \quad \text{for } t < 0$$
$$x_i = kt \quad \text{for } t > 0$$

The indicated temperature lags behind the actual temperature, and it may be important in practice to know the size of the lag.

248 Dynamics of Mechanical Systems

(iii) The thermometer is used to measure a temperature which varies periodically with time. A convenient idealization for such a disturbance is a sinusoidal or *harmonic* input for which $x_i = X_i \cos \omega t$. In this case, interest centres on the amplitude of the response and its phase shift relative to the input.

These three types of input, in particular the first and third, are the ones generally used to study the response of a system, whether this be by mathematics or experiment. As we shall see, mathematically it is always much easier to solve the equations for a harmonic input than it is for a step input. Experimentally, the choice of input depends on circumstances. Referring back to the thermometer, clearly it might be difficult to arrange for a sinusoidal variation in the temperature of a bath of liquid, whereas there is no difficulty in subjecting the thermometer to a step input. With electrical systems, on the other hand, it is generally very easy to provide a harmonic excitation and there are well established methods for measuring the amplitude and phase of the response. Consequently, the step input is less likely to be used.

There is one important difference of attitude involved between the use of the step and harmonic input. In the case of the step input we are interested to see how long it takes for the system to settle down to a steady response, and the way in which it settles down. With the harmonic excitation, no attention is paid to what happens immediately after the excitation is switched on. Initially, the response usually varies somewhat irregularly with time and it is only after several cycles that it settles down to a regular periodic variation (a sinusoidal variation in the case of a linear system). No attempt is made to measure the response until the system has settled down. This difference in attitude has experimental consequences.

When the vibrational properties of an elastic structure, e.g., an engine mounting, are being investigated it is generally difficult to arrange for a step change of force to be applied to a structure, but a harmonically varying force can readily be applied by a commercial shaker made specifically for the job. The frequencies which are of interest are generally of the order of several cycles per second, so that no great inconvenience is caused by having to wait a little time for the system to settle down under the influence of the exciting force. If we turn to a different example, that of an aircraft whose response in flight to a particular control is being tested, we can visualize that it may be very inconvenient, and expensive, to wait for the aircraft to settle down to a steady response to a harmonic displacement of the particular control, and that it would be preferable to consider a step input.

Two other types of standard input should be mentioned here, first the *impulse* where the system is subjected to an intense disturbance for a very short interval of time, second the random input. We shall study the impulse response below where it is developed by the use of the principle of superposition from the step response. The response to a *random* input can only be deduced by an excursion into statistical theory and will not be considered.

LINEAR SYSTEMS I First order systems 249

6.9 Step response of a first order system
The general solution of the differential equation
$$(1 + TD)x_o = x_i$$
where
$$x_i = 0 \quad \text{for } t < 0$$
$$x_i = X_i \quad \text{for } t > 0$$
and
$$x_o = 0 \quad \text{at } t = 0$$
consists of two parts:

1 the *complementary function*, which is the solution of the equation
$$(1 + TD)x_o = 0$$
namely
$$x_o = A\,e^{-t/T}$$
where A is an arbitrary constant; and

2 the *particular integral*, which is the simplest solution which can be found to the equation
$$(1 + TD)x_o = X_i$$
the simplest solution being
$$x_o = X_i$$
The general solution of the equation is therefore
$$x_o = A\,e^{-t/T} + X_i$$
The arbitrary constant A is determined by inserting any known pair of values for x_o and t. Usually, when time is the independent variable the only known conditions are those which occur at $t = 0$, when the input starts. In this case, we have specified that at $t = 0$, $x_o = 0$, so that $A = -X_i$, and the response is
$$x_o = X_i(1 - e^{-t/T}) \tag{6.5}$$
Graphs of the input and response are given in fig. 6.8. The time constant T is seen to have simple and useful physical interpretation: differentiating x_o with respect to t we have
$$\frac{dx_o}{dt} = \frac{X_i}{T}e^{-t/T}$$
At the origin
$$\left(\frac{dx_o}{dt}\right)_{t=0} = \frac{X_i}{T}$$
so that the tangent to the response curve at the origin intersects the line which represents the input, and eventual response, at $t = T$. The time constant T is therefore a measure of the rate of response of the system. If T is small the

250 Dynamics of Mechanical Systems

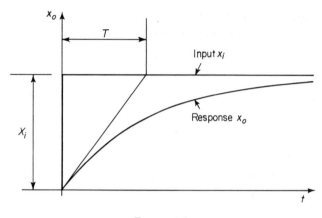

FIGURE 6.8

system responds quickly, if T is large the system is sluggish. We obtain different impressions of the response with different values of T. The responses over a period of 3 seconds corresponding to $T = 0.1$ second, 1 second, and 10 seconds, are as given in fig. 6.9. If the element is operating in a context

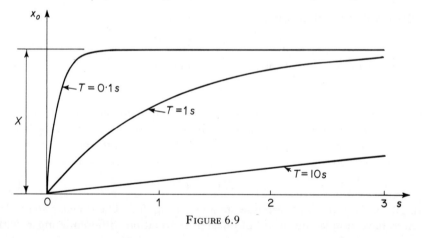

FIGURE 6.9

where other happenings occur in seconds, rather than tenths of seconds or tens of seconds, then the response curve corresponding to $T = 1$ second can be regarded as being typical of a first order system. On the same time scale the rate of response for the system with $T = 0.1$ second is so rapid that we might be inclined to say that for all practical purposes $x_o = X_i$, that is to say the system is behaving as a simple proportional system for which the coefficient of proportionality is 1. The system with $T = 10$ seconds appears to have a response curve which rises linearly so that

$$\frac{dx_o}{dt} = \frac{X_i}{T}$$

or
$$x_o = \int \frac{X_i}{T} \, dt$$

and it behaves like an integrating device. This brings us back to the earlier discussion of idealizations and the goodness of theory. These examples show that even if we know that a system might properly be regarded as a system for which the equation

$$(1 + T D)x_o = x_i$$

applies, we may well be justified in appropriate circumstances of writing instead either $x_o = x_i$ or $T D x_o = x_i$.

All the curves in fig. 6.9 are coalesced into a single curve if we take as abscissa t/T instead of t. In drawing these curves we have presupposed that they are being compared on the basis of the same step input. This restriction is removed if we take as ordinate x_o/X instead of x_o. The response curve corresponding to $(1 + T D)x_o = x_i$ in fig. 6.10 is therefore a universal response curve which is applicable to all systems of this particular type. The use of dimensionless coordinates in this way is an extremely useful and powerful means of simplifying the presentation of data, and will be used in respect of other systems. Figure 6.10

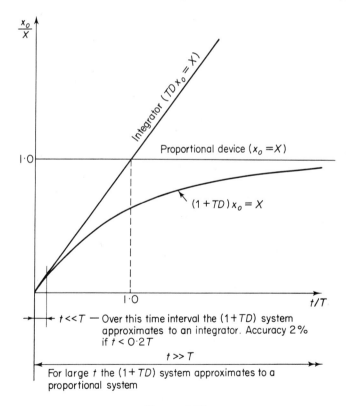

FIGURE 6.10

also shows the dimensionless response curves for a proportional element, and an integrator.

6.10 Integration by analogue computer

An analogue computer provides a convenient way of assembling a system whose behaviour will be governed by a given equation, and so obtaining an analogue of some other system whose behaviour is under investigation. By studying the analogue, rather than the original system, it is possible to investigate very quickly the effect of varying parameters in a way which might call for major structural alterations in the prototype. It is also possible to adjust the time scale and either slow the response down in order to study it more easily, or to speed things up and so carry out tests much more rapidly than would otherwise be possible. Integration is a basic operation in this process, and since we have referred to integrating elements it is of interest to note how integration is carried out in an electrical analogue computer.

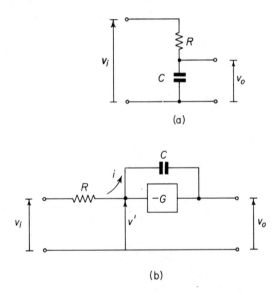

FIGURE 6.11

One way in which integration could be achieved is by use of the circuit in fig. 6.11(a) for which, as we have seen earlier, the differential equation is

$$(1 + RC\,D)v_o = v_i$$

Provided that we consider the response only over a time period which is much less than the time constant RC, this circuit operates as an integrating circuit. There are, however, two serious objections to its use in this way: first, the limitation of the time interval to $0 \cdot 2T$ or less is very inconvenient; second, the circuit behaves in accordance with the given equation only if no current is

LINEAR SYSTEMS I First order systems 253

taken from the output terminals. Bearing in mind that we shall want to use the circuit element in conjunction with other, perhaps similar, elements, this second limitation is also very inconvenient.

The standard way of integrating in an analogue computer makes use of an electronic amplifier, fig. 6.11(b). Assuming that the amplifier draws no current from its input signal, so that the current through R equals that through C, we have

$$v_i - v' = iR$$

and

$$v' - v_o = \frac{i}{CD}$$

where the voltages are as indicated in the figure, also if the gain of the amplifier is $-G$

$$v' = -\frac{v_o}{G}$$

On eliminating i and v' from these equations we obtain

$$\frac{1}{G} + \left(1 + \frac{1}{G}\right) RC\, Dv_o = -v_i$$

Usually $G = 10^4$ to 10^5, so that with little error we can write

$$(1 + GT\, D)v_o = -Gv_i$$

where $T = RC$.

For a step change V_i in V_i, the solution of this differential equation is

$$v_o = -GV_i(1 - e^{-t/GT})$$

So the introduction of the amplifier effectively increases the time constant by a factor G and the integration can proceed for G times as long and still achieve the same accuracy as the circuit in fig. 6.11(a). On expanding the exponential term we have

$$v_o = -GV_i \left\{ \frac{t}{GT} + \frac{1}{2}\left(\frac{t}{GT}\right)^2 + \cdots \right\}$$

$$= -\frac{V_i}{T}t - \frac{1}{2G}\left(\frac{t}{T}\right)^2 - \cdots$$

$$\approx -\frac{v_i}{T}t$$

6.11 Superposition

The principle of superposition can be used to determine the response of a system to any transient input once the response to a step input is known. Thus the response to the input shown in fig. 6.12 can be calculated by adding the

9*

responses of two suitably phased step inputs. For a reason which will appear subsequently, let one step of magnitude $X/2$ be applied at time $t = -\tau/2$ and

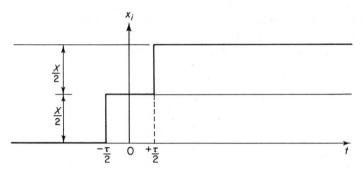

FIGURE 6.12

the other applied at $t = +\tau/2$. At any subsequent time t the first step will have been applied for a time $t + \tau/2$ so that the response due to it is

$$x_o = \frac{X}{2}(1 - e^{-(t+0\cdot5\tau)/T})$$

The response to the second step is

$$x_o = \frac{X}{2}(1 - e^{-(t-0\cdot5\tau)/T})$$

For $t > \tau$ the total response is

$$x_o = \frac{X}{2}[2 - e^{-(t+0\cdot5\tau)/T} - e^{-(t-0\cdot5\tau)/T}]$$

$$= \frac{X}{2}[2 - e^{-t/T}(e^{-\tau/2T} + e^{+\tau/2T})]$$

$$= X\left[1 - e^{-t/T}\left(1 + \frac{1}{2}\left(\frac{\tau}{2T}\right)^2 + \cdots\right)\right]$$

Provided that $\tau < 0\cdot28T$, the coefficient of $e^{-t/T}$ will differ from 1 by less than 1 per cent, and we can reasonably write

$$x_o = X(1 - e^{-t/T})$$

which is the response due to a unit step applied at $t = 0$. Even if $\tau = T$, the coefficient of $e^{-t/T}$ is still only $1\cdot127$. We can generalize from this result and conclude that, provided the area under a 'step' input curve equals the area under an equivalent ideal step, its actual shape does not matter very much provided also that the total movement is carried out in a time interval which is small compared with the time constant of the system. This conclusion is very important because an ideal step is impossible to achieve in practice.

As a second example let us consider the effect of applying a step at $t = 0$ and then an equal and opposite step at $t = \tau$, fig. 6.13. This example provides

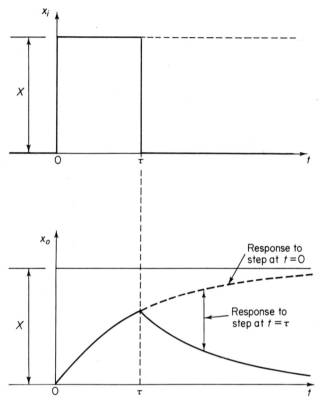

FIGURE 6.13

the basis of a simple experiment on superposition using as the system a mercury thermometer whose bulb has been lagged with a suitable insulator (e.g., a layer of epoxy resin) to increase the time constant to several seconds. The thermometer is placed in a bath of water at air temperature and allowed to reach a steady reading, and then plunged quickly into a bath of hot water. This corresponds to a step input, the response to which can be determined by noting the increase in the thermometer reading with time. When the response has been thus determined we can predict what the readings would have been had the thermometer been replaced in the cold bath at any time. A second experiment which does involve the two moves of the thermometer can then be made to verify the prediction. It should be appreciated that the value of this prediction does not depend on the behaviour of the thermometer conforming to an equation of the type $(1 + T\mathrm{D})x_o = x_i$, though on the basis of the analysis in section 6.6 we would expect it to do so.

To simplify the mathematical determination of the response to this input and to bring it into line with our previous example we will consider the input to be symmetrical with respect to the origin $t = 0$, fig. 6.14. The response for $t > \tau/2$ is

$$x_o = X(1 - e^{-[(t+\tau/2)/(T)]}) - X(1 - e^{-[(t+\tau/2)/(T)]})$$

which reduces to

$$x_o = X e^{-t/T} \left[\frac{\tau}{T} + \frac{1}{3}\left(\frac{\tau}{2T}\right)^3 + \cdots \right]$$

$$\approx \frac{X\tau}{T} e^{-t/T} \tag{6.6}$$

with an error of less than 1 per cent if $\tau < 0.34T$.

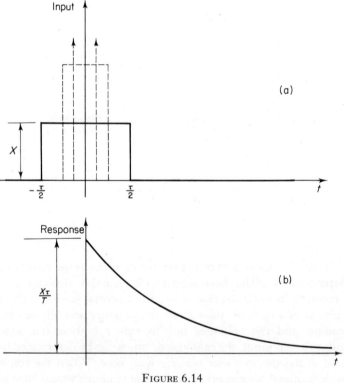

FIGURE 6.14

The interesting thing to note here is that X and τ do not matter individually, it is their product which counts, i.e., the area under the input. We could therefore double X and reduce τ by one half and obtain the same response (for $t > \tau/2$). Carrying on this argument, we have eventually an input of infinite amplitude applied for an infinitesimal time, but such that their product is the finite quantity $X\tau$. In the realm of classical mechanics we regard an infinite force applied for an infinitesimal time such that their product is a finite quantity as an *impulse*. The same word is used in the more general context in which this analysis is framed and so $(X\tau/T)\,\mathrm{e}^{-t/T}$ is the response of a first order system to an impulse $X\tau$. This response is plotted in fig. 6.14(b).

6.12 Relationships between impulse, step, and ramp responses

There is a more direct way of obtaining the impulse response from the step response than that used in the last section. If we take the differential equation

$$(1 + T\mathrm{D})x_o = x_i$$

and differentiate it with respect to t, we have

$$(1 + T\mathrm{D})\,\mathrm{D}x_o = \mathrm{D}x_i$$

so that the response to $\mathrm{D}x_i$ is $\mathrm{D}x_o$. It follows that, if an input and its response are differentiated with respect to time, the differentiated input and response will still correspond.

We have already seen that we can, with little error, replace an ideal step input by something which approximates to it as in fig. 6.15. Whereas there is an obvious difficulty in differentiating an ideal step, there is no difficulty in differentiating the approximate step. On doing this we have an impulsive input the strength of which is $(X/\tau)\tau = X$. It follows that if we differentiate the response to a step input of unit amplitude we shall obtain the response to unit impulse.

The response in the case of a system for which $(1 + T\mathrm{D})x_o = x_i$ to a unit step is

$$x_o = (1 - \mathrm{e}^{-t/T})$$

On differentiating the right-hand side with respect to t, we have as the response to a unit impulse $\mathrm{e}^{-t/T}/T$, as obtained at the end of the last section.

Although we have taken a first order system as an example, the general argument is applicable to a system of any order.

If we integrated a unit step input and its response, again taking the first order system as an example, we have as our integrated input a ramp, fig. 6.16. The response, obtained by integrating $(1 - \mathrm{e}^{-t/T})$, is

$$(x_o)_{\mathrm{ramp}} = \int_0^t (1 - \mathrm{e}^{-t/T})\,\mathrm{d}t$$
$$= \left[t + T\mathrm{e}^{-t/T}\right]_0^t$$
$$= t - T(1 - \mathrm{e}^{-t/T}) \qquad (6.7)$$

The response in this case lags behind the input by an amount which increases with time and approaches the limiting value T as t tends to infinity.

258 Dynamics of Mechanical Systems

It is by no means essential to have a response expressed as an equation in order to apply the above results. If, for example, the response of a system to an impulse has been determined experimentally, its step response may be obtained by numerical integration.

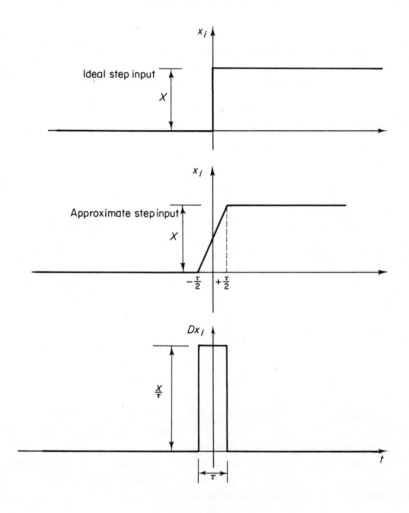

FIGURE 6.15

The impulse and ramp response can be used in application of the principle of superposition to build up the response to other inputs, just as we have used the step input. It is purely a matter of personal choice as to whether the response to an irregular transient input, fig. 6.17, is regarded as a series of steps or as a series of impulses.

LINEAR SYSTEMS I First order systems

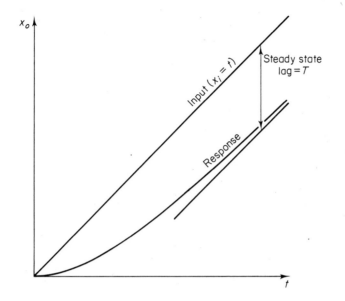

Figure 6.16

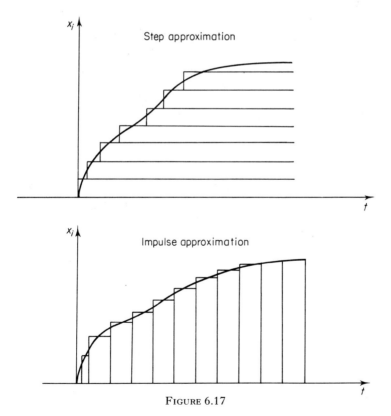

Figure 6.17

6.13 Harmonic response of a first order system (1)

The response of a linear system to a periodic input can be obtained by expressing the input as a Fourier series

$$x_i = a_1 \sin \omega t + a_2 \sin 2\omega t + \cdots$$
$$+ b_1 \cos \omega t + b_2 \cos 2\omega t + \cdots$$
$$= A_1 \sin(\omega t + \phi_1) + A_2 \sin(2\omega t + \phi_2) + \cdots$$

determining the response to each component of the input and adding the responses together. The basic problem is therefore to determine the response to a sinusoidal input.

Referring to the first order system, we have to obtain the solution of the equation

$$(1 + TD)x_o = X_i \cos \omega t$$

The complete solution of this equation is comprised of two parts; the complementary function which in this case is

$$x_o = A\,e^{-t/T},$$

and the particular integral. As t becomes large the complementary function tends to zero, and we are left with the particular integral which represents the steady periodic response to the input — the *steady-state response*. Usually interest is focused on the steady-state response, rather than the transient motion which occurs immediately after the input has been initiated, and so the complementary function is ignored.

For the rest of this book we shall, when faced with the problem of determining the particular integral of a differential equation due to a harmonic input, assume that there is a solution of the form

$$x_o = X_o \cos(\omega t - \phi) \tag{6.8}$$

and proceed by the techniques which we shall develop to determine the amplitude X_o and the phase angle ϕ. We shall be justified in doing this if, at this stage, we show formally that any differential equation of the form

$$P(D)x_o = X_i \cos \omega t$$

where $P(D)$ is a polynomial in D, has a particular integral of the assumed form. To do this we follow the standard method of evaluating a particular integral and rewrite the equation in the form

$$x_o = \frac{X_i}{P(D)} \cos \omega t$$

Substituting $\mathcal{R}\,e^{j\omega t}$ (the real part of $e^{j\omega t}$) for $\cos \omega t$ we have

$$x_o = \mathcal{R}\,\frac{X_i}{P(D)} e^{j\omega t}$$

LINEAR SYSTEMS I First order systems

By a standard theorem of linear differential equations we can now write

$$x_o = \mathscr{R} \frac{X_i}{P(j\omega)} e^{j\omega t}$$

Now
$$P(D) = (a_o + a_1 D + a_2 D^2 + \cdots + a_n D^n)$$

therefore
$$P(j\omega) = (a_o + a_1 j\omega + a_2(j\omega)^2 + \cdots)$$
$$= (a_o - a_2 \omega^2 + a_4 \omega^4 - \cdots)$$
$$+ j(a_1 \omega - a_3 \omega^3 + \cdots)$$
$$= \alpha + j\beta$$

Hence
$$x_o = \mathscr{R} \frac{X_i}{\alpha + j\beta} e^{j\omega t}$$
$$= \mathscr{R} \frac{X_i}{\alpha^2 + \beta^2}(\alpha - j\beta) e^{j\omega t}$$
$$= \frac{X_i}{\alpha^2 + \beta^2}(\alpha \cos \omega t + \beta \sin \omega t)$$
$$= \frac{X_i}{(\alpha^2 + \beta^2)^{\frac{1}{2}}} \cos(\omega t - \phi)$$
$$= X_o \cos(\omega t - \phi)$$

where $X_o = X_i/(\alpha^2 + \beta^2)^{\frac{1}{2}}$ and $\phi = \tan^{-1}(\beta/\alpha)$. In a specific case, α and β, and hence X_o and ϕ, emerge from the calculation with definite values. However, having demonstrated that $x_o = X_o \cos(\omega t - \phi)$ is a valid solution for differential equations of the type which we shall be concerned with, we shall henceforth assume this solution and determine X_o and ϕ by more direct means.

6.14 Vector diagram solution of $(1 + T D)x_o = X_i \cos \omega t$

A harmonically varying quantity, such as $x_1 = A \cos \omega t$, can be represented by the projection on a suitable datum of a line OA of length A rotating with angular speed ω, fig. 6.18(a). In this case, the datum corresponds to the position of the line when $t = 0$. A variable $x_2 = B \cos(\omega t + \alpha)$ is similarly represented by a rotating line OB of length B, but starting when $t = 0$ at an angle α to the datum, fig. 6.18(b).

The question now arises as to how we represent the sum $x_1 + x_2$ of these two quantities. In the particular case when $\alpha = 0$ there is no difficulty, since

$x_1 + x_2 = (A + B)\cos \omega t$, and clearly may be represented by a rotating line of length $A + B$. If $\alpha = \pi/2$ we have

$$x_1 + x_2 = A \cos \omega t + B \cos \left(\omega t + \frac{\pi}{2}\right)$$

$$= A \cos \omega t - B \sin \omega t$$

$$= (A^2 + B^2)^{\frac{1}{2}} \cos(\omega t + \psi)$$

where $\psi = \tan^{-1} B/A$. The sum $x_1 + x_2$ is therefore represented by a rotating line OC of length $(A^2 + B^2)^{\frac{1}{2}}$, which is at an angle ψ in advance of OA, fig. 6.18(c). It follows that to obtain $x_1 + x_2$ we can treat the representative

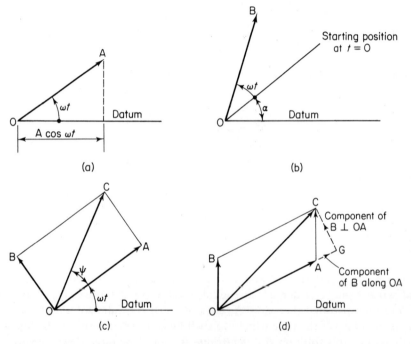

FIGURE 6.18

lines as vectors and apply the normal rule for vector addition. It also follows that we can resolve a vector OC, say, into two components at right-angles to each other, and hence that the vector rule for addition applies, even if OA and OB are not at right-angles to each other, fig. 6.18(d).

To solve the equation

$$(1 + TD)x_o = X_i \cos \omega t$$

we assume that there is a solution of the form

$$x_o = X_o \cos(\omega t - \phi)$$

LINEAR SYSTEMS I First order systems

so that the equation becomes

$$(1 + TD)X_o \cos(\omega t - \phi) = X_i \cos \omega t$$

The two terms on the left-hand side of this equation can be represented by rotating vectors the sum of which must be identical with the single vector which represents the right-hand side of this equation. Now if

$$x_o = X_o \cos(\omega t - \phi)$$

then

$$\frac{dx_o}{dt} = -\omega X_o \sin(\omega t - \phi)$$

$$= \omega X_o \cos\left(\omega t - \phi + \frac{\pi}{2}\right)$$

The effect of differentiating a function is to rotate its representative vector forward through $\pi/2$ and to multiply its length by ω, fig. 6.19(a). In fig. 6.19(b) the representative vectors for x_o and $T D x_o$ have been added to give the representative vector X_i. The angle which X_i makes with the datum is ωt; and the angle between X_o and the datum is $(\omega t - \phi)$, so that the angle between X_i and X_o is the phase angle ϕ.

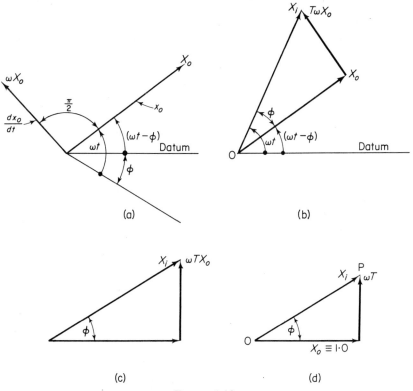

FIGURE 6.19

As t increases, the triangle in fig. 6.19(b) rotates about O without change of shape. As we are interested in the shape of the triangle rather than its orientation at any particular instant it is usual to draw X_o horizontal and omit the datum as in fig. 6.19(c). The relationship between the amplitude is obtained from the geometry of the figure

$$X_i^2 = X_o^2 + (\omega T X_o)^2$$

so that
$$X_o = \frac{X_i}{(1 + \omega^2 T^2)^{\frac{1}{2}}} \tag{6.9}$$

We also have

$$\tan \phi = \frac{\omega T X_o}{X_o}$$

so that
$$\phi = \tan^{-1} \omega T \tag{6.10}$$

These simple calculations demonstrate the use of vector diagrams to obtain general expressions for amplitude and phase angle. It is also possible to determine the amplitude and phase angle in a particular case by drawing the diagram to scale. We cannot do this immediately because the amplitude X_o is unknown, instead we draw the diagram assuming X_o to be of unit amplitude and so find the corresponding amplitude X_i of the input. The actual amplitude of the response for an input of specified amplitude is then obtained by direct proportion.

6.15 Representation of harmonic response data

It is of considerable interest to know how the amplitude and phase of the response of a system vary with the frequency of the input. The amplitude and phase relationships for a first order system are defined by eqns. (6.9) and (6.10). As the amplitude of the response is directly proportional to the amplitude of the input it is usual to think in terms of the *dynamic magnification* $M = X_o/X_i$ rather than the response to an input of particular amplitude. The variation of the dynamic magnification, and of the phase angle with frequency, are shown in fig. 6.20. If eqns. (6.9) and (6.10) are inspected it will be seen that ω and T always occur together in a product. It is logical, therefore, to consider ωT to be the independent variable rather than ω by itself. In this way the equations are each represented by a single curve rather than a family of curves corresponding to different values of T.

We will defer discussion of the way in which the amplitude and frequency vary with frequency, and consider first alternative ways of presenting the same information.

In fig. 6.19(a) we have a diagram which for a particular frequency gives the amplitude of the input required to cause unit amplitude of the response, and the phase angle. If ω is varied the terminal point P of the X_i vector traces a locus which is a straight line, fig. 6.21, on which the variation in ω can be

LINEAR SYSTEMS I First order systems

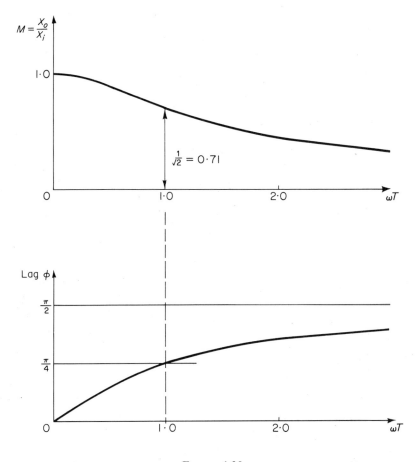

FIGURE 6.20

noted. The way in which the dynamic magnification and phase angle vary with frequency can be seen very quickly from this diagram, which is simpler to draw than the curves in fig. 6.20. The diagram is known as the *inverse harmonic response locus*. As we have seen, it has evolved naturally out of the vector diagram method of solving a differential equation. Effectively, we assume a unit amplitude for the response and then by means of the diagram determine the corresponding amplitude of the input. Whilst this is a perfectly rational means of calculation it does mean that fig. 6.21 does not give, directly, the information which is of most interest, namely the way in which the amplitude of the response varies when the input is of unit amplitude; to obtain the dynamic magnification from fig. 6.21 we have to take the reciprocal of the length OP. A more convenient representation would be one in which the dynamic magnification is given directly by keeping OP $\equiv X_i$ fixed with unit length and tracing the locus of Q, the terminal point of the X_o vector, as ω

266 Dynamics of Mechanical Systems

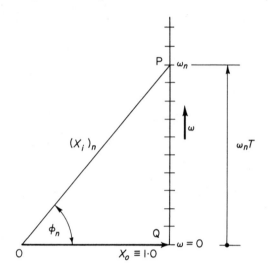

FIGURE 6.21

varies. The resultant locus is the semi-circle shown in fig. 6.22. The semi-circular shape is explained by noting that for a particular frequency ω_n the triangle OPQ in figs. 6.21 and 6.22 must be geometrically similar to each other, since the phase angle ϕ_n must be the same in both diagrams and X_o/X_i must be the same in both. It follows that in fig. 6.22 the angle OQP is a right-angle, and that consequently the locus of Q is a semi-circle. Figure 6.22 is known as the *direct harmonic response locus* for a first order system. It is also referred to as the *polar plot* for the system (polar coordinates X_o, ϕ). In an experimental investigation of the behaviour of a system it is usual to keep the amplitude of the input constant and to measure the amplitude of the response and the phase angle as the frequency is changed. The direct locus therefore provides a natural way of plotting experimental data, and for this reason it is more generally used than the inverse plot.

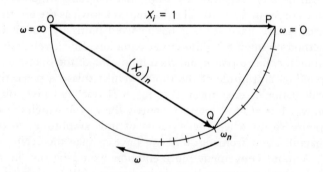

FIGURE 6.22

LINEAR SYSTEMS I First order systems

When the direct locus is to be determined from the governing equation there are two possible approaches. One, which has already been described in relation to the first order system, involves drawing the indirect locus first and deducing the direct locus from it. It is, of course, not necessary to prove by geometrical argument that the locus has a particular shape; X_i (for $X_o = 1$) and ϕ are measured on the inverse locus for a convenient series of frequencies, X_o (for $X_i = 1$) is then calculated for each frequency and plotted, using its corresponding phase angle, in polar form. This method is quite convenient if the governing differential equation of the system is of low order and of the form $F(D)x_o = x_i$. When this is not the case it is better to proceed by direct calculation in the manner described in the next section.

6.16 Use of the Argand diagram
The differential equation

$$P(D)x_o = X_i \cos \omega t$$

may be written in the form

$$P(D)x_o = \mathscr{R} X_i\, e^{j\omega t}$$

On substituting the assumed solution

$$x_o = X_o \cos(\omega t - \phi) = \mathscr{R} X_o\, e^{j(\omega t - \phi)}$$

we have

$$P(D) \mathscr{R} X_o\, e^{j(\omega t - \phi)} = \mathscr{R} X_i\, e^{j\omega t}$$

This equation for the real parts of the exponential functions will obviously be satisfied if

$$P(D) X_o\, e^{j(\omega t - \phi)} = X_i\, e^{j\omega t}$$

is satisfied.

On evaluating the differential coefficients this equation becomes

$$P(j\omega) X_o\, e^{j(\omega t - \phi)} = X_i\, e^{j\omega t}$$

and on cancelling the common factor $e^{j\omega t}$, and dividing by $P(j\omega)$ we have, for $X_i = 1$,

$$X_o\, e^{-j\phi} = \frac{1}{P(j\omega)} \tag{6.11}$$

Provided that this equation is satisfied, our assumed solution of the original differential equation is valid. In other words, the assumed solution is valid provided that X_o and ϕ have the values which are given as a solution of this last equation.

Now $X_o\, e^{-j\phi}$ is a complex number which is represented in the Argand diagram by a point, whose position is defined by a radius vector of length X_o which is at an angle $-\phi$ to the real axis. In the case of a first order system we would therefore expect a diagram which is identical with fig. 6.22. Let us verify that eqn. (6.11) does indeed lead to this diagram:

For a first order system

$$P(D) \equiv (1 + TD)$$

so that

$$P(j\omega) \equiv (1 + j\omega t)$$

and

$$\frac{1}{P(j\omega)} \equiv \frac{1}{(1 + j\omega t)}$$

$$\equiv \frac{1 - j\omega T}{(1 + \omega^2 T^2)} X_i$$

A typical point Q in the Argand diagram is therefore located as shown in fig. 6.23. If we now identify the real (Re) and imaginary (Im) axes by x and y, respectively, we have

$$x = \frac{1}{1 + \omega^2 T^2}; \quad y = \frac{-\omega T}{1 + \omega^2 T^2} \tag{6.12}$$

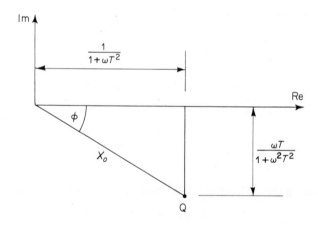

FIGURE 6.23

The locus of Q can now be plotted by evaluating x and y for a series of values of ω which are then marked on the locus as shown in fig. 6.22. The equation of the locus of Q is obtained by eliminating ω from these two equations, and is found to be

$$(x - \tfrac{1}{2})^2 + y^2 = (\tfrac{1}{2})^2$$

This is the equation of a circle of radius $\tfrac{1}{2}$ and centred on the point $(\tfrac{1}{2}, 0)$. Only the semi-circle below the real (or x-) axis is of interest, the upper half relating to negative values of ω.

The inverse locus is obtained by assuming $x_o = 1 \cdot \cos \omega t$ and
$$x_i = X_i \cos(\omega t + \phi)$$
so that we have
$$X_i e^{j\phi} = F(j\omega)$$
In the case of the first order system
$$X_i e^{j\phi} = (1 + j\omega T)$$
which is readily identified with the inverse locus in fig. 6.21.

It may seem somewhat perverse to refer to $1/P(j\omega)$ as the direct locus and $P(j\omega)$ as the inverse locus, but it must be remembered that the basic object of this work is to predict the response of a system to an input, rather than the reverse.

The use of the Argand diagram extends to more complicated systems for which the differential equation is of the form
$$P(D)x_o = Q(D)x_i$$
By rephrasing the analysis leading up to eqn. (6.11) it can be shown that for this more general case
$$X_o e^{-j\phi} = \frac{Q(j\omega)}{P(j\omega)} \qquad (6.13)$$

6.17 Harmonic response of first order systems (2)

Having considered how the response of a first order system to a harmonic input can be determined and described, let us now look at the consequences of that behaviour. We have seen that for an input $X_i \cos \omega t$, the amplitude of the response of a first order system is
$$X_o = \frac{X_i}{(1 + \omega^2 T^2)^{\frac{1}{2}}}$$
and its phase lag is
$$\phi = \tan^{-1} \omega T$$
The amplitude of the response decreases with frequency, whilst the phase lag increases.

The attenuation property of a first order system is sometimes a nuisance but is sometimes used to advantage, as, for example, in the use of the electrical circuit shown in fig. 6.24(a) as a low-pass filter (see p. 244 for analysis). The purpose of this element is to reduce spurious high frequency variations, known as noise, which may affect the input voltage signal v_i. Physically, the condenser C acts as a short-circuit for high frequency components of the input signal without affecting the main low frequency component. Pneumatic and hydraulic components which act as low-pass filters are also given in fig. 6.24.

The hydraulic relay in fig. 6.7(d) is an interesting example of a case where an extraneous high frequency component may be deliberately injected into the input signal in the knowledge that it will not affect the output. This happens when, as is sometimes done, the spool is made to vibrate at high frequency to prevent it sticking.

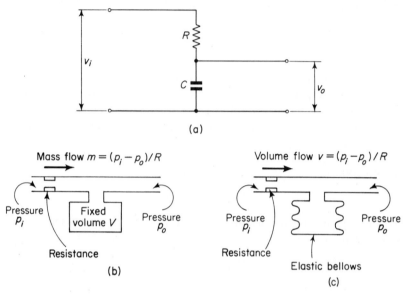

FIGURE 6.24

When a first order element is required to approximate in its behaviour to a proportional element the attenuation and phase shift are both undesirable sources of distortion. We can regard any periodic input as being composed of a series of harmonics. Each harmonic will suffer a different amount of attenuation and phase shift so that the response is bound to be a poor replica of the input unless the frequency of the highest harmonic is small compared with T^{-1}. In this case, $X_o \approx X_i$ for all harmonics, which are consequently kept in their correct proportions. It might appear that distortion will still occur due to the different phase shifts of the harmonics; in fact, this does not happen. The response to the nth harmonic is

$$X_{o(n)} = X_{i(n)} \cos n\omega(t + \phi_n)$$

where ω is the frequency of the fundamental component. If

$$n\omega \ll T^{-1}$$

we have

$$\phi_n \approx n\omega T$$

so that

$$X_{o(n)} = X_{i(n)} \cos n\omega(t + T)$$

This means that all the harmonics suffer the same delay T, and are kept in the correct phase in relation to each other.

It should be borne in mind that at high frequencies errors may arise because

LINEAR SYSTEMS I First order systems

of the idealizations which have been made in setting up the differential equation of the element. Provided that this does not happen, the high frequency ($\omega \gg T^{-1}$) behaviour of a first order element approximates to that of an integrating element with $X_o = X_i/\omega T$ and $\phi = \pi/2$. Practical application of this is in vibration measurement.

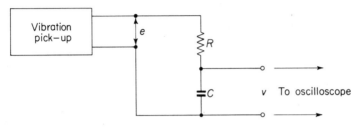

FIGURE 6.25

The amplitude of vibration of a piece of machinery may be determined by attaching to it a vibration pick-up. This consists essentially of a coil and a magnet, of which one remains stationary while the other moves with the vibrating member. The relative motion of the coil and the magnet induces an e.m.f. in the coil proportional to the relative velocity. The e.m.f. can easily be fed directly to the plates of an oscilloscope and so measured. A direct measurement of the amplitude of the motion is obtained if an integrating circuit is interposed between the pick-up and the oscilloscope, fig. 6.25. If the displacement being measured is

$$x = X \sin \omega t$$

then the e.m.f. generated is

$$c = kX\omega \cos \omega t$$

The exact relationship between e and v is

$$(1 + TD)v = kX\omega \cos \omega t$$

but provided $\omega \gg T^{-1}$ the first term in the left-hand side is negligible compared with the second so that we have

$$T Dv = kX\omega \cos \omega t$$

or

$$v = \frac{kX}{T} \frac{\omega}{D} \cos \omega t$$

$$= \frac{kX}{T} \sin \omega t$$

$$= \frac{k}{T} x$$

272 Dynamics of Mechanical Systems

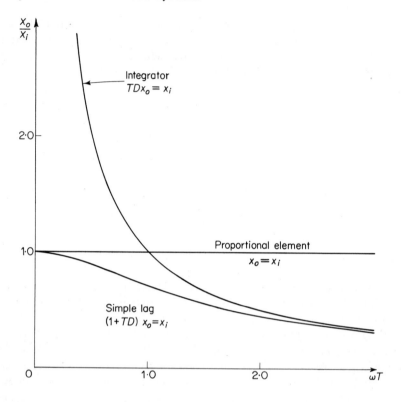

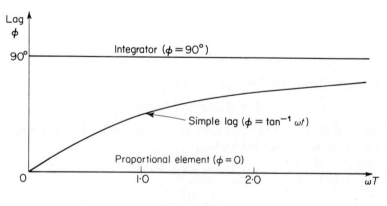

FIGURE 6.26

A comparison of the frequency response for proportional, integrating, and simple lag elements is given in fig. 6.26.

7

LINEAR SYSTEMS
II Second order systems

7.1 Introduction
We define a second order system as a system whose behaviour is governed by an ordinary second order differential equation, namely an equation of the form

$$(a_2 D^2 + a_1 D + a_o)x_o = (b_2 D^2 + b_1 D + b_o)x_i(t) \qquad (7.1)$$

The point has already been made in the last chapter that a particular type of differential equation may represent many diverse physical situations, and it need hardly be made again. In this chapter, we shall consider only two types of system: we shall take the behaviour of a simple servo-mechanism as our first example, and the mechanical vibrations of a spring-mounted mass as our second example.

7.2 Equation of motion for a simple position control system (servo-mechanism)
The object of the system which we are to study is to control the position of a mass. It does not matter whether the mass is the reflector of a radio telescope, a radar antenna, the rudder of a large ship, the tool-head of an automatic milling machine or the pen of an automatic recording device, the basic principle used and the mathematical equations are the same. We shall take as our example one for which it is easy to visualize just what is happening.

Figure 7.1 shows a rotor of moment of inertia J driven by a motor. The rotation θ_o of the rotor from a datum position is required to be equal to the angle θ_i through which a human operator turns a control wheel. The error ε in the position of the rotor is determined mechanically by making θ_i and θ_o the inputs to a differential gear. The output shaft of the differential controls the power supply to the motor. If $\varepsilon = \theta_i - \theta_o$ is zero, the rotor is in its required position, and it need not be moved further. If an error exists, the motor applies a torque to the rotor causing it to turn and reduce ε. This type of system is referred to as being *error-activated*, it is also power amplifying in that the power input by the operator is negligible compared with the power developed by the motor, and it is capable of responding to an arbitrarily varying input. A system which does all this is referred to as a *servo-system* ('servo' = Greek for slave). If, as in this case, the output is mechanical, it is a *servo-mechanism*. Examples of servo-systems with non-mechanical outputs are temperature and voltage control systems.

In order to keep the analysis as simple as possible we will assume that the torque output for the motor is proportional to the rotation ε of the motor

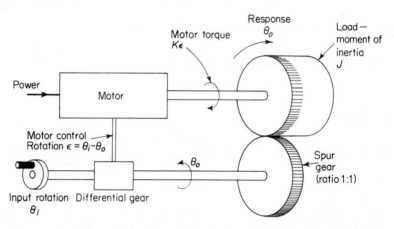

FIGURE 7.1

control, and is independent of the motor speed. We will leave for the moment consideration of what type of motor, if any, will operate in this way. The motor torque is absorbed partly in accelerating the load and partly in overcoming friction. We will assume that the friction component is proportional to speed. The equation of motion is thus

$$\underbrace{K(\theta_i - \theta_o)}_{\text{Motor torque}} = \underbrace{J\frac{d^2\theta_o}{dt^2}}_{\text{Inertia torque}} + \underbrace{\lambda\frac{d\theta_o}{dt}}_{\text{Friction torque}}$$

On rearranging, this becomes

$$J\frac{d^2\theta_o}{dt^2} + \lambda\frac{d\theta_o}{dt} + K\theta_o = K\theta_i(t) \tag{7.2}$$

which can be seen to be of the same basic form as eqn. (7.1). In terms of the operator D, we can write this equation as

$$\left(\frac{J}{K}D^2 + \frac{\lambda}{K}D + 1\right)\theta_o = \theta_i(t)$$

The speed with which the system is capable of responding to an input depends primarily on the power available to overcome the inertia of the load, so that K/J is a measure of the speed of response of the system, or conversely J/K is a measure of the time taken to respond. On inspecting the differential equation we see that J/K must have the dimensions of (time)^2 so that it is convenient to put $J/K = T^2$, where T is defined as the *time constant* of the system. It must be realized that this definition of the time constant is quite independent of the definition used when referring to a first order system. There is no equivalence between the two except that they both, in one way or another, provide a measure of the time taken by the system to respond to an input.

It is convenient to pair the D of the middle term on the left-hand side of the equation with a T, and so introducing a dimensionless damping factor $c = \lambda/2\sqrt{JK}$, so that the final form of the equation is

$$(T^2 D^2 + 2cT D + 1)\theta_o = \theta_i(t) \qquad (7.3)$$

For the moment we shall treat this system merely as an example of a second order system, and leave any assessment of its performance in the light of its function until the next chapter where the general theory of control systems is elaborated. The reader might, however, like to compare the way in which this system works with the way in which the hydraulic relay in fig. 6.7(d) operates, and to reflect on why this system leads to a second order differential equation whilst the relay leads to a first order equation.

7.3 Equation of motion for a spring-mounted mass

As our other example of a second order system we will consider the motion of a mass supported on a spring. In real life the system might be an engine or other piece of machinery which has been set on an elastic footing to prevent vibrations being transmitted to the supporting structure. Alternatively, it might be an instrument panel in an aircraft spring-mounted to prevent vibrations being transmitted to it. In the idealized situation we consider a rigid mass m supported on a light spring of stiffness s which is in parallel with a damping device, as shown in fig. 7.2. The damping device may be a physical entity, as for example the shock absorbers on an automobile, or it may represent damping which is inherent in the system, e.g., windage due to the motion, friction in the guides, hysteresis in the material of the spring. We will assume that, whatever its origin, the damping force is proportional to the velocity across the damping device; this is known as *viscous damping*. This form of damping is assumed because it is the one form of damping which leads to a differential equation of

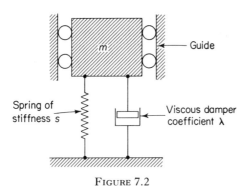

FIGURE 7.2

the form which we have stated above. Other forms of damping, Coulomb friction where the resisting force is independent of the velocity, and velocity squared damping which arises from turbulent flow of liquid through pipes, lead to non-linear differential equations which require special methods of solution. The justification for our assumption of viscous damping comes from the fact that it generally leads to predictions of overall behaviour which accord well with experience.

To obtain the equation of motion for the system we simply write down the equilibrium equation for the forces which act on the mass. First, however, we must decide on the origin for the variable x which we are to use to define the motion of the mass. There are various possibilities: we could, for example, let x be simply the height of the mass above ground, or the compression in the spring. It proves to be most convenient to let x be the displacement of the mass from its position of statical equilibrium, where the compression x_g of the spring is then simply that due to the weight and equals mg/s. The general set of forces acting on the mass are shown in fig. 7.3, they are the exciting force

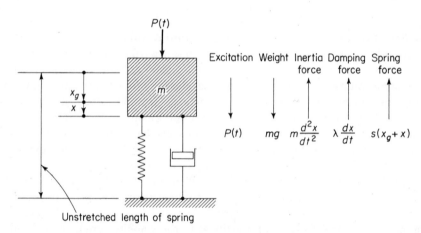

FIGURE 7.3

$P(t)$ which is an arbitrary function of time, the weight mg, the spring force $s(x + x_g)$, the damping force $\lambda\, dx/dt$, and the inertia force $m\, d^2x/dt^2$. The equilibrium equation, or equation of motion, is

$$m\frac{d^2x}{dt^2} + \lambda\frac{dx}{dt} + s(x + x_g) = mg + P(t)$$

As
$$sx_g = mg$$

the equation simplifies to

$$m\frac{d^2x}{dt^2} + \lambda\frac{dx}{dt} + sx = P(t) \tag{7.4}$$

The equation is of precisely the same form as eqn. (7.2).

Equation (7.4) can be regarded as the starting point for vibration theory and will be studied in this particular context in a later chapter. For the rest of this chapter we shall consider the general behaviour of systems governed by equations of the type

$$(T^2 D^2 + 2cT D + 1)\theta_o = \theta_i(t)$$

It is, however, helpful to have a particular system in mind, and for this purpose we will refer to the control system of fig. 7.1 as a typical example.

7.4 Solution of the equation of motion — the complementary function

The general solution of the equation

$$(T^2 D^2 + 2cT D + 1)\theta_o = \theta_i(t)$$

is comprised of two parts:

1 the particular integral, which depends on the excitation; and
2 the complementary function, which is the solution of the equation

$$(T^2 D^2 + 2cT D + 1)\theta_o = 0 \tag{7.5}$$

and is independent of the input $\theta_i(t)$.

The complementary function represents the free motion of the system following a disturbance from its equilibrium state, which for our elementary control system is $\theta_o = \theta_i = 0$, and $d\theta_o/dt = 0$.

To solve eqn. (7.5) we substitute an assumed solution $\theta_o = e^{\alpha t}$. This yields the auxiliary equation

$$\alpha^2 T^2 + 2c\alpha T + 1 = 0 \tag{7.6}$$

for which the solution is

$$\alpha = [-c \pm \sqrt{(c^2 - 1)}]/T \tag{7.7}$$

Our assumed solution $\theta_o = e^{\alpha t}$ is valid provided that α has one or other of these two values. Accordingly, the general form for the solution of eqn. (7.5) is

$$(\theta_o)_{CF} = A_1 e^{\alpha_1 t} + A_2 e^{\alpha_2 t} \tag{7.8}$$

where α_1 and α_2 are the two roots of eqn. (7.6).

The complete solution of eqn. (7.3) is therefore

$$\theta_o = A_1 e^{\alpha_1 t} + A_2 e^{\alpha_2 t} + particular\ integral \tag{7.9}$$

where the constants A_1 and A_2 are determined by specifying the values of θ_o and $d\theta_o/dt$ at $t = 0$.

The character of the complementary function depends on the value of the damping coefficient c.

Overdamped system $c > 1$

It can be seen by inspection of eqn. (7.7) that if $c > 1$ then both α_1 and α_2 are negative so that both components of the complementary function decay exponentially to zero as t tends to infinity. Such a system is said to be over-damped. In practice, an overdamped position control system would usually be regarded as too sluggish for satisfactory operation.

Underdamped system $c < 1$

If $c < 1$ then α_1 and α_2 are both complex, i.e.,

$$\left.\begin{array}{c}\alpha_1 \\ \alpha_2\end{array}\right\} = [-c \pm j(1 - c^2)]/T$$

$$= -c/T \pm j\beta$$

where
$$\beta = \sqrt{(1 - c^2)}/T \qquad (7.10)$$

The complementary function is now

$$(\theta_o)_{CF} = (A_1 e^{j\beta t} + A_2 e^{-j\beta t}) e^{-ct/T}$$

Now
$$e^{j\beta t} = \cos \beta t + j \sin \beta t$$

and
$$e^{-j\beta t} = \cos \beta t - j \sin \beta t$$

so that
$$(\theta_o)_{CF} = (a_1 \cos \beta t + a_2 \sin \beta t) e^{-ct/T} \qquad (7.11)$$

where a_1 and a_2 are now the constants to be determined by the initial values of θ_o and $d\theta_o/dt$. The relationships between these constants and the original ones A_1 and A_2 are of no interest.

We can, alternatively, coalesce the two trigonometric terms and write

$$(\theta_o)_{CF} = A e^{-ct/T} \cos (\beta t + \psi) \qquad (7.12)$$

where A and ψ are the two constants of integration. The latter form of $(\theta_o)_{CF}$ makes it clear that if $c < 1$, the complementary function is a damped sine wave whose amplitude tends to zero as t tends to infinity.

Critical damping $c = 1$

This case marks the boundary between the non-oscillatory complementary function when $c > 1$ and the oscillatory complementary function when $c < 1$. It can be seen from eqn. (7.7) that for $c = 1$

$$\alpha_1 = \alpha_2 = -1/T$$

so that
$$(\theta_o)_{CF} = (A_1 + A_2 t) e^{-t/T} \qquad (7.13)$$

which is non-oscillatory, but still tends to zero as t tends to infinity.

Whatever the value of c (assuming it to be non-zero and positive) the complementary function becomes negligible for large t, by which we mean $ct/T \ll 1$, and we are left with

$$\theta_o = \text{particular integral}$$

Accordingly we identify the complementary function with the transient response of a system to a disturbance or excitation, and the particular integral with its steady-state response.

Control systems are almost invariably designed so that c is at least 0·6. This is so that the response will settle down reasonably quickly to the desired steady state. In the context of mechanical vibrations, on the other hand, the amount of damping will usually be very small (i.e., $c \ll 1$) unless it has been deliberately introduced. It is easy to show from eqn. (7.12) that the amplitude of the transient response is reduced to 0·2 per cent of its initial value after $1/c$ cycles. The effect of c on the rate of decay of the transient vibrations is therefore strong. The frequency β of the transient vibrations is, however, relatively insensitive to c. Equation (7.10) defines $\beta = \sqrt{(1 - c^2)}/T$ so that provided c is not too large $\beta \approx 1/T$; if, for example, $c = 0\cdot 4$, which is fairly large in the context of mechanical vibrations, $\beta = 0\cdot 92/T$.

7.5 Step response

Following on the approach developed in respect of first order systems we will now investigate the response of a second order system to a step input, i.e.,

$$\theta_i(t) = 0 \quad \text{for } t < 0$$
$$\theta_i(t) = \theta_i \quad \text{for } t > 0$$

(fig. 7.4) in the knowledge that once the step input has been determined we can obtain the impulse and ramp responses by differentiation and integration respectively, and other cases by use of the principle of superposition.

For $t > 0$ the equation of motion is

$$(T^2 D^2 + 2cT D + 1)\theta_o = \theta_i \tag{7.14}$$

The particular integral is

$$(\theta_o)_{\text{PI}} = \theta_i \tag{7.15}$$

since this is the simplest expression for θ_o which satisfies the equation. Assuming that the system is underdamped ($c < 1$), the solution is

$$\theta_o = A\, e^{-ct/T} \cos(\beta t + \psi) + \theta_i \tag{7.16}$$

If we assume that at $t = 0$ the system was at rest in its equilibrium position, we have at $t = 0$

$$\theta_o = 0 = A \cos \psi + \theta_i \tag{7.17}$$

To obtain the second initial condition we will consider the mechanics involved. Referring to the position control system, it can be seen that the torque applied by the motor immediately after a step input must be finite in

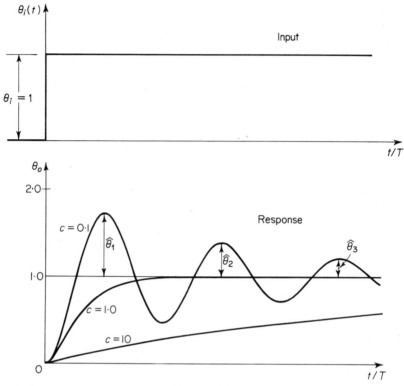

FIGURE 7.4

magnitude and therefore cannot impart an instantaneous angular velocity to the rotor. Hence, at $t = 0$, $d\theta_o/dt = 0$. In general,

$$\frac{d\theta_o}{dt} = -A\,e^{-ct/T}[c\cos(\beta t + \psi)/T + \beta \sin(\beta t + \psi)]$$

so that at $t = 0$, we have

$$0 = -A[c\cos\psi/T + \beta \sin\psi]$$

As eqn. (7.17) does not allow the possibility $A = 0$

$$\tan\psi = -c/\beta T = -c/\sqrt{(1-c^2)}$$

or

$$\psi = -\tan^{-1}\frac{c}{\sqrt{(1-c^2)}}$$

Now

$$\cos\psi = \frac{1}{\sqrt{(1 + \tan^2\psi)}} = \sqrt{(1-c^2)}$$

Hence from eqn. (7.17)

$$A = -\theta_i/\sqrt{(1-c^2)}$$

Substituting for A in eqn. (7.16) we obtain

$$\theta_o = \theta_i\left[1 - \frac{e^{-ct/T}}{\sqrt{(1-c^2)}}\cos(\beta t + \psi)\right] \qquad (7.18)$$

where $\quad \psi = -\tan^{-1}\dfrac{c}{\sqrt{(1-c^2)}}$

as the response to a step input of magnitude θ_i. The response when $c = 1$ can be obtained by inserting the initial conditions in

$$\theta_o = (A_1 + A_2 t)e^{-t/T} + \theta_i$$

or by treating $c = 1$ as a limiting case of eqn. (7.18). The response of a critically damped system to a step input is

$$\theta_o = \theta_i[1 - e^{-t/T}(1 + t/T)] \qquad (7.19)$$

For an overdamped system, $c > 1$,

$$\theta_o = \theta_i\left[1 - \frac{e^{-ct/T}}{\sqrt{(c^2-1)}}\sinh(\beta^* t + \psi^*)\right] \qquad (7.20)$$

where $\quad \beta^* = \sqrt{(c^2-1)}/T \quad \text{and} \quad \psi^* = \tanh^{-1}\dfrac{\sqrt{(c^2-1)}}{c}$

Equations (7.18), (7.19), and (7.20) have been plotted for an appropriate range of values of c in fig. 7.4. These graphs make quite clear the influence of the damping coefficient c on the rate at which the system approaches its steady-state condition.

With a position control system, the need to adjust the damping coefficient to some optimum is clear; if c is too large the system is sluggish, but if c is small the response seriously overshoots its steady-state value and the subsequent oscillations take several cycles to die away. As the amount of the first overshoot is easily measured it provides a convenient criterion whereby to judge the performance of a position control system. It also enables c to be determined experimentally for a real system. Differentiation of eqn. (7.18) with respect to time gives

$$\frac{d\theta_o}{dt} = \theta_i\frac{e^{-t/T}}{(1-c^2)}\beta\sin(\beta t + \psi) + c\cos(\beta t + \psi)/T$$

On substituting for ψ, and rearranging, this becomes

$$\frac{d\theta_o}{dt} = \frac{\theta_i}{T}\frac{e^{-ct/T}}{(1-c^2)}\sin\beta t \qquad (7.21)$$

which is zero, so that x is a maximum or minimum, when $\beta t = 0, \pi, 2\pi, 3\pi, \ldots$ The first overshoot occurs when $\beta t = \pi$. On substituting this value for t in eqn. (7.18) we have

$$(\theta_o)_{\max} = \theta_i[1 + e^{-\pi c/(1-c^2)}] \qquad (7.22)$$

Expressed as a fraction of the steady-state response the overshoot is

$$\frac{\hat{\theta}_1}{\theta_i} = e^{-\pi c/(1-c^2)} \tag{7.23}$$

This equation does not always provide a convenient means of determining the damping coefficient because of practical difficulties in applying a step input. An approximation to a step input is permissible provided that the time taken to effect the step change in θ_i is small compared with the time constant T. If there is any doubt about this the evaluation of c should not be based on eqn. (7.23) alone. Instead, successive overshoots $\hat{\theta}_1$, $\hat{\theta}_2$, $\hat{\theta}_3$, etc. should be measured. These occur when $t/T = \pi, 3\pi, 5\pi$, etc. respectively. Now

$$\hat{\theta}_1 = \theta_i \, e^{-\pi c/(1-c^2)}$$
$$\hat{\theta}_2 = \theta_i \, e^{-3\pi c/(1-c^2)}$$
$$\hat{\theta}_3 = \theta_i \, e^{-5\pi c/(1-c^2)}$$

etc.

Hence
$$\frac{\hat{\theta}_1}{\hat{\theta}_2} = \frac{\hat{\theta}_2}{\hat{\theta}_3} = \cdots = e^{2\pi c/(1-c^2)}$$

and in general
$$\log_e (\hat{\theta}_i/\hat{\theta}_{i-1}) = 2\pi c/(1-c^2) \tag{7.24}$$

Thus a graph of $\log_e \hat{\theta}_i$ against the number of swings i should be a straight line whose slope is $2\pi c/(1-c^2)$. The straightness of the line demonstrates the validity of the assumption of viscous damping. The quantity $2\pi c/(1-c^2)$ is known as the *logarithmic decrement*.

7.6 Other transient responses

The responses to a unit impulse and to a unit ramp input can be obtained respectively by differentiating and integrating with respect to t, the response to a unit step.

Responses to non-standard inputs can be obtained by applying the principle of superposition, and using either the impulse or step response as a basis in the manner described in the last chapter. The process is tedious, though fairly easily programmed for evaluation by computer. Transient forcing of lightly damped systems is more easily investigated by the phase-plane technique, which will be described later.

7.7 Harmonic response

The general solution of the differential equation

$$(T^2 D^2 + 2cT D + 1)\theta_o = \begin{cases} 0 & \text{for } t < 0 \\ \Theta_i \cos \omega t & \text{for } t > 0 \end{cases} \tag{7.25}$$

is
$$\theta_o = A \, e^{-ct/T} \cos (\beta t + \psi) + \text{particular integral}$$

We know from experience that the steady-state response to a harmonic excitation is sinusoidal, and at the same frequency as the excitation, so that the particular integral is of the form $\Theta_o \cos(\omega t - \phi)$. It will be our aim later to determine expressions for Θ_o and ϕ. For the moment we note that, unless the forcing frequency happens to equal the frequency β of free damped vibrations, the motion must be rather complicated with a damped sine wave superimposed on a constant amplitude sine wave of different frequency. This general motion is, perhaps fortunately, of little interest, and attention is focused almost exclusively on the steady-state response, which is represented by the particular integral. It is in any event unlikely that, in a practical case, the excitation will suddenly be switched on at full amplitude and frequency; normally, it will build up from zero. This is an added reason for not devoting too much attention to the transient response to a harmonic excitation.

7.8 The particular integral of $(T^2 D^2 + 2cT D + 1)\theta_o = \Theta_i \cos \omega t$

There are well-known formal methods of determining the particular integral to a linear differential equation to a harmonic excitation. However, as we have already shown earlier that the particular must be sinusoidal in form and at the same frequency as the excitation, we will assume a solution

$$\theta_o = \Theta_o \cos(\omega t - \phi) \qquad (7.26)$$

and will proceed directly to evaluating Θ_o and ϕ using the vector diagram method which was developed in the last chapter. The relevant diagram is drawn in fig. 7.5.

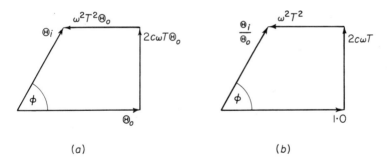

FIGURE 7.5

The third term on the left-hand side of the differential equation is represented by the vector Θ_o. In accordance with the work of the last chapter, $d\theta_o/dt$ must be represented by a vector of length $\omega\Theta_o$ turned 90 degrees in the anticlockwise direction from Θ_o. The central term is therefore represented by the vector $2c\omega T\Theta_o$. The second differentiation in the first term means a further rotation through 90 degrees, and an extra multiplication by ω so that this term is represented by the vector labelled $\omega^2 T^2 \Theta_o$. The resultant of these vectors must be identical to the vector Θ_i, which represents the input $\Theta_i \cos \omega t$.

The phase angle ϕ, by which the response lags behind the input, is the angle between Θ_o-vector and the Θ_i-vector.

Following the line of argument used in relation to the first order system, fig. 7.5(a) has been redrawn in fig. 7.5(b) with all the vectors divided by Θ_o. This diagram can be drawn to scale to determine Θ_o and ϕ numerically for a given system, or it can be used to obtain convenient general expressions. By inspection

$$\tan \phi = \frac{2c\omega T}{1 - \omega^2 T^2} \qquad (7.27)$$

Using Pythagoras' theorem,

$$\left(\frac{\Theta_i}{\Theta_o}\right)^2 = (1 - \omega^2 T^2)^2 + (2c\omega T)^2$$

so that

$$\Theta_o = \frac{\Theta_i}{[(1 - \omega^2 T^2)^2 + 4c^2\omega^2 T^2]^{\frac{1}{2}}} \qquad (7.28)$$

If ω is varied, the terminal point P of the Θ_o/Θ_i vector traces the inverse harmonic response locus, which is a parabola, because one coordinate is proportional to ωT and the other is proportional to $(\omega T)^2$, fig. 7.6(a). The 'aspect ratio' of the parabola depends on the value of c, fig. 7.6(b) (page 285).

7.9 The direct harmonic response locus

In order to plot the direct harmonic response locus we first substitute $j\omega$ for D in the differential equation

$$(T^2 D^2 + 2cT D + 1)\theta_o = \Theta_i \cos \omega t$$

and so obtain

$$[(j\omega)^2 T^2 + 2(j\omega)cT + 1]\Theta_o = \Theta_i$$

The direct harmonic response locus is defined by the complex expression

$$\frac{\Theta_o}{\Theta_i} = \frac{1}{(j\omega)^2 T^2 + 2(j\omega)cT + 1}$$

$$= \frac{1}{(1 - \omega^2 T^2) + 2jc\omega T} \qquad (7.29)$$

It is by no means easy to visualize what the shape of the locus will be as ω (or ωT) is varied, whilst c is kept constant. We could, of course, simply evaluate the real and imaginary parts of Θ_o/Θ_i for a suitable range of ωT and suitable values for c. This would, however, be a lengthy computation. The curves in fig. 7.7 have in fact been drawn accurately as the result of such a calculation. We will consider how the general shape could have been deduced, and hence the curves sketched without extensive calculation.

LINEAR SYSTEMS II Second order systems

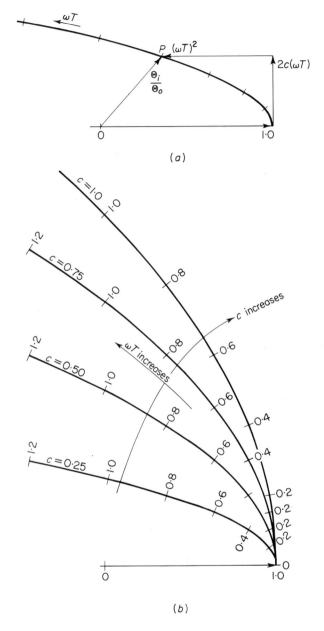

FIGURE 7.6

First, we note that for

$$\omega = 0, \quad \Theta_o/\Theta_i = 1 + j0$$

for

$$\omega = \infty, \quad \Theta_o/\Theta_i = 0$$

Hence, all loci, irrespective of the value of c, start with $\omega = 0$, at the point $1 + j0$ and terminate with $\omega = \infty$ at the origin. We now consider what happens as ω increases from zero, and tends to infinity.

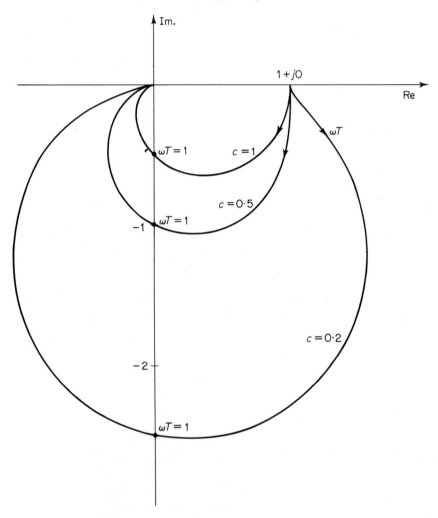

FIGURE 7.7

In the region where ω is small (i.e., $\omega T \ll 1$), we can approximate Θ_o/Θ_i by a binomial expansion, thus

$$\Theta_o/\Theta_i = 1 - (-\omega^2 T^2 + 2jc\omega T) + (-\omega^2 T^2 + 2jc\omega T)^2 - \cdots$$
$$= 1 + (\omega^2 T^2 - 2jc\omega T) + (\omega^4 T^4 - 4jc\omega^3 T^3 - 4c^2\omega^2 T^2) - \cdots$$
$$\approx 1 + (1 - 4c^2)\omega^2 T^2 - 2jc\omega T$$

LINEAR SYSTEMS II Second order systems

This approximate equation shows that for very small values of ωT, where $\omega^2 T^2$ is infinitesimal, the locus must be parallel to the imaginary axis, and proceeding for increasing ω in the negative direction.

As ω is increased to the extent that the $\omega^2 T^2$ term makes a measurable contribution, we see that the real part tends to increase if $c < 0.5$ and decrease if $c > 0.5$.

For large values of ωT

$$\Theta_o/\Theta_i \approx -1/\omega^2 T^2$$

so that as ω is decreased from infinity the loci are traced in the negative direction along the real axis. Rather less approximately we have

$$\frac{\Theta_o}{\Theta_i} = \frac{-1}{(\omega T)^2} \frac{1}{1 - 2jc/\omega T}$$

$$= \frac{-1}{(\omega T)^2}(1 + 2jc/\omega T)$$

so that as ω is reduced further the loci turn down from the negative real axis in the third quadrant of the Argand diagram. The rate of 'turn-down' increases with c.

The next thing to investigate is whether there are any values of ωT, other than 0 or ∞, which cause either the real or the imaginary parts of the denominator to be zero. Clearly, the imaginary part never goes to zero in this example except when $\omega = 0$ or ∞. The real part, on the other hand, is zero if $\omega T = 1$ so that Θ_o/Θ_i is then $1/2jc$. The loci therefore cross the imaginary axis at the points $0 - j/2c$. We now have sufficient information to deduce that the general shape of the loci must be as in fig. 7.7.

7.10 Variation of amplitude and phase with frequency

The variations in amplitude and phase angle with frequency can be computed directly from eqns. (7.27) and (7.28) respectively, or they can be deduced by measurements on either fig. 7.6 or fig. 7.7. The general shape of the curves can be deduced immediately from these diagrams.

We note that at zero frequency, the dynamic magnification Θ_o/Θ_i is 1·0, whatever the value of c. If c is large then Θ_o/Θ_i decreases continuously as ω increases and tends to zero as ω tends to infinity. Also if c is large we note a steady change in the phase lag ϕ from zero at $\omega = 0$ up to 180 degrees as ω tends to infinity. One notable feature of the phase curves is that whatever the value of c, the phase shift at $\omega T = 1$ is 90 degrees.

If c is small we observe that the dynamic magnification increases as ω is increased from zero, slowly at first, but then very rapidly as ωT approaches unity. As ω increases further there is a rapid fall-off in the dynamic magnification which, again, tends to zero as ω tends to infinity. The phase angle grows very slowly with ω initially, increases rapidly to 90 degrees as ωT approaches unity, and continues to increase rapidly as ωT becomes greater than unity. The

288 Dynamics of Mechanical Systems

rate of increase then slows down considerably with the 180-degree phase lag approached asymptotically as ω tends to infinity. The actual graphs of dynamic magnification and phase lag against frequency are given in fig. 7.8.

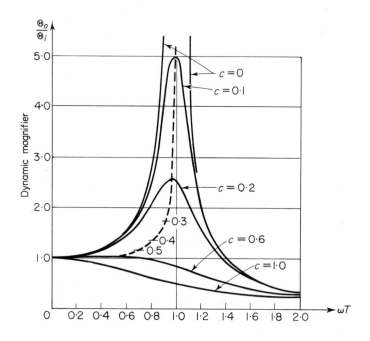

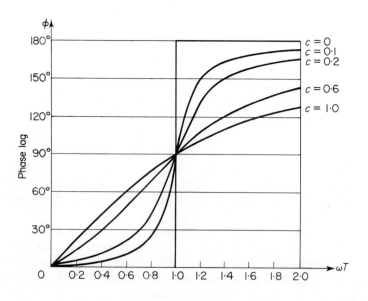

FIGURE 7.8

LINEAR SYSTEMS II Second order systems

The large amplitude response which occurs in lightly damped systems for $\omega T \approx 1$ is known as *resonance*. The physical explanation of this phenomenon is most readily understood in relation to the spring-mounted mass. If the terms of eqn. (7.2) are identified with the vectors in fig. 7.5 it will be seen that resonance occurs when the P-vector which represents the exciting form is parallel to, i.e., in phase with, the vector which represents the damping force. As the damping force vector is simply the velocity vector multiplied by ω, we can say that at resonance the velocity is in phase with the excitation. This means that the power input into the system must be relatively large, and the resulting vibrations correspondingly large in amplitude.

Figure 7.8 shows that the resonance peaks do not occur precisely when $\omega T = 1$. The exact location of the maxima can be obtained by differentiating eqn. (7.28). It is, in fact, slightly easier to differentiate

$$\left(\frac{\Theta_i}{\Theta_o}\right)^2 = (1 - \omega^2 T^2)^2 + 4c^2\omega^2 T^2$$

and to say that Θ_o is a maximum when this is a minimum

$$\frac{d}{d(\omega T)}\left(\frac{\Theta_i}{\Theta_o}\right)^2 = -4\omega T(1 - \omega^2 T^2) + 8c^2 \omega T$$

$$= 0 \text{ for minimum}$$

Hence Θ_o/Θ_i is a maximum when $\omega T = 0$ or $\sqrt{(1 - 2c^2)}$. It can be seen that all the curves have zero slope at $\omega T = 0$. When $\omega T = \sqrt{(1 - 2c^2)}$,

$$\Theta_o/\Theta_i = 2c\sqrt{(1 - c^2)^{-1}} \tag{7.30}$$

The locus of the maxima represented by this equation has been drawn in fig. 7.8.

7.11 Comparison with first order systems

Figure 7.9 shows two similar systems. In one a light platform is supported by a spring in parallel with a damper, in the other a mass is supported by a spring and a damper. The equation of motion for the first system is

$$\lambda \frac{dx}{dt} + sx = P(t)$$

or
$$(T_1 D + 1) = P(t)/s \tag{7.31}$$

For the second system the equation is

$$m\frac{d^2 x}{dt^2} + \lambda \frac{dx}{dt} + sx = P(t)$$

or
$$(T_2^2 D^2 + 2cT_2 D + 1)x = P(t)/s \tag{7.32}$$

Now a massless platform is something which does not exist in the real world so that eqn. (7.32) is bound to give a more accurate representation of the

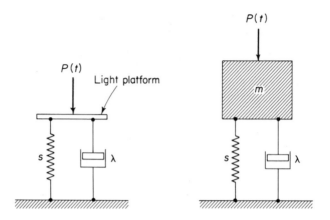

FIGURE 7.9

behaviour of this type of system than eqn. (7.31). Nevertheless, our intuition tells us that there must be situations in which the mass is small enough to be neglected — the problem is to know how small. A comparison of the responses to a step input of first and second order systems respectively, figs. 6.8 and 7.4, makes it clear that the controlling parameter is c. It would be quite unrealistic to represent a second order system by a first order equation if $c < 1$, because the second order response is oscillatory whilst the first order response cannot be oscillatory.

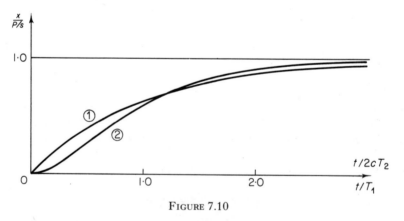

FIGURE 7.10

In fig. 7.10, the response to a step input of a second order system, for which $c = 1$, is compared with the response of a first order system. If the inertia term is simply dropped from eqn. (7.32) it becomes

$$(2cT_2 D + 1) = P(t)/s$$

which is the equation for a first order system where the time constant T_1 is $2cT_2$. It follows that, in order to make a direct comparison between the two

LINEAR SYSTEMS II Second order systems 291

responses, eqn. (7.31) must be plotted to a base of t/T_1 and eqn. (7.31) must be plotted to a base of $t/2cT_2$.

It can be seen that the two curves lie very close to each other except at the origin where the inertia term is dominant in eqn. (7.32), and non-existent in eqn. (7.31).

Figure 7.11 compares the responses of the two systems to a harmonic excitation, $P(t) = P \cos \omega t$.

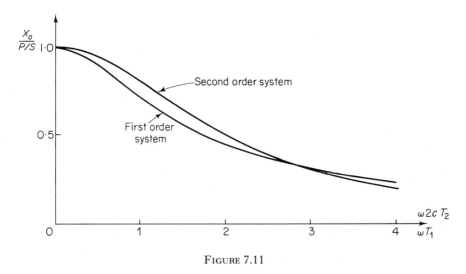

FIGURE 7.11

It can be deduced from figs. 7.10 and 7.11 that it is reasonable to ignore the inertia effects in a second order system if the system is more than critically damped. If similar comparisons are made for a system for which $c = 2$, there is found to be very little difference indeed between the curves for the first and second order systems.

8

AUTOMATIC CONTROL

8.1 Introduction
In the engineering context, the purpose of a control system is to control the operation of a piece of plant or machinery: the purpose of an engine governor is to prevent or limit the change in speed which might otherwise be the result of a change in load; in a domestic water heating system a thermostat ensures a constant supply of water at the right temperature irrespective, within limits, of the amount drawn off; the automatic pilot of an aircraft ensures that the plane maintains its proper course despite variations in wind speed and direction; the rudder of a ship is required to move in response to a rotation of the helm. It is the purpose of this chapter to study the way in which such systems operate.

The same basic principles apply in other contexts: the human body, for example, incorporates many control systems. We soon feel ill if our body temperature departs from its usual value; things are liable to go wrong if our blood pressure is too high or too low; the iris controls the amount of light falling on the retina, and so on. A set of control systems come into operation when we use our hands. The same hands that can crack a walnut can pick up an egg without crushing it; we can place our hands where we want them and the speed with which we move them is under control.

In order to identify the basic principles which underly the operation of all these systems, we might start by asking two questions. First, do all the systems to which we have referred have the same sort of purpose? Second, do they all function basically in the same way?

When we study the various purposes we can distinguish two classes. With one the purpose is to maintain the output of the system constant and to offset the effect of external agencies which tend to alter the output. Members of this class of system are known as *regulators* (or in appropriate circumstances *process controllers*): thermostatic devices, voltage regulators, speed regulators, and so on all come into this class of operation.

The other class of system is typically a *position control system*, whose purpose is to cause the system to respond in the required way to a continuously and arbitrarily varying input. The steering mechanism of a ship would be an example of such a system. Here, the emphasis is on the ability of a system to do as it is told rather than simply its ability to withstand disturbing effects, though in practice this is also important.

Mathematically, there is little to distinguish these two classes of system, and we shall not make particular reference to this distinction again.

AUTOMATIC CONTROL

A much more basic classification of control systems is revealed when we consider, and try to answer, the question as to whether all control systems function basically in the same way?

We find that in some systems the mode of operation is completely sequential, as indicated by fig. 8.1. An input signal, which is proportional to the desired

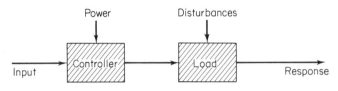

FIGURE 8.1

response, is applied to a controller which as a result takes appropriate action to elicit the desired response. If the system has been properly calibrated the response will indeed correspond with the input, provided that it is operated under the same conditions as those under which it was calibrated. Any change in these conditions constitutes a disturbance, which leads to a lack of correspondence between input and response. As an example of open-loop control we might think of a boy sailing his model boat on a pond. In order to send the boat to his father on the other side of the pond the boy sets the sails and rudder, points the boat in what he hopes is the right direction and gives it a push. If the boy has judged conditions correctly at the time of launching, the boat will reach its appointed destination provided that there are no disturbances to upset the arrangements. The disturbances might be a change of wind strength, or collision with another boat or inquisitive duck. The boy starts events moving but thereafter has no control over them. His more sophisticated friend with his radio-controlled boat, on the other hand, on observing a difference between the actual course of the boat and the desired course can change the rudder setting in order to correct the error. In this second example the boy is more than simply the agent who sends the boat on its way, he is part of the system which ensures that the boat makes its destination. This system is represented diagrammatically in fig. 8.2. The feedback is in this example purely visual; the boy observes what is happening and mentally compares it with what he wants to happen. If the boat is on its correct course no corrective action is called for.

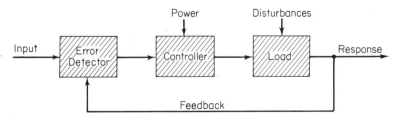

FIGURE 8.2

294 Dynamics of Mechanical Systems

If the boat is off course the controller, which is really a subsystem composed of the radio-link and the rudder mechanism, causes the rudder to turn appropriately. The load represents the boat, or rather the dynamics of the boat which determine the relationship between rudder setting and course sailed.

Many systems which in themselves are open loop are made effectively closed-loop by the intervention of a human operator. We have an *automatic control system* when the human feedback link is replaced by a mechanical, electrical, hydraulic, or some other inanimate connection. It is the automatic closed-loop control system to which we will now turn our attention.

8.2 Elements in series

In the last chapter we took a simple position control system as an example of a second order system. There was no particular difficulty in deducing the equation of motion, and many simple systems can be analysed in the same way, by simply writing down an appropriate set of equations, and eliminating intermediate variables. We will now consider how this process can be formalized, taking as an example the hydraulic device in fig. 8.3.

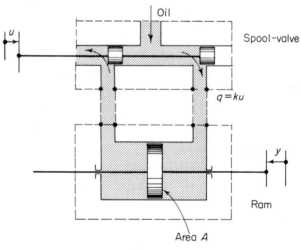

FIGURE 8.3

The device consists of two elements, a spool-valve and a ram. The equation which describes the behaviour of the spool-valve is

$$q = ku \tag{8.1}$$

where q is the oil-flow through the valve, u the valve opening, and k is a constant. The flow q can be regarded as the response of the spool-valve to an input u and eqn. (8.1) can be expressed in the alternative form

$$\frac{q}{u} \equiv \frac{\text{response}}{\text{input}} = k \tag{8.1(a)}$$

For the ram we have

$$A\,Dy = q \qquad (8.2)$$

or

$$\frac{y}{q} \equiv \frac{\text{response}}{\text{input}} = \frac{1}{A\,D} \qquad (8.2(a))$$

The separation of the operator D ($\equiv d/dt$) from the variable on which it operates in eqn. (8.2(a)) is at first a little difficult to accept. Equation (8.2(a)) is, however, simply a convenient way of writing eqn. (8.2). If, in general, we have

$$\frac{x_o}{x_i} \equiv \frac{\text{response}}{\text{input}} = \frac{P(D)}{Q(D)} \equiv F(D) \qquad (8.3)$$

where $P(D)$ and $Q(D)$ are polynomials in D, it simply means that the equation for the element (or system) in question is

$$Q(D)x_o = P(D)x_i$$

The quantity $F(D) = P(D)/Q(D)$ is known as the *transfer operator* of the element in question. It is convenient to represent an element symbolically by a box or block, fig. 8.4(a), in which the transfer operator is written.

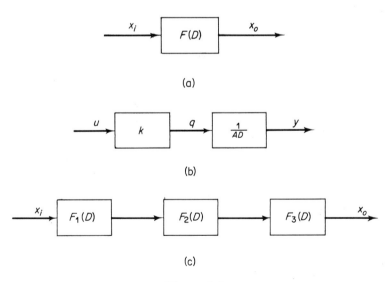

FIGURE 8.4

Following this line of thought we can represent the spool-valve and ram of fig. 8.3 by two elements in series fig. 8.4(b). The transfer operator for the spool-valve is the simple constant k; for the ram it is $1/A$ D. The overall transfer operator for the two elements is the product of the two individual transfer operators. That is

$$\frac{y}{u} = \frac{k}{A\,D} \tag{8.3(a)}$$

so that
$$A\,Dy = ku$$

is the overall equation for the device.

This may seem a somewhat roundabout way of eliminating q from eqns. (8.1) and (8.2), for that is in effect all that we have done, but the argument can be extended to deal with any number of elements in series, fig. 8.4(c). The overall transfer operator for a series of elements is the product of the transfer operators for the individual elements, i.e.,

$$x_o/x_i = F_1(D).F_2(D).F_3(D)\cdots F_n(D) \tag{8.4}$$

This equation is based on an important assumption; it is assumed that there is no *interaction* between the elements. This means we assume that no element is influenced by the behaviour of an element which comes after it. In the case of the hydraulic spool-valve and ram, we assume that the flow as given by eqn. (8.2) is quite uninfluenced by the way in which the ram operates, or indeed whether the ram is there or not. In practice, this means that the pressure drop across the ram due to whatever load it is driving must be negligible compared with the pressure drop across the orifices of the spool-valve. If this condition is not met, the valve and ram must be considered as a single unit rather than two elements in series. Figure 8.5 gives an example of two elements in series

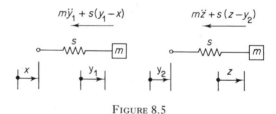

FIGURE 8.5

where the effect of interaction must be considered, the intention being to connect the two elements together by attaching the right-hand spring to the left-hand mass.

For the left-hand element, the relationship between the displacement x of the free end of the spring and the displacement y_1 is determined by equating the forces which act on the mass. We have

$$(m\,D^2 + s)x = sy_1$$

or
$$\frac{y_1}{x} = \frac{1}{1 + \dfrac{m}{s}D^2} \tag{8.5}$$

Similarly, for the right-hand element

$$\frac{z}{y_2} = \frac{1}{1 + \dfrac{m}{s} D^2} \quad (8.6)$$

If there were no interaction between these elements we could obtain the overall transfer operator by multiplying the right-hand sides of eqns. (8.5) and (8.6) together. We may not do this, however, because the force in the right-hand spring affects the motion of the left-hand mass. Instead, we must consider the system as a whole, fig. 8.6.

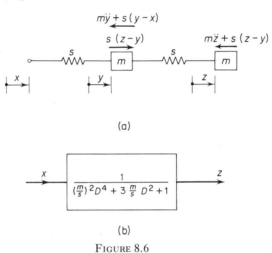

FIGURE 8.6

The equations of motion for the two masses are

$$m D^2 y + s(y - z) - s(z - y) = 0$$
$$m D^2 z + s(z - y) = 0$$

If z is eliminated we find

$$\frac{x}{z} = \frac{1}{\left(\dfrac{m}{s}\right)^2 D^4 + 3 \dfrac{m}{s} D^2 + 1} \quad (8.7)$$

It would be misleading to represent this system diagrammatically by two blocks in series. We will adopt the convention that only non-interacting elements are entitled to separate representation in a block diagram. The system of fig. 8.6(a) is therefore represented by a single block, as in fig. 8.6(b).

This example illustrates one of the main differences between the analysis of control systems and vibration analysis. In control theory we rely heavily on the concept of a string of non-intersecting elements, in vibration theory there is almost always considerable interaction between elements.

298 Dynamics of Mechanical Systems

In practice, there is almost always some interaction between elements which are coupled together. The decision on whether to treat elements as non-interacting or otherwise is a matter of judgement. It is the same sort of judgement as that involved in deciding whether to treat bodies as rigid or flexible, heavy or massless, and so on.

8.3 Closing the loop

As they stand, the spool-valve and ram in fig. 8.3 constitute an open-loop system, with the input ε controlling the ram velocity dy/dt. With a human operator the system can operate as a closed-loop control system because the operator can observe the response y and control it by manipulating ε. We can convert the system into an automatic position control system by adding a mechanical *feedback* link ABC, as in fig. 8.7. It may well be advantageous in

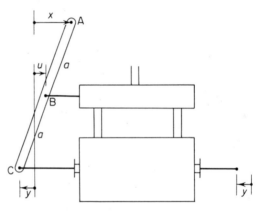

FIGURE 8.7

particular circumstances not to have AB = BC, but the explanation and algebra are simplified by taking this special case. The way in which the system functions following a step displacement x from its zero position is as follows:

initially the lever simply pivots about end C and the orifices of the spool-valve are opened due to the resulting displacement of B. The oil flowing to the ram causes a displacement of C in the direction which causes B to be carried back to its zero position. The oil-flow continues until B is back in its zero position when $y = x$.

From the geometry we have

$$u = (x - y)/2 \qquad (8.8)$$

This equation can be represented diagrammatically as in fig. 8.8. This diagram is combined in fig. 8.9 with the corresponding diagram, fig. 8.4(b), for the valve and ram to form the *block diagram* for the whole system. The block diagram shows how the non-interacting elements of the system are inter-

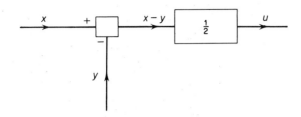

FIGURE 8.8

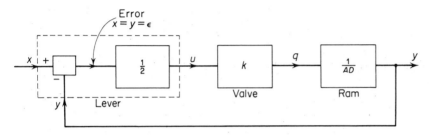

FIGURE 8.9

connected. With simple systems the block diagram is, for the person who draws it, little more than a convenient way of keeping a note of the relevant transfer operators, and a means of explaining the way in which the system operates to somebody else. The value of the block diagram to its author increases with the complexity of the system by providing a convenient intermediate step between drawings of the actual hardware and the mathematical analysis.

Because of the simple multiplication rule for the transfer operators of elements in series, the formulation of the equation of motion for the whole system involves only two equations:

$$\frac{y}{\varepsilon} = \frac{1}{2} \times k \times \frac{1}{AD} = \frac{k}{2AD} \qquad (8.9)$$

and
$$\varepsilon = x - y \qquad (8.10)$$

On eliminating ε from these two equations we have

$$\left(1 + \frac{2A}{k} D\right) y = x \qquad (8.11)$$

as the equation for the system.

The quantity ε which is defined in fig. 8.9 as the difference $x - y$ is known as the *error*. It is the difference between the input and the response and by convention is taken as

$$(\text{error}) = (\text{input}) - (\text{response}) \qquad (8.12)$$
$$\varepsilon \quad = \quad x \quad - \quad y$$

300 Dynamics of Mechanical Systems

rather than the other way round. This usage grew up with the study of position control systems, where it is natural to think of the response as lagging behind the input and to call the error in this case positive. It is less suitable when dealing with, say, a temperature control system for which it would be natural to regard the error as positive when the output is too hot.

The hydraulic relay in fig. 8.7 operates in essentially the same way as the relay shown in fig. 6.7(d), and (except for the factor 2) the same differential equation holds for both devices. It is to be noted that the relay in fig. 6.7(d) is a feedback system, and, except for the omission of the transfer operator of $\frac{1}{2}$, can be represented by the block diagram in fig. 8.9 even though there is no actual piece of hardware which can be identified as the feedback link. This feedback is inherent in the mode of operation.

We have already, in chapter 6, examined in detail the response of a first order system to different types of input, and there is nothing more to be noted in regard to first order systems as such.

8.4 Remote position control system

We used a remote position control system as an example of a second order system in the last chapter. We have now to study this system as a control system rather than as a particular example of a device which operates according to a certain type of mathematical equation. Figure 8.10 shows schematically a

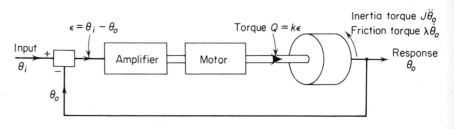

FIGURE 8.10

system which is basically the same as that shown in fig. 7.1. There is a rotor whose angular position is to be controlled by a motor. As the control lever on the motor is likely to require more force to move it than is available from the error signal $\varepsilon = \theta_i - \theta_o$, we have shown an amplifier between the differencing device and the motor. There are many ways of measuring the error in a control system, the simplest is perhaps the lever which we used in the hydraulic relay. Figure 7.1 shows an alternative arrangement which could be used for a compact position control system. Obviously this method would not be suitable if the place where θ_i is set is a long distance from the mass under control. In this situation the 'error detection' would almost certainly be done electrically. A potentiometer brush attached to the input would pick up a voltage proportional to θ_i, a similar potentiometer at the output would give a voltage proportional to θ_o. The potential difference then gives $\theta_i - \theta_o$ as a voltage which can be fed straight into an electronic amplifier.

Assuming that there are no time delays in the operation of the motor, and that the motor torque Q is independent of speed, we have

$$\frac{Q}{\varepsilon} = k$$

For the load

$$Q = (J\,\mathrm{D}^2 + \lambda\,\mathrm{D})\theta_o$$

so that

$$\frac{\theta_o}{Q} = \frac{1}{J\,\mathrm{D}^2 + \lambda\,\mathrm{D}}$$

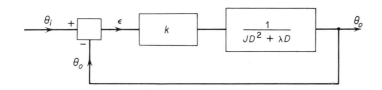

FIGURE 8.11

A block diagram for the system is accordingly as shown in fig. 8.11, and to obtain the equation of motion we eliminate ε from

$$\frac{\theta_o}{\varepsilon} = \frac{k}{J\,\mathrm{D}^2 + \mathrm{D}} \tag{8.13}$$

and

$$\varepsilon = \theta_i - \theta_o \tag{8.14}$$

to give

$$\left(\frac{J}{k}\mathrm{D}^2 + \frac{\lambda}{k}\mathrm{D} + 1\right)\theta_o = \theta_i \tag{8.15}$$

or

$$(T^2\,\mathrm{D}^2 + 2cT\,\mathrm{D} + 1)\theta_o = \theta_i \tag{8.16}$$

where

$$T = \sqrt{J/k} \quad \text{and} \quad c = \lambda/2\sqrt{(Jk)}$$

A position control system would normally be designed for a damping coefficient $c = 0\cdot 6$ to $0\cdot 7$. This gives a reasonably lively response with a fairly quick decay of the transient motion. The overshoot in response to a step input is 5 to 10 per cent, with the time to the first peak of $4\cdot 0T$ to $4\cdot 4T$, as in fig. 8.12.

The time constant T is determined by the speed of response which is required, and the time taken to reach the first peak is a measure of this. The moment of inertia J of the load is known, and normally the motor chosen will be the smallest (and hence the cheapest) which will have sufficient power to drive the load in response to a step input of specified maximum size. The amplifier can then be chosen to give the required value of k.

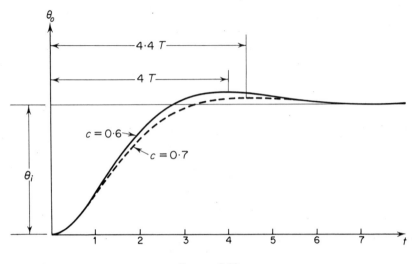

FIGURE 8.12

The problem now is to arrange for the correct amount of damping λ to give the desired damping factor of $c \approx 0.6$. So far we have thought of the damping term as being due to viscous friction inherent in or deliberately introduced into the system. The same effect may be produced by other means. As an example, we will consider a position control system in which the motor is an electrical d.c. motor with separate excitation. The output torque is controlled primarily by the armature current which we can regard as being proportional to the error signal, but is also dependent upon the output speed $d\theta_o/dt$. The idealized speed torque characteristics for a d.c. motor are as in fig. 8.13. The output torque is

$$Q = \alpha I_a - \beta \frac{d\theta_o}{dt}$$

If I_a is proportional to ε,

$$Q = k\varepsilon - \beta \frac{d\theta_o}{dt}$$

But

$$Q = J \frac{d^2\theta_o}{dt^2} + \lambda \frac{d\theta_o}{dt}$$

so that on eliminating Q

$$\frac{\theta_o}{\varepsilon} = \frac{k}{J D^2 + (\lambda + \beta)D} \qquad (8.17)$$

If the torque is independent of speed then $\beta = 0$ and eqn. (8.17) is the same as eqn. (8.13). A comparison of these two equations shows that the effect of the speed dependence of the motor torque is the same as viscous damping. It is,

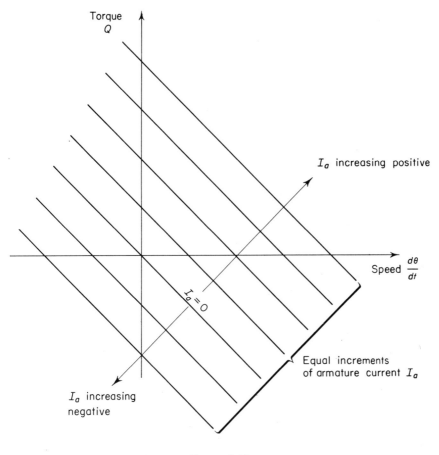

FIGURE 8.13

however, unlikely that a motor would be designed to give a specific amount of damping in this way, and we are left with the problem of introducing just the right amount.

If there is insufficient effective damping in the system the obvious thing to do is to introduce more. In a low power system, when, for example, the inertia load is the pen of a recording instrument, this might well be the simplest and cheapest solution. It is merely necessary to mount an aluminium disc on the shaft and allow it to rotate between the jaws of a permanent magnet. For larger systems this method would be very unsatisfactory. All the power absorbed by the brake must be produced by the motor, and a system which generates power merely to dissipate it in heat is very wasteful. The heat produced may itself be embarrassing in that special cooling facilities may be required to dispose of it. There is a further objection to this method if the system is required to respond to a constantly changing input. Suppose, for example, that $\theta_i = \Omega t$.

From eqns. (8.13) and (8.14)

$$\frac{\theta_i - \varepsilon}{\varepsilon} = \frac{k}{J D^2 + \lambda D}$$

so that
$$(J D^2 + \lambda D + k)\varepsilon = (J D^2 + \lambda D)\theta_i$$

If $\theta_i = \Omega t$ we have

$$(J D^2 + D + k)\varepsilon = \lambda \Omega \qquad (8.18)$$

The particular integral of this equation is

$$\varepsilon = \lambda \Omega / k \qquad (8.19)$$

This is the steady-state error, and clearly if we increase λ we increase this error.

In order to find a way of introducing extra damping without causing the difficulties which have been mentioned let us think for a moment about how the system works. The system measures the error ε which in turn causes the load to be moved in the direction which reduces ε, the magnitude of the corrective torque being directly proportional to the error. We would surely achieve better control if the system were constructed to take notice of any tendency for the error to build up, as well as taking notice of the error which actually exists. We can do this by the use of *derivative control*, which provides a motor torque with two components, one proportional to ε, the other proportional to $d\varepsilon/dt$.

8.5 Derivative control

Ideally we introduce into the system an element for which the transfer operator is $(1 + T_d D)$, as in fig. 8.14.

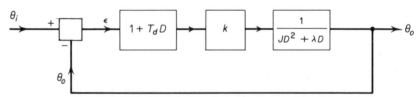

FIGURE 8.14

We then have

$$\frac{\theta_o}{\varepsilon} = \frac{k(1 + T_d D)}{J D^2 + \lambda D} \qquad (8.20)$$

and
$$\varepsilon = \theta_i - \theta_o$$

On eliminating ε and re-arranging, these equations become

$$\left[\frac{J}{k} D^2 + \left(\frac{\lambda}{k} + T_d\right) D + 1\right]\theta_o = (1 + T_d D)\theta_i \qquad (8.21)$$

AUTOMATIC CONTROL 305

It can be seen that with the introduction of derivative action the damping coefficient is effectively increased, with $c = (\lambda + kT_d)/2\sqrt{(kJ)}$. The power input into the derivative element is very small compared with the power output of the motor so that the question of wastage does not arise. Another advantage of this method of damping the transient motion is that it does not increase the steady-state error for $\theta_i = \Omega t$.

We now have

$$\frac{\theta_o}{\varepsilon} = \frac{\theta_i - \varepsilon}{\varepsilon} = \frac{k(1 + T_d D)}{JD^2 + \lambda D}$$

or
$$\left[\frac{J}{k}D^2 + \left(\frac{\lambda}{k} + T_d\right)D + 1\right]\varepsilon = \left(\frac{J}{k}D^2 + \frac{\lambda}{k}D\right)\theta_i \qquad (8.22)$$

With $\theta_i = \Omega t$, the right-hand side of this equation becomes $\lambda\Omega/k$. The particular integral, and hence the steady-state error is $\varepsilon = \lambda\Omega/k$ as before.

The D on the right-hand side of eqn. (8.21) causes some difficulty in deducing the response to a step input. The difficulty can be circumvented by using the principle of superposition.

First, we find the general solution of

$$\left[\frac{J}{k}D^2 + \left(\frac{\lambda}{k} + T_d\right)D + 1\right]\theta_o = \theta_i(t)$$

where $\theta_i(t)$ is a unit step, the solution is a standard one and has been given in the preceding chapter, eqn. (7.18). On differentiating this response with respect to t we obtain the response due to $D\theta_i(t)$, which is the response due to a unit impulse and is also standard, see exercise 7.9. The second response is then simply multiplied by T_d and added to the first to give the complete response of the system to a step input. Figure 8.15 compares the response to a

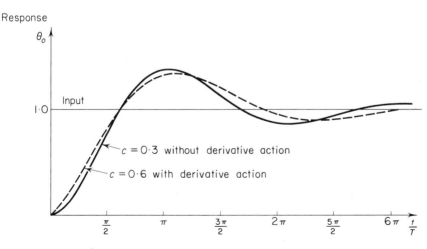

FIGURE 8.15

step input of an unmodified system for which $c = 0\cdot3$ with the response of a system to which derivative control has been added to make $c = 0\cdot6$. It can be seen that the sluggishness introduced by the addition of more damping is offset by the impulsive start ($d\theta_o/dt \neq 0$ at $t = 0$). This comparison is, however, more of academic than practical interest because derivative control as envisaged here is an ideal which cannot be achieved, and we have to be content with devices which do the job approximately. An electrical circuit which is used in practice to achieve an approximation to derivative control is shown in fig. 8.16.

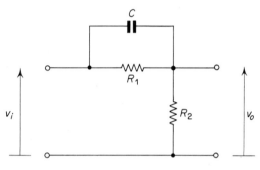

FIGURE 8.16

Assuming that the output takes negligible current, we have on equating the sum of the currents through the condenser and R_1 to the current through R_2

$$C D(v_i - v_o) + (v_i - v_o)/R_1 = v_o/R_2$$

$$\frac{v_o}{v_i} = \frac{1 + R_1 C D}{1 + \frac{R_1}{R_2} + R_1 C D}$$

$$= \frac{\alpha(1 + T_d D)}{(1 + \alpha T_d D)} \qquad (8.23)$$

where
$$\alpha = \frac{R_2}{R_1 + R_2}$$

α is less than $1\cdot0$ and can be made as small as we please, so that provided changes in v_o and v_i occur slowly enough for the derivative terms to be small we have approximately

$$\frac{v_o}{v_o} = \alpha(1 + T_d D)$$

which is of the required form.

The reader might reasonably regard the neglect of the D-operator in the denominator with some suspicion, so let us see if this element produces the desired effect, even though its transfer operator is not quite of the form which we envisaged. The block diagram is now as in fig. 8.17. The equation of motion is obtained from

$$\frac{\theta_o}{\varepsilon} = \frac{k\alpha(1 + T_d D)}{(1 + \alpha T_d D)(J D^2 + \lambda D)}$$

$$\varepsilon = \theta_i - \theta_o$$

and is found to be

$$\left[\frac{JT_d}{k} D^3 + \left(\frac{\alpha T_d}{k} + \frac{J}{\alpha k}\right) D^2 + \underline{\left(\frac{\lambda}{\alpha k} + T_d\right) D + 1}\right]\theta_o$$

$$= (1 + T_d D)\theta_i \qquad (8.24)$$

If this equation is compared with eqn. (8.21), which was obtained for an ideal derivative element, we notice two things: first, the order of the differential

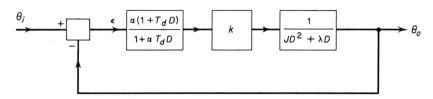

FIGURE 8.17

equation has been increased from second order to third order. Second, if we compare the underlined terms in eqn. (8.24) with the associated terms in eqn. (8.21), we note that k has been replaced by αk, where $\alpha < 1$. Now $J/k = T^2$ and T is a measure of the speed of response of the system, and it is unlikely that we would be prepared to tolerate an increase in T by a factor $1/\alpha$. Accordingly, when such an element is introduced we would normally increase k by a factor $1/\alpha$ in order to maintain the speed of response. If this is done eqn. (8.24) becomes

$$\left[\frac{\alpha JT_d}{k} D^3 + \left(\frac{\alpha^2 T_d}{k} + \frac{J}{k}\right) D^2 + \left(\frac{\lambda}{k} + T_d\right) D + 1\right]\theta_o$$

$$= (1 + T_d D)\theta_i \qquad (8.24(a))$$

In order to show that the response of a system governed by eqn. (8.24(a)) differs little from one governed by eqn. (8.21) we must consider a specific example. Let the object of the derivative element be to increase c from 0·3 to 0·6, and let us assume that $\alpha = 0·2$. Equation (8.21) becomes

$$(T^2 D^2 + 1·2T D + 1)\theta_o = (1 + 0·6T D)\theta_i$$

for which the complementary function is

$$\theta_o = A e^{-0·6t/T} \sin(0·8t/T + \psi) \qquad (8.21(a))$$

Equation (8.24(a)) becomes

$$(0 \cdot 12 T^3 \, D^3 + 1 \cdot 072 T^2 \, D^2 + 1 \cdot 20 T \, D + 1)\theta_o = (1 + 0 \cdot 6 T \, D)\theta_i$$

for which the complementary function is

$$\theta_o = A_1 \, e^{-7 \cdot 82 t/T} + \underline{A_2 \, e^{-0 \cdot 573 t/T} \sin(0 \cdot 86 t/T + \psi)} \qquad (8.24(b))$$

The underlined term in eqn. (8.24(b)) is very little different from the response as given by eqn. (8.21(a)), and the additional term decays very rapidly. With the approximate derivative device the main effect on the response to a step input is to annul the effect of the impulse arising from the D on the right-hand side of the equation and to cause $d\theta_o/dt$ to be zero at $t = 0$.

Derivative control is used a great deal in practical control systems but it does have disadvantages. One which we have seen is the attenuation of the error signal which required an increase in the amplification used. This is not too serious; a much more troublesome feature arises from the fact that the feedback signal often contains *noise*, i.e., high-frequency oscillations which are imposed on the main signal. A derivative device accentuates the noise component. Suppose, for example, that

$$\varepsilon = v + v_n \sin \omega t$$

where $v_n \sin \omega t$ is the noise component, probably quite small.

$$(1 + T_d \, D)\varepsilon = v + v_n \sin \omega t + \omega T_d v_n \cos \omega t$$

If $\omega \gg 1/T_d$ this noise content increases considerably so that the signal passed on by the derivative element is much larger than is intended. If the amplitude of the input signal to an element is larger than it is designed to accept, small changes due to variations in the main feedback signal will be undetected. An alternative method of introducing extra damping without introducing the noise problem uses *velocity feedback*.

8.6 Velocity feedback

Velocity feedback means that a signal proportional to the output velocity is fed back in addition to a signal proportional to the output displacement. The velocity signal is obtained by attaching a small electric generator with fixed excitation to the output shaft so that a voltage is generated proportional to the output velocity. The block diagram for the system is then as shown in fig. 8.18. The governing differential equation obtained from

$$\frac{\theta_o}{\varepsilon'} = \frac{k}{J \, D^2 + \lambda \, D}$$

and

$$\varepsilon' = (\theta_i - \theta_o) - T_v \, D\theta_o$$

is found to be

$$\left[\frac{J}{k} D^2 + \left(\frac{\lambda}{k} + T_v \right) D + 1 \right] \theta_o = \theta_i \qquad (8.25)$$

AUTOMATIC CONTROL 309

The effect of velocity feedback is seen to be an increase in the effective damping in the system. As the power output from the generator is virtually zero the damping is achieved with very little dissipation of energy. The disadvantage of velocity feedback, apart from the need to provide a generator, is

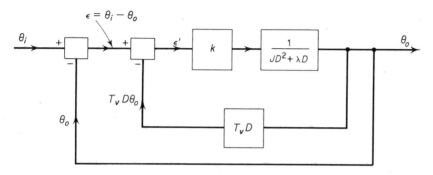

FIGURE 8.18

that it results in an increased velocity lag. On substituting $\theta_i = \Omega t$ and $\theta_o = \theta_i - \varepsilon$ in eqn. (8.25), the steady-state error in ε is found to be $(\lambda/k + T_v)\Omega$. If the increased velocity lag cannot be tolerated it can be combated by further modification of the error signal, using what is known as integral control, section 8.8.

8.7 Speed regulation of an engine

The purpose of a position control system is to control the position of a heavy object in response to a signal which may be varied in an arbitrary manner. The purpose of a regulator system is to ensure that the output of a system remains constant in spite of disturbances.

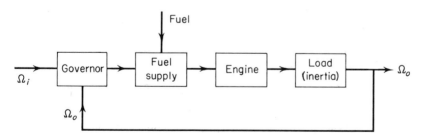

FIGURE 8.19

As a particular example of a regulator system we will take the speed control of an engine which drives an electric generator, which is required to produce power at a fixed frequency. The torque needed to drive the generator at constant speed varies with the load increment. It is the purpose of the engine governor to control the supply of fuel to the engine so as to prevent variations in the load causing a variation in speed. Obviously, a sudden increase in the torque against which an engine is operating is bound to result at least momentarily in a reduction of engine speed. What interests us is the form that this variation takes, the time taken for transient variations to decay, and the steady-state reduction in speed. We approach this analysis in the same way as we did the analysis of the position control system, first by deriving the transfer operators for the individual elements and then connecting the elements together.

The engine and the load

The engine and load (the generator) must be considered together because of the interdependence of engine torque and speed. The total torque which the engine is required to develop is

$$Q = J\frac{d\Omega_o}{dt} + Q_o \tag{8.26}$$

where J is the total moment of inertia of the engine and load, Ω_o is the motor speed and Q_o is the load torque.

The torque developed by the engine depends on the fuel supplied, H, and the speed in the way indicated in fig. 8.20. There is no point in attempting to find algebraic expressions for this family of curves, we will simply write

$$H = f(\Omega_o, Q) \tag{8.27}$$

Now we are not particularly interested in the net torque Q developed by the engine, but in the way in which the speed and fuel consumption are related to the load torque Q_o. That is to say, we want to eliminate Q between eqns. (8.26) and (8.27). Obviously we cannot do this with eqn. (8.27) as it stands. In order

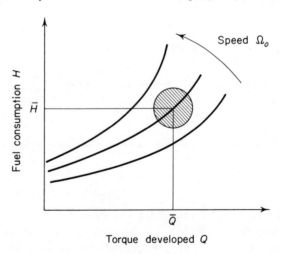

FIGURE 8.20

AUTOMATIC CONTROL 311

to express eqn. (8.27) in a less elegant but more manageable form we take account of the purpose of the control system, which is to prevent the speed varying by a large amount from its desired value $\Omega_i = \bar{\Omega}$. We assume that our system achieves this purpose, and that the relevant variables vary by only small amounts from their standard values, that is to say the operating state always lies in the small shaded area in fig. 8.20. Denoting these small variations by δH, $\delta \Omega_o$, and δQ we have

$$\delta H = \left(\frac{\partial f}{\partial \Omega_o}\right) \delta \Omega_o + \left(\frac{\partial f}{\partial Q}\right) \delta Q \tag{8.28}$$

where the partial differential coefficients are evaluated at the point $\Omega_o = \bar{\Omega}$, $Q = \bar{Q}$ in fig. 8.20. On replacing Q by $\bar{Q} + \delta Q$, etc., eqn. (8.26) becomes

$$(\bar{Q} + \delta Q) = J\frac{d}{dt}(\bar{\Omega} + \delta \Omega_o) + (\bar{Q}_o + \delta Q_o) \tag{8.29}$$

If we now let $\delta \Omega_o = \omega_o$, $\delta Q_o = q_o$, etc.

$$\left(\frac{\partial f}{\partial \Omega_o}\right)_{\substack{\Omega=\bar{\Omega}\\Q=\bar{Q}}} = f_\omega \quad \text{and} \quad \left(\frac{\partial f}{\partial Q}\right)_{\substack{\Omega=\bar{\Omega}\\Q=\bar{Q}}} = f_q$$

and recognize that $\bar{Q} = \bar{Q}_o$, eqns. (8.28) and (8.29) become

$$h = f_\omega \omega_o + f_q q$$

and

$$q = J\frac{d\omega_o}{dt} + q_o$$

On eliminating q these equations yield

$$h = f_\omega(1 + TD)\omega_o + f_q q_o \tag{8.30}$$

where

$$T = f_q J / f_\omega$$

In an ungoverned system, $h = 0$ because the fuel supply is fixed at its standard value. The steady-state increase in speed due to a torque q_o is

$$\omega_o = -\frac{f_q}{f_\omega} q_o \tag{8.31}$$

The governor
An ideal governor would detect, instantaneously, a difference between the actual speed and the required speed, and give

$$h = k_g(\omega_i - \omega_o) \tag{8.32}$$

The block diagram for the whole system [as represented by eqns. (8.30) and (8.32)] is then as shown in fig. 8.21.

The equation for the system is then

$$k_g(\omega_i - \omega_o) = h = f_\omega(1 + TD)\omega_o + f_q q_o$$

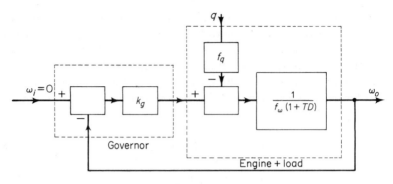

FIGURE 8.21

Assuming that the engine is required to run at its standard speed ($\omega_i = 0$) we have

$$\left(1 + \frac{f_\omega}{k_g + f_\omega} TD\right)\omega_o = -\frac{f_q}{k_g + f_\omega} q_o \tag{8.33}$$

The effect of the governor is two-fold:

1 it increases the speed of response of the engine by reducing the time constant from T to $Tf_\omega/(f_\omega + k_g)$; and
2 it reduces the steady-state speed reduction due to $q_o = $ constant from $q_o f_q/f_\omega$ to $q_o f_q/(k_g + f_\omega)$.

These general conclusions as to the effect of a governor hold, even though eqn. (8.32) may represent a gross simplification of the governor equation. However, as this simplification masks other aspects of the system behaviour which must be considered, we will now derive a more realistic governor equation, taking as an example the governor shown in fig. 8.22. The whole device rotates about the vertical axis at a speed equal to, or proportional to, the engine speed Ω_o. Any increase in Ω_o causes the flyballs to move to a larger radius against the resistance of the central spring, and in so doing causes a reduction in the rate at which fuel is supplied. The algebra is reduced if we assume that the arms BC, B'C' are vertical when the engine runs steadily at its standard speed $\bar{\Omega}$. Taking moments about B for one of the arms we find that the force in the spring is then $2md\bar{\Omega}^2 b/a = \bar{P}$.

The equation of motion for the flyballs when the engine is not running steadily is obtained by taking moments about B for the system of forces shown in fig. 8.22(b). We have

$$mb^2 \frac{d^2\theta}{dt^2} - m(d + b\sin\theta)\Omega_o^2 b\cos\theta + \tfrac{1}{2}(\bar{P} + sa\sin\theta)a\cos\theta = 0 \tag{8.34}$$

assuming that the weight of the ball is negligible compared with the spring and inertia forces. The relationship between the displacement x of the fuel control and the angle θ is

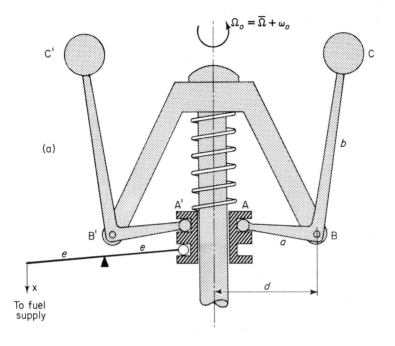

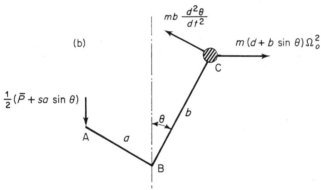

FIGURE 8.22

$$x = a \sin \theta \qquad (8.35)$$

The operation of the governor as defined by eqns. (8.34) and (8.35) is non-linear for we cannot reduce the input/response relationship to an equation of the form

$$\frac{x}{\Omega_o} = F(D)$$

This difficulty is overcome by adopting again the philosophy introduced earlier in this section: we assume that all variables change by only small amounts from the steady-state values which they have when $\Omega_i = \Omega_o = \bar{\Omega}$. This means that in eqn. (8.34) we put $\Omega_o = \bar{\Omega} + \omega_o$, and we assume θ to be a small angle, so that ignoring products of small quantities as being of second order, eqns. (8.34) and (8.35) become

$$mb^2 \frac{d^2\theta}{dt^2} - mb(d\bar{\Omega}^2 + b\bar{\Omega}^2\theta + 2d\bar{\Omega}\omega_o) + \frac{a}{2}(\bar{P} + sa\theta) = 0 \qquad (8.34(a))$$

and $\qquad\qquad\qquad\qquad\qquad\qquad\qquad\qquad x = a\theta \qquad (8.35(a))$

On substituting $\bar{P} = 2md\bar{\Omega}^2 b/a$ and eliminating θ these equations become

$$\frac{d^2x}{dt^2} + (sa^2/2mb^2 - \bar{\Omega}^2)x = 2\frac{ad}{b}\bar{\Omega}\omega_o \qquad (8.36)$$

Equation 8.36 shows that the system is unstable if $s < 2mb^2\bar{\Omega}^2/a^2$, so it is reasonable to conclude that, for satisfactory operation, the spring stiffness must be substantially greater than this critical value. On the other hand, we note that the steady-state value of x, for constant ω_o, decreases as s is increased, so that if the spring is made too stiff the governor will be insensitive. The problem of achieving an acceptable margin of stability without sacrificing too much sensitivity is really the central problem of control theory.

Putting

$$(sa^2/2mb^2 - \bar{\Omega}^2) = 1/T_g^2 \quad \text{and} \quad 2ad\bar{\Omega}/b = k_1/T_g^2$$

eqn. (8.36) becomes

$$(T_g^2 D^2 + 1)x = k_1\omega_o \qquad (8.37)$$

The complementary function of this equation represents an undamped simple harmonic motion. A dash-pot attached to the sleeve of the governor will cause these transient oscillations to die away so that eqn. (8.37) becomes

$$(T_g^2 D^2 + 2c_g T_g D + 1)x = k_1\omega_o$$

If the governor output x, corresponding to an increase ω_o in the output speed, causes a decrease h in the fuel supplied, we have

$$(T_g^2 D^2 + 2c_g T_g D + 1)h = -k_g\omega_o \qquad (8.38)$$

The block diagram for the whole system is now as shown in fig. 8.23. The equation of motion for the complete system is obtained by eliminating h from eqns. (8.38) and (8.30), and is found to be

$$[T_g^2 T D^3 + (2c_g T + T_g)T_g D^2 + (2c_g T_g + T) D + (1 + k_g/f_\omega)]\omega_o$$

$$= -\frac{f_q}{f_\omega}(T_g^2 D^2 + 2c_g T_g D + 1)q_o \qquad (8.39)$$

The steady-state speed change ω_o due to a constant load torque q_o is

$$\omega_o = -\frac{f_g}{f_\omega + k_g} q_o$$

as before, and is unaffected by the dynamics of the governor itself. The way in which the system approaches the steady state does depend on the dynamics of

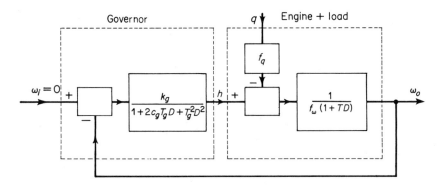

FIGURE 8.23

the governor. The simple exponential approach to the steady state, which is implied by eqn. (8.33), will be complicated by a superimposed harmonic component which component decays to zero if the system is stable. Under certain circumstances, which will be investigated in a later section, the oscillations may build up. It will be sufficient for the moment to observe that the system tends to an unstable condition as the governor sensitivity k_g is increased; it will be recalled that we have already noted the same tendency in the governor treated as an isolated element. If the steady-state drop in speed due to a constant load torque q_o is greater than can be tolerated, the obvious solution is to increase the governor sensitivity k_g. The possibility that the system may thereby become unstable must be realized. An alternative solution uses *integral control*.

8.8 Integral control

Integral control, like derivative control, is achieved by modification of the error signal. In our study of the speed regulation of an engine we have postulated an increase in the fuel supply proportional to the error detected by the governor. If the fuel supply is further increased by an amount proportional to the time integral of the error, we have *proportional plus integral action*, which, for conciseness, will simply be referred to as integral control. Ideally, we would wish to insert between the governor and the fuel control a device for which the transfer operator is

$$\left(1 + \frac{1}{T_i D}\right) \qquad (8.40)$$

The block diagram for the whole system is now as in fig. 8.24, and the relevant system equations are

$$h = -\frac{k_g\left(1 + \dfrac{1}{T_i D}\right)\omega_o}{1 + 2c_g T_g D + T_g^2 D^2}$$

and
$$h = f_\omega(1 + TD)\omega_o + f_q q_o$$

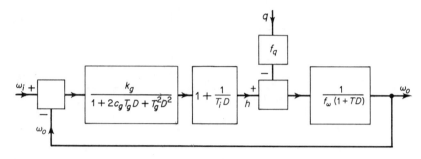

FIGURE 8.24

Equating these two expressions for h, and rationalizing, we have

$$k_g(1 + T_i D)\omega_o + f_\omega(1 + TD)(1 + 2c_g T_g D + T_g^2 D^2)T_i D\omega_o$$
$$= -f_q(1 + 2c_g T_g D + T_g^2 D^2)T_i D q_o \quad (8.41)$$

If q_o is constant, ω_o is constant in the steady state, and this equation reduces to

$$-k_g \omega_o = 0$$

so that the engine runs at the standard speed. We can argue from the mechanics of the situation that if the system is stable then ultimately ω_o must be zero, for if it were to settle down to a non-zero value the integral term would grow indefinitely with time, and this would be contrary to our assumption of a stable system.

As with derivative control, practical devices which are used to obtain integral control do not generally have a transfer operator of the ideal form, eqn. (8.40). One method of obtaining approximate integral control is shown in fig. 8.25.

The device is a hydraulic relay; the governor output is connected to the spool-valve, and the ram is connected to the fuel-supply valve. It should be noted that, as the governor cannot itself exert much effort it may be necessary to introduce a relay in this position, irrespective of whether or not integral control is desired.

If simple proportional control is satisfactory, a hydraulic relay of the type shown in fig. 8.3 may be used. The device in fig. 8.25 has a sleeve introduced between the cylinder and piston of the spool-valve. When the relay is inactive

AUTOMATIC CONTROL 317

the sleeve is held in the neutral position by the centring spring, stiffness s_1, with its ports aligned with those of the cylinder. The sleeve is connected to the piston-rod of the ram by means of a dash-pot and lever as shown. To simplify the algebra, it will be assumed that the lever ratio is 1:1. In order to understand the way in which the device operates, consider the effect of a sudden displacement x_i of the spool-valve from its central position. Oil is immediately admitted to the ram through the lower port. As the force in the centring spring is initially

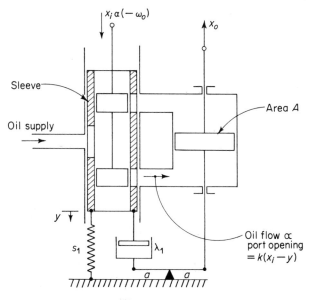

FIGURE 8.25

zero, there can be no force across the dash-pot so that the sleeve starts to move down with the same speed as the piston moves up. If there were no centring spring this motion would continue until $x_o = x_i$ when the port in the sleeve would be covered by the piston of the spool-valve and the flow of oil would cease. The spring, however, does not allow the sleeve to cut off completely the oil supplied to the ram. An equilibrium state is reached in which the spring force $s_1 y$ is balanced by the dash-pot force $\lambda_1 \, Dx_o$. If the spring is relatively soft $y \approx x_i$, so that $\lambda_1 \, Dx_o \approx s_1 x_i$. The total movement of the ram therefore consists of two components, one equal to x_i and the other equal to the time integral of x_i.

The 'exact' transfer operator is obtained as follows:

The flow equation is

$$q = k(x_i - y) = A \, Dx_o$$

or

$$x_i - y = T_2 \, Dx_o \qquad (8.42)$$

where $T_2 = A/k$.

11*

Equating the spring force to the dash-pot force (and ignoring inertia effects)

$$\lambda_1 \frac{d}{dt}(x_o - y) = s_1 y$$

so that
$$y = \frac{T_1 D x_o}{1 + T_1 D} \tag{8.43}$$

where $T_1 = \lambda_1/s_1$.

Eliminating y between eqns. (8.42) and (8.43) we find

$$\frac{x_o}{x_i} = \frac{1 + T_1 D}{D(T_1 + T_2 + T_1 T_2 D)}$$

This transfer operator can be reduced to the ideal form [eqn. (8.40)] by assuming first that $T_2 \ll T_1$ and then assuming that all dynamic effects take place sufficiently slowly for $T_2 D$ to be negligible compared with unity. However, instead of arguing that the transfer operator which we have obtained approximates to the one ideal we want, it is better simply to verify that it achieves the desired effect.

We now have

$$h = \frac{-k_g(1 + T_1 D)\omega_o}{D(T_1 + T_2 + T_1 T_2 D)(1 + 2c_g T_g D + T_g^2 D^2)}$$

and $\quad h = f_\omega(1 + T D)\omega_o + f_q q_o$

from which we obtain

$k_g(1 + T_1 D)\omega_o$
$+ f_\omega(1 + T D)(T_1 + T_2 + T_1 T_2 D)(1 + 2c_g T_g D + T_g^2 D^2)D\omega_o$
$= -f_q(T_1 + T_2 + T_1 T_2 D)(1 + 2c_g T_g D + T_g^2 D^2)D q_o \tag{8.44}$

If q_o is constant, ω_o is constant in the steady state so that the particular integral is $\omega_o = 0$, as required. This argument takes it for granted that the complementary function tends to zero as t tends to infinity. If this does not happen the system is unstable and never settles down to a steady state. One has only to glance at eqn. (8.44) to dismiss the idea of determining the complementary function as a general expression in terms of the various parameters. This is not, in general, possible for a differential equation of order higher than two, and eqn. (8.44) is a fifth order equation! It would be a fairly straightforward task to determine the complementary function if all the parameters were given numerical values, but not a task to be undertaken lightly. Fortunately, there are easier ways of determining whether or not a system is stable.

8.9(a) Stability — Routh's criterion
The differential equation

$$(a_0 + a_1 D)x_o = x_i$$

has the general solution

$$x_o = A\,e^{\lambda t} + \text{particular integral}$$

where $\lambda = -a_0/a_1$. The system is stable provided that λ is negative, i.e., provided that a_0 and a_1 both have the same sign.

In general, the stability of a system whose motion is governed by a differential equation of the type

$$P(D)x_o = Q(D)x_i$$

where $P(D)$ and $Q(D)$ are polynomials in the operator D ($\equiv d/dt$), depends on the complementary function which is the general solution of the equation

$$P(D)x_o = 0$$

In order to solve this equation we substitute for x_o an assumed solution of the form $x_o = e^{\lambda t}$. This yields the auxiliary equation

$$P(\lambda) = 0$$

The assumed solution is valid provided that the auxiliary equation is satisfied, that is to say provided $\lambda = \lambda_1, \lambda_2, \lambda_3$, etc. where $\lambda_1, \lambda_2, \lambda_3$ are the roots of the auxiliary equation. The complementary function is thus

$$(x_o)_{CF} = A_1\,e^{\lambda_1 t} + A_2\,e^{\lambda_2 t} + A_3\,e^{\lambda_3 t} + \cdots$$

The roots λ_1, λ_2, etc. may be real or complex. The sole criterion for the system to be stable is that all the roots λ_1, λ_2, etc. must have negative real parts.

As we have seen, in the case of a first order system, this criterion is satisfied if a_0 and a_1 both have the same sign.

A second order system, for which

$$P(D) \equiv a_0 + a_1 D + a_2 D^2$$

is stable provided that a_0, a_1, and a_2 all have the same sign.

A third order system, with

$$P(D) = a_0 + a_1 D + a_2 D^2 + a_3 D^3$$

must also have all the coefficients as positive if it is to be stable. There is a further requirement that

$$a_1 a_2 > a_0 a_3 \tag{8.45}$$

This condition, together with the requirement that a_0, a_1, a_2, and a_3 must all be positive, is known as *Routh's criterion*.

The requirement that all the coefficients should be positive applies to all systems, whatever the order of their differential equations. We shall not be concerned with the additional criteria which apply to fourth and higher order equations.

A proof of the stability criteria for second and third order systems is given in section 8.9(b). It is simply an exercise in algebra which can be ignored by readers who are prepared to accept the statements given above.

As an example, we will apply Routh's criterion to eqn. (8.39), which was derived as the equation of motion for an engine/governor system. The equation is

$$[T_g^2 T \, D^3 + (2c_g T + T_g)T_g \, D^2 + (2c_g T_g + T)D + (1 + k_g/f_\omega)]\omega_o$$
$$= -\frac{f_q}{f_\omega}(T_g^2 \, D^2 + 2c_g T_g \, D + 1)q_o \quad (8.39(a))$$

We note first that all the coefficients of D on the left-hand side of this equation have the same sign. Routh's criterion

$$a_1 a_2 > a_0 a_3$$

becomes

$$(2c_g T_g + T).(2c_g T + T_g)T_g > (1 + k_g/f_\omega)T_g^2 T$$

This simplifies to

$$2cg\left(\frac{T_g}{T} + 2cg + \frac{T}{T_g}\right) > k_g/f_\omega$$

The system is stable provided this inequality is satisfied. The point made earlier, when eqn. (8.39) was derived, that the system will become unstable if the governor sensitivity k_g is made too large is therefore confirmed by Routh's criterion.

With third order systems Routh's criterion provides a quick and simple means of determining the way in which the main parameters of the system affect the overall stability of the system. Unfortunately, it is, in general, not sufficient to know that a system is stable and that the transient effects due to the complementary function will decay with time. For a satisfactory performance it will usually be necessary for the transient effects to decay quickly, and Routh's criterion does not give any indication of this aspect of performance. Even though $a_1 a_2 \gg a_0 a_3$, the system may still have an unacceptable transient response.

When extended forms of Routh's criterion (Routh-Hurwitz criteria) are applied to higher order systems it is often found that the resulting expressions are so complicated that it becomes difficult to isolate the influence of a particular parameter. The usefulness of these purely algebraic criteria is therefore somewhat restricted. A more fruitful approach is given in section 8.10.

8.9(b) Stability — proof of Routh's criterion
Second order system
The auxiliary equation for a second order system is

$$a_0 + a_1 \lambda + a_2 \lambda^2 = 0$$

so that the condition for stability is that the roots

$$\left.\begin{matrix}\lambda_1\\ \lambda_2\end{matrix}\right\} = \frac{-a_1 \pm \sqrt{(a_1{}^2 - 4a_0a_2)}}{2a_0}$$

shall both be negative if they are real, or shall have negative real parts if they are complex.

We can assume, without loss of generality, that a_0 is positive. Now if a_2 is negative

$$|(a_1{}^2 - 4a_0a_2)^{\frac{1}{2}}| > |a_1|$$

so that irrespective of the sign of a_1, λ_1 and λ_2 must be of different sign. If a_2 is positive and $4a_0a_2 < a_1{}^2$, then

$$|(a_1{}^2 - 4a_0a_2)^{\frac{1}{2}}| < |a_1|$$

so that λ_1 and λ_2 are both positive if a_1 is negative. Finally, if a_2 is positive and $4a_0a_2 > a_1{}^2$, then the roots are

$$\left.\begin{matrix}\lambda_1\\ \lambda_2\end{matrix}\right\} = \frac{-a_1 \pm j\sqrt{(4a_0a_2 - a_1{}^2)}}{2a_0}$$

$$= -\alpha \pm j\beta \quad \text{(say)}$$

and the complementary function is

$$x_o = A\,e^{-\alpha t}\sin(\beta t + \psi)$$

which decays provided $\alpha = a_1/2a_0$ is positive, i.e., provided a_1 is positive.

A second order system is therefore stable, provided all the coefficients of D have the same sign (positive for convenience).

Third order system

With a third order system the auxiliary equation is

$$f(\lambda) = a_0 + a_1\lambda + a_2\lambda^2 + a_3\lambda^3 = 0$$

Depending upon the coefficients a, this equation either has three real roots or has one real root and a complex pair of roots. There must be at least one real root, let it be λ_1. This means that $(\lambda - \lambda_1)$ is a factor of $f(\lambda)$. To obtain the associated quadratic factor let us divide $f(\lambda)$ by $(\lambda - \lambda_1)$ thus:

$$\begin{array}{r}
a_3\lambda^2 + (a_2 + a_3\lambda_1)\lambda + (a_1 + a_2\lambda_1 + a_3\lambda_1{}^2) \\
\lambda - \lambda_1 \overline{)\, a_3\lambda^3 + a_2\lambda^2 + a_1\lambda + a_0 } \\
\underline{a_3\lambda^3 - a_3\lambda_1\lambda^2} \\
(a_2 + a_3\lambda_1)\lambda^2 \\
\underline{(a_2 + a_3\lambda_1)\lambda^2 - (a_2 + a_3\lambda_1)\lambda_1\lambda} \\
(a_1 + a_2\lambda_1 + a_3\lambda_1{}^2)\lambda + a_0 \\
\underline{(a_1 + a_2\lambda_1 + a_3\lambda_1{}^2)(\lambda - \lambda_1)} \\
(a_0 + a_1\lambda_1 + a_2\lambda_1{}^2 + a_3\lambda_1{}^3)
\end{array}$$

The remainder
$$(a_0 + a_1\lambda_1 + a_2\lambda_1^2 + a_3\lambda_1^3) = f(\lambda_1)$$
is zero because λ_1 is a root of $f(\lambda) = 0$. For stability
$$\lambda_1 < 0 \qquad (i)$$
Furthermore, if the quadratic factor is to represent a pair of stable roots
$$a_3 > 0 \qquad (ii)$$
$$(a_2 + a_3\lambda_1) > 0 \qquad (iii)$$
$$(a_1 + a_2\lambda_1 + a_3\lambda_1^2) > 0 \qquad (iv)$$

As
$$a_0 + a_1\lambda_1 + a_2\lambda_1^2 + a_3\lambda_1^3 = 0$$
we can express (iv) more compactly as
$$-a_0/\lambda_1 > 0 \qquad (iv(a))$$

From (iii)
$$\lambda_1 > -a_2/a_3$$

As $a_3 > 0$, a_2 must also be positive if λ_1 is to be negative. From (iv)
$$\lambda_1 > -a_1/(a_2 + a_3\lambda_1)$$

As
$$(a_2 + a_3\lambda_1) > 0$$

a_1 must be positive if λ_1 is to be negative. Condition (iv(a)) shows that a_0 must also be positive if λ_1 is to be negative.

It follows, that for $\lambda_1 < 0$, all the coefficients a must be positive (or all negative).

Having all the coefficients a of the same sign, and hence $\lambda_1 < 0$, is not by itself a sufficient condition for stability. Conditions (i), (ii), and (iv(a)), and hence (iv), are automatically satisfied, but condition (iii) is not; we might for example have $a_2 = a_3 = 1$, and $\lambda_1 = -2$.

Given that a_2 and a_3 are positive, condition (iii) tells us that
$$\lambda_1 > -a_2/a_3 \qquad (iii(a))$$

Now
$$a_0 + a_1\lambda_1 + a_2\lambda_1^2 + a_3\lambda_1^3 = 0$$
or
$$(a_0 + a_1\lambda_1) + \lambda_1^2(a_2 + a_3\lambda_1) = 0$$

This expression tells us that if (iii) is to be satisfied we must also have
$$(a_0 + a_1\lambda_1) < 0$$
or
$$\lambda_1 < -a_0/a_1$$

Taking this in conjunction with (iii(a)) we have
$$-a_2/a_3 < \lambda_1 < -a_0/a_1$$

It follows that
$$a_2/a_3 > a_0/a_1$$
or
$$a_1 a_2 > a_0 a_3 \qquad \text{(iii(b))}$$

The necessary and sufficient conditions for the stability of a third order system are:

1 All the coefficients a must have the same sign.

2 $$a_1 a_2 > a_0 a_3$$

Taken together, these conditions are known as *Routh's criterion*.

These conditions have been extended to cover higher order systems; the extended forms are known as the *Routh-Hurwitz* criteria. For a fourth order system these are:

1 All the coefficients a must have the same sign.

2 $$a_1 a_2 a_3 > a_0 a_3^2 + a_1^2 a_4 \qquad (8.46)$$

8.10 Stability — Nyquist's criterion (I)

One of the limitations of Routh's criterion as a means of determining the stability of a system is that it gives no indication of the degree of stability achieved, i.e., whether transient oscillations die quickly or slowly. So far we have used harmonic response loci (direct and inverse) simply as a graphical means of determining and representing the behaviour of a system. One of the most useful properties of these loci is that theory also enables us to determine readily, not only whether a system is stable, but also how stable. Figure 8.26

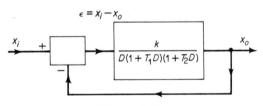

FIGURE 8.26

is an example of the simplest type of system for which it is possible to make the point; physically, the system might be a simple servomechanism with a simple exponential delay in the amplifier.

The relevant equations are
$$\frac{x_o}{\varepsilon} = \frac{K}{D(1 + T_1 D)(1 + T_2 D)}$$
and
$$\varepsilon = x_i - x_o$$

On eliminating ε and rearranging, we find
$$[T_1 T_2 D^3 + (T_1 + T_2)D^2 + D + K]x_o = K x_i \qquad (8.47)$$

324 Dynamics of Mechanical Systems

The inverse transfer operator is therefore

$$\frac{x_i}{x_o} = 1 + \frac{1}{K}[D + (T_1 + T_2)D^2 + T_1 T_2 D^3] \qquad (8.48)$$

and its associated vector diagram is as in fig. 8.27(a). As ω is varied the terminal point P of the x_i/x_o vector traces the inverse harmonic response locus. The

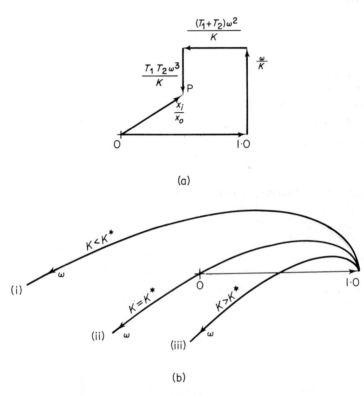

FIGURE 8.27

shape of the locus depends upon the parameters T_1, T_2, and K. The effect of varying K can readily be visualized since the magnitude of the three vectors involved are all affected in the same way. K also happens to be the parameter which is most readily varied in practice, simply by varying the gain of the amplifier. Figure 8.27(b) shows loci for three values of K, curves (i), (ii), and (iii). It will be noticed that curve (ii) is somewhat special in that it passes through the origin of the diagram, curve (i) is for a smaller value of K, and curve (iii) is for a larger value. At the point in the locus which coincides with the origin ($x_i/x_o = 0$), x_i must be zero but x_o need not be. Physically, this means that the system is capable of producing an output without any input. As this point corresponds to a particular value of ω, the output is an oscillation at that frequency. Under these circumstances, power is supplied, via the amplifier, at a rate which is just sufficient to balance the rate at which it is dissipated. If we increase the rate at which power is supplied by increasing K, it is reasonable to suppose that the amplitude of the output will increase, and that the system will be unstable. Conversely, if K is decreased so as to reduce the power supplied, the oscillations would be expected to die out. Thinking of these loci in relation to the axes in an Argand diagram, we deduce that *a system is stable if the inverse harmonic response locus cuts the real axis to the left of the origin and unstable if it cuts the real axis to the right of the origin.*

This test for stability is known as *Nyquist's criterion*. It is a criterion which can be expressed in a number of different forms and we shall meet others later. It must be realized that we have deduced this criterion by the use of a plausible argument rather than rigorous analysis. A formal proof shows that the criterion as stated above is not always valid. Fortunately, the exceptions are sufficiently unusual and esoteric for us to ignore them. As the associated mathematics is appreciably more advanced than that which we have used so far, or will use in the rest of this book, the reader is referred elsewhere for a formal proof of Nyquist's criterion.[1]

Nyquist's criterion does no more than Routh's criterion, in fact it is equivalent to it. We can see from fig. 8.28 that, at the point where the locus cuts the real axis, the two vertical sides of the vector diagram are equal in length, and

$$T_1 T_2 \omega^3 = \omega$$

or

$$\omega^2 = 1/T_1 T_2$$

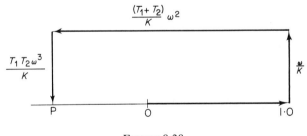

FIGURE 8.28

For stability, P must, in accordance with Nyquist's criterion, be to the left of O, that is

$$\frac{(T_1 + T_2)}{K} \omega^2 > 1$$

On substituting for ω^2, the condition becomes

$$(T_1 + T_2) > K T_1 T_2$$

which is the result obtained when Routh's criterion is applied to eqn. (8.47).

In order to estimate the degree of stability we must look at the shape of the locus. Consider, for example, a system for which the differential equation is

$$(25D^3 + 25D^2 + 2D + 1)x_o = x_i$$

The product of the two inner coefficients $a_1 a_2$ is 50, and the product of the two outer coefficients $a_0 a_3$ is 25. With $a_1 a_2 = 2a_0 a_3$ the sensitivity, or gain, of the system could be doubled before the system becomes unstable, so that it would appear from this test that the system is quite satisfactory. In fact, it is unlikely that such a system would be acceptable in practice. It can be seen from fig. 8.29(a) that although the harmonic response locus cuts the real axis well to the left of the origin at $(-1, 0)$, it also passes very close to the origin. At the point of nearest approach to the origin x_i/x_o is small, and x_o/x_i is correspondingly large. This means that there must be a pronounced resonance, at

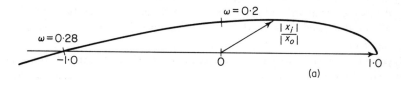

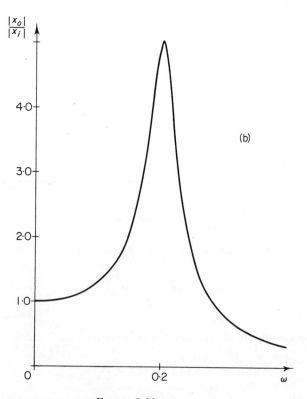

FIGURE 8.29

the corresponding frequency, as shown by the graph of dynamic magnification against frequency, fig. 8.29(b). Thinking of the behaviour of second order systems we can identify a pronounced resonance peak with small damping, and hence with a relatively slow rate of decay of transient vibrations. This switch in our attention from the steady-state harmonic response, which is what we have plotted in fig. 8.29(a), to the transient response which follows a disturbance is crucial. The fact that we can make this switch provides one of the reasons why the harmonic response loci are so valuable.

As with Nyquist's criterion, we have proceeded to this point by plausible argument rather than rigorous mathematics. The mathematical relationship between the transient response and steady-state harmonic response is explained in the appendix to this chapter, where it is also shown how a fairly accurate estimate of the rate of decay of transient oscillations may be obtained. In a later section we shall give simple rules which normally ensure an acceptable transient response.

The inverse harmonic response locus provides a simple and convenient means of investigating the stability of second order systems. Its use in this way has been presented as a logical development of the vector diagram method of solving differential equations. There is no reason in principle why we should not extend these ideas to fourth and higher order equations (the derivation of Routh's criterion for a fourth order equation is a simple exercise), but in practice this does not prove to be very convenient. It may be noted in passing that, even with a third order system, the drawing of the locus is complicated if there are D terms operating on the input function, that is, on what is conventionally the right-hand side of the differential equation. The main objection to a straightforward extension of this method to higher order systems is that it generally becomes difficult to isolate the effect a particular parameter or component has on the overall performance. This means that if the performance is unsatisfactory it is not easy to decide what alterations should be made to the system to remedy the deficiencies. It will be remembered that the same difficulty arises in relation to the application of Routh's criterion to higher order systems.

In order to meet this difficulty we forget, for a moment, about the overall performance of the system and go back to consider how the transfer operators for separate components of a control system are combined to form its equation of motion.

8.11 Open-loop harmonic response loci
Elements in series
We have shown in chapter 7 that the harmonic response locus and its inverse can be regarded as loci plotted in the Argand diagram. In the case of the direct harmonic response locus we plot

$$\frac{x_o}{x_i} = F(j\omega)$$

for varying ω, and in the inverse diagram we plot

$$\frac{x_i}{x_o} = \frac{1}{F(j\omega)}$$

where $F(j\omega)$ is the transfer function for the element in question. $F(j\omega)$, it will be remembered, is obtained by substituting $j\omega$ for D in the transfer operator.

We have already seen, eqn. (8.4), that when several non-interacting elements are in series the overall transfer operator is the product of the individual transfer operators.

$$F(D) = F_1(D).F_2(D).F_3(D)\ldots$$

It follows that the overall transfer function is the product of the individual transfer functions

$$F(j\omega) = F_1(j\omega).F_2(j\omega).F_3(j\omega)\ldots \qquad (8.49)$$

At any particular value ω_n of ω, $F_1(j\omega_n)$ and $F_2(j\omega_n)$ are complex numbers and are represented by two points in the Argand diagram, fig. 8.30, each of which

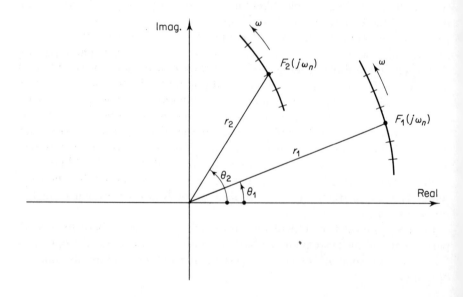

FIGURE 8.30

is defined by its modulus r, and argument θ. The product is obtained by multiplying moduli and adding arguments, thus

$$F_1(j\omega_n).F_2(j\omega_n) \equiv (r_1\, e^{j\theta_1}).(r_2\, e^{j\theta_2})$$
$$= r_1 r_2\, e^{j(\theta_1 + \theta_2)} \qquad (8.50)$$

The combined harmonic locus is obtained by repeating this calculation for a series of values of ω. As an example we can consider an integrating element in series with a simple exponential delay

$$F_1(j\omega) \equiv \frac{K}{j\omega}$$

$$F_2(j\omega) \equiv \frac{1}{(1 + j\omega T_1)}$$

In fig. 8.31, $F_1(j\omega)$ is a line drawn along the negative imaginary axis with ω increasing to infinity at the origin. $F_2(j\omega)$ is a semi-circle as shown in chapter 6.

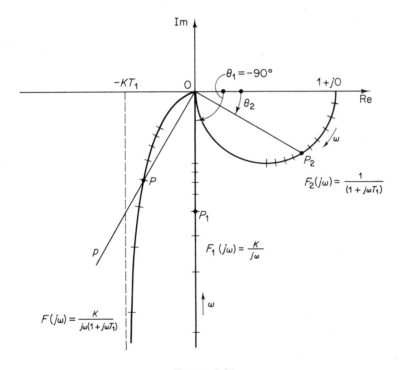

FIGURE 8.31

If P_1 and P_2 correspond to the same value of ω, the associated point P on the combined locus is obtained as follows:

first draw the ray Op in the direction obtained by adding the arguments. In this example, $\theta_1 = -90$ degrees for all values of ω so that Op is at 90 degrees (in the negative direction) to OP_2. The length OP is then set off equal to the product of the lengths OP_1, OP_2. If this process is repeated for a series of values of ω the shape of the resulting locus is found to be as shown. The reader may find it instructive to draw this diagram accurately for, say,

$T_1 = K^{-1} = 1$ second. It should be noted that, although the phase angle for the combined curve tends to -90 degrees as ω tends to zero, the locus does not tend to lie along the imaginary axis. Instead, it approaches an asymptote which is parallel to the imaginary axis and crosses the real axis at the point $-KT_1$. This can be shown by expanding $F(j\omega)$ as a binomial series

$$F(j\omega) = \frac{K}{j\omega}\left[1 - j\omega T_1 + (j\omega T_1)^2 \ldots\right]$$

$$= \frac{K}{j\omega} - KT_1 + j\omega KT_1^2 \ldots$$

$$\approx -KT_1 + \frac{K}{j\omega}$$

as $\omega \to 0$.

This last piece of analysis is really a side issue: the method which we have described of combining harmonic response loci is essentially a graphical construction which relates to the loci themselves, and is not really concerned with the algebraic expressions from which they may have been derived. Indeed, it is not even necessary to know the transfer operator of an element once its harmonic response locus has been drawn. This means that we can take an existing element, determine its harmonic response by direct measurement and predict what the overall performance will be when it is combined with another element whose individual performance may likewise have been determined experimentally, or could equally well have been calculated.

The same reasoning applies for the inverse loci, for which precisely the same method of combination applies is illustrated in fig. 8.32. There is no difficulty in continuing the process, whether in relation to the direct or inverse loci, to take account of further elements in series. We have only to establish a means of incorporating a feedback link into the system, and we have developed an obviously extremely powerful design technique. It will prove convenient to illustrate this with the system which we studied in the previous section, so let us add an element for which the transfer operator is $(1 + T_2 D)^{-1}$ to the system represented by figs. 8.31 and 8.32. The reader should satisfy himself that the resultant direct and inverse loci are as in fig. 8.33. There are various features to be noted: both loci are tangential to the imaginary axis at the origin, the direct locus approaches an asymptote as $\omega \to 0$, and the distances from the origin to the points at which the loci cut the negative real axis are reciprocals of each other.

The two harmonic response loci in fig. 8.33 are drawn for x_o/ε and ε/x_o, respectively. We could, alternatively, regard the direct harmonic response locus as a plot to show the variations of x_o when ε is taken as a vector of unit length directed along the real axis, fig. 8.34(a), and the inverse locus as showing the variation in ε when x_o is represented by a unit vector along the real axis,

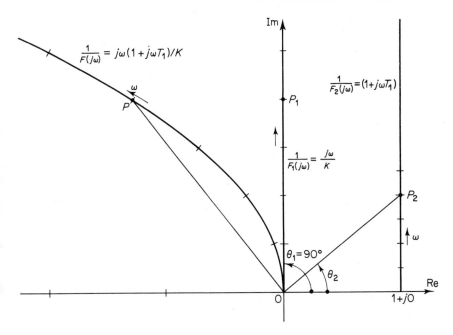

FIGURE 8.32

fig. 8.34(b). The angle ϕ_p in each diagram is the angle by which the output x_o lags behind the input ε when the frequency is ω_p.

So far we have considered the overall response of a simple series of elements. In order to change this series of elements into a closed-loop system we must add a feedback link and a differencing element, fig. 8.35, so that now

$$\varepsilon = x_i - x_o \tag{8.51}$$

x_i being the input to the new system and x_o being its response. The *closed-loop* relationship between x_o and x_i can be determined from *open-loop* response curves in fig. 8.33 by adding the vector triangles represented by eqn. (8.51). In fig. 8.35(a), the direct locus, we have drawn the vector $-\varepsilon$ from the origin to the point $(-1 + j0)$. Then, given

$$x_o = -\varepsilon + x_i$$

x_i is represented by a vector drawn from the $(-1 + j0)$ point to the locus. In fig. 8.35(b), the inverse locus, the vector $-x_o$ is drawn from the origin to the $(-1 + j0)$ point and again, as

$$\varepsilon = -x_o + x_i$$

x_i is represented by a vector drawn from $(-1 + j0)$ to the locus.

The variation in x_o/x_i with frequency can readily be obtained from either diagram by measuring x_o and x_i at a series of frequencies. The phase angle ψ, by which the output lags behind the input, is measured as shown.

332 Dynamics of Mechanical Systems

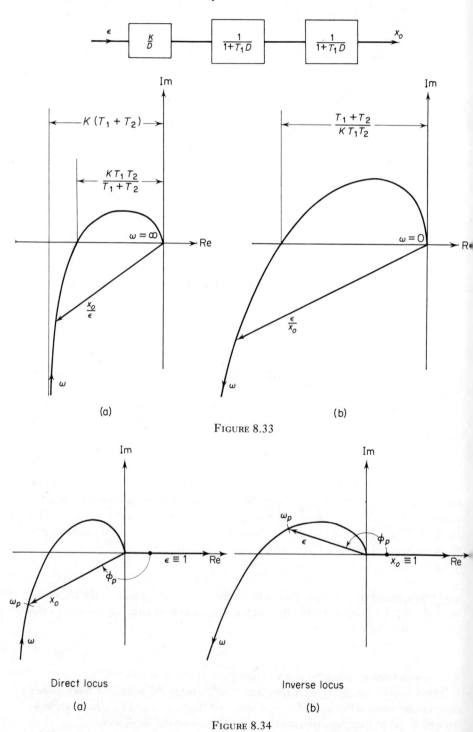

FIGURE 8.33

Direct locus
(a)

Inverse locus
(b)

FIGURE 8.34

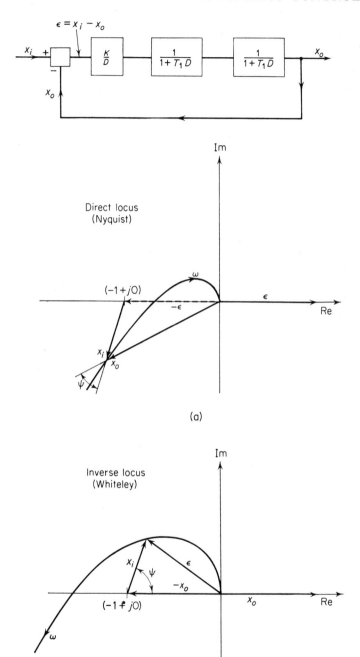

Figure 8.35

334 Dynamics of Mechanical Systems

The two diagrams, figs. 8.35(a) and (b), are respectively the direct and inverse open-loop harmonic response loci, and are known respectively as the Nyquist and Whiteley diagrams. Although, as we have seen, both enable x_o/x_i to be determined, they show essentially the relationship between x_o and ε, and they are referred to as open-loop loci, because the feedback link need not exist. The Nyquist diagram is the one which is most favoured in practice because it is the natural one to use when plotting experimental results. It involves, for each element, plotting the phase and amplitude of the output in response to an input signal of unit amplitude. The physical counterpart of the Whiteley diagram, on the other hand, involves varying the amplitude of the input signal so as to produce unit output. Nevertheless, the Whiteley diagram has one very valuable feature which will be described later.

8.12 Relationship between the open-loop and closed-loop harmonic response loci

The inverse-loop harmonic response locus (Whiteley diagram) shows the way in which ε/x_o varies with ω, i.e.,

$$\frac{\varepsilon}{x_o} = \frac{1}{F(j\omega)} \qquad (8.52)$$

Now

$$\varepsilon = x_i - x_o$$

so that

$$\frac{x_i}{x_o} = 1 + \frac{1}{F(j\omega)} \qquad (8.53)$$

Equations (8.52) and (8.53) show that the closed-loop harmonic response locus may be obtained from the open-loop locus by adding unity to the real part of the complex coordinate of each point in the locus. If the coordinate axes are considered to be fixed, the effect is that of moving the open-loop locus bodily a unit distance to the right. A simpler solution is to move the imaginary axis a unit distance to the left so as to transfer the origin to the $(-1 + j0)$ point in the open-loop diagram. This transformation is made clear in fig. 8.35(b), which shows x_i as being measured from $(-1 + j0)$.

Provided that there are no D-operators on the right-hand side of the governing differential equation, the vector locus for the equation is identical with the inverse closed-loop harmonic response locus, except possibly for a change of scale. Consequently, the inverse open-loop harmonic response locus in fig. 8.35(b) must, for a given system, be identical in shape to the vector locus constructed in fig. 8.27. To demonstrate this we merely replace $F(j\omega)$ in eqn. (8.53) by

$$F(D) \equiv \frac{K}{D(1 + T_1 D)(1 + T_2 D)} \qquad (8.54)$$

AUTOMATIC CONTROL 335

and verify that the resulting relationship

$$\frac{x_i}{x_o} = 1 + \frac{1}{K}[D + (T_1 + T_2)D^2 + T_1 T_2 D^3]$$

is identical with eqn. (8.48), which defines the vector locus.

This identity between the inverse open-loop locus and the inverse closed-loop locus enables us to deduce immediately that, for the closed-loop system to be stable, the open-loop locus must cut the negative real axis to the left of the point $(-1 + j0)$.

The direct open-loop harmonic response locus (Nyquist diagram) shows the way in which x_o/ε varies with ω, i.e.,

$$\frac{x_o}{\varepsilon} = F(j\omega) \tag{8.55}$$

Substituting

$$\varepsilon = x_i - x_o$$

we find

$$\frac{x_o}{x_i} = \frac{F(j\omega)}{1 + F(j\omega)} \tag{8.56}$$

On comparing eqns. (8.55) and (8.56) we deduce that the relationship between the direct open-loop and closed-loop loci does not have a simple geometrical interpretation, such as was found with the inverse loci.

8.13 Nyquist's criterion (II)

The most common form for the open loop transfer operator of a system is

$$F(D) = \frac{KQ(D)}{D^s . P(D)}$$

where $Q(D)$ and $P(D)$ are polynomials in D:

$$Q(D) = 1 + q_1 D + q_2 D^2 + \cdots$$
$$P(D) = 1 + p_1 D + p_2 D^2 + \cdots$$

the particular example which we have been using in earlier sections [eqn. (8.54)] is of this type. The quantity K is defined as the *gain* of the system, and the effect of varying K is to cause the Nyquist locus to expand or contract bodily, with the opposite effect in the Whiteley diagram.

Provided that the Nyquist locus crosses the negative real axis it is possible to choose K so that the point of cross-over is at $(-1 + j0)$. The same value of K must cause the Whiteley locus also to cross over at $(-1 + j0)$, because when this happens, see fig. 8.35(a) and (b), the input x_i is of zero amplitude, and the output x_o is of unit amplitude, in both diagrams. The fact that we have zero input and a non-zero output means that the system is in a state of self-sustained oscillation. The frequency of this oscillation is given by the value of ω at which the loci cut the real axes. The argument by which we now develop

Nyquist's criterion is identical with that used in section 8.10 in respect of the inverse closed-loop harmonic response locus:

a reduction in K means that the power supplied to the system is reduced so that there is insufficient energy to sustain the oscillations, which consequently die out. Conversely, an increase in K causes the oscillations to grow and the system is unstable. The effect of reducing K on the harmonic response loci is to cause the Nyquist locus to cross the real axis to the right of $(-1 + j0)$, and the Whiteley locus to cross it to the left of $(-1 + j0)$.

Accordingly Nyquist's criterion is:

A system is stable if the Nyquist locus crosses the negative real axis to the right of $(-1 + j0)$, and unstable if the cross-over is to the left of $(-1 + j0)$. Alternatively, a system is stable if the Whiteley locus crosses the negative real axis to the left of $(-1 + j0)$, and unstable if the cross-over is to the right of $(-1 + j0)$, see fig. 8.36.

These rules require qualification if the loci cut the negative real axis more than once as in fig. 8.37. A general way to test the stability of a system is to imagine oneself walking along the locus (direct or inverse) in the direction of ω increasing. If the point $(-1 + j0)$ is on the left as one passes it, the system is stable; if it is on the right, the system is unstable.

In the case of the closed-loop loci the same rule holds, except that the point to consider is not $(-1 + j0)$, but the origin.

Exceptions to these rules occur when the open-loop system is unstable. This situation is sufficiently rare for us to omit further reference to it.

8.14 Design criteria

The crudest design criterion which we can apply to a system is simply to require that it shall be stable. If the harmonic response locus does not cross the negative real axis the system is inherently stable, and in theory the gain K could be increased indefinitely. If the locus does cross the negative real axis but the system is nevertheless operating stably the factor by which the gain may be increased before the system becomes unstable is known as the *gain margin*. This quantity is defined with reference to the Nyquist diagram as the reciprocal of the distance of the point of cross-over from the origin, fig. 8.38(a). A reasonable gain margin is 2·5 to 10.

The gain margin is also defined, differently, as the fraction by which the gain falls short of the value which just causes instability. On this definition, the gain margin is represented by the distance indicated by the asterisk in fig. 8.38(a). This double meaning is unfortunate, but should not lead to error as on the first definition the gain margin must be greater than one, and on the second it must be less than one. On both definitions, a large gain margin means that a large increase in the gain can be tolerated.

An acceptable gain margin does not by itself ensure a satisfactory performance. The locus in fig. 8.38(b) has a large gain margin, but only a small increase

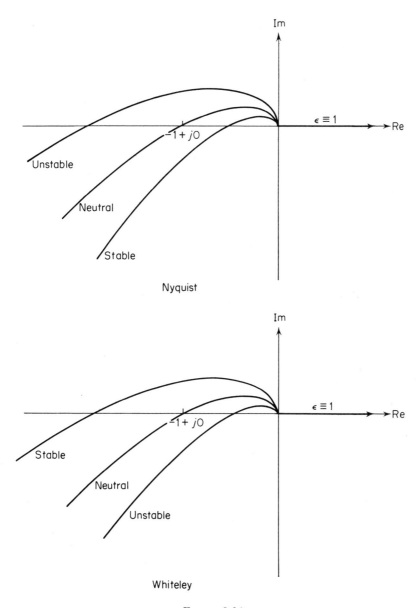

FIGURE 8.36

in the phase lag is needed to cause the locus to pass on the other side of $(-1 + j0)$. To avoid this situation, systems are required to operate with an adequate *phase margin*. The phase margin is defined as the increase in the open-loop phase lag, without change of amplitude which will just cause a system to become unstable, fig. 8.38(c). Normal design values for the phase margin are 30 to 45 degrees.

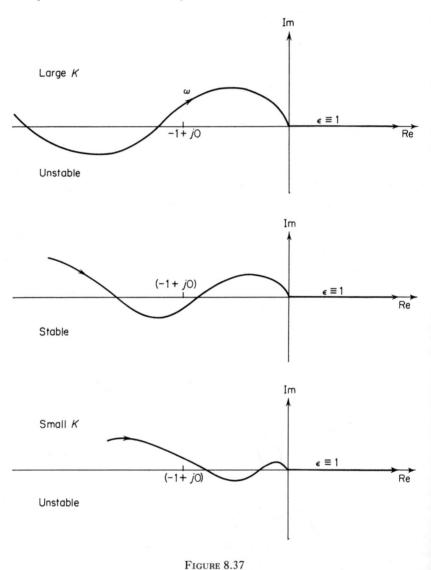

FIGURE 8.37

Adequate phase and gain margins generally ensure a satisfactory performance of a system. The phase margin in particular usually prevents the locus approaching $(-1 + j0)$ too close and so prevents excessive resonance. The shape of the Nyquist locus may, however, be such as to make the gain and phase margins alone an insufficient protection against excessive resonance, as in fig. 8.39. Experience shows that if $|x_o/x_i|_{max}$ lies in the range 1·3 to 1·5 the transient characteristics will be about right, reasonably lively but without too long a decay time.

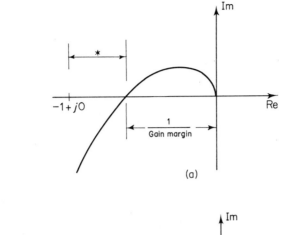

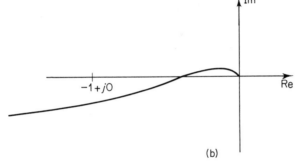

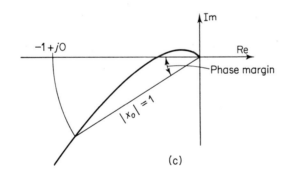

Figure 8.38

8.15 Constant M and ϕ contours

So far, we have referred somewhat loosely to the need to prevent a harmonic response locus passing 'too close' to $(-1 + j0)$. The question naturally arises as to whether we can draw, as it were, a fence around $(-1 + j0)$, outside which a locus must lie if $|x_o/x_i|_{max}$ is to be limited to some specified value M.

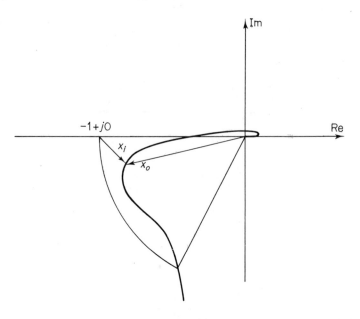

FIGURE 8.39

In the Whiteley diagram $|x_o| = 1$, and at any given frequency $|x_i|$ is the distance from $(-1 + j0)$ to the corresponding point on the locus. It follows that for a specified value of M the locus must not, at any frequency, be at a distance less than M^{-1} from $(-1 + j0)$, and hence that the M-contour is a circle centred at $(-1 + j0)$ and of radius M^{-1}. In fig. 8.40 the two loci both touch the same M-contour and therefore operate with the same maximum dynamic magnification. It can be seen that in both cases, an increase in the gain K causes the locus to cross the given M-circle, and hence an increase in the dynamic magnification of the system.

In the Nyquist diagram the M-contours are again circles. At any point on the harmonic response locus $|x_i|$ is the distance from $(-1 + j0)$ and $|x_o|$ is the distance from the origin. An M-contour is defined by the relationship $|x_o| = M|x_i|$. If the coordinates of a point P on the contour are (r, s), fig. 8.41, we have

$$|x_o|^2 = r^2 + s^2$$
$$|x_i|^2 = (r + 1)^2 + s^2$$

hence $\quad (r + 1)^2 + s^2 = (r^2 + s^2)/M^2$

This equation can be rearranged to the form

$$\left(r + \frac{M^2}{M^2 - 1}\right)^2 + s^2 = \frac{M^2}{(M^2 - 1)^2}$$

and is seen to be the equation of a circle with its centre at $[-M^2/(M^2 - 1), 0]$ and of radius $M/|M^2 - 1|$.

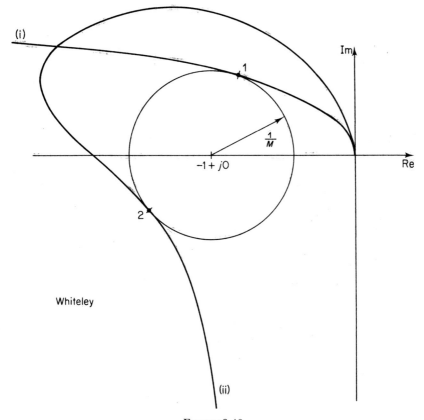

FIGURE 8.40

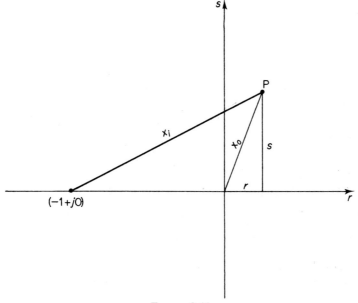

FIGURE 8.41

342 Dynamics of Mechanical Systems

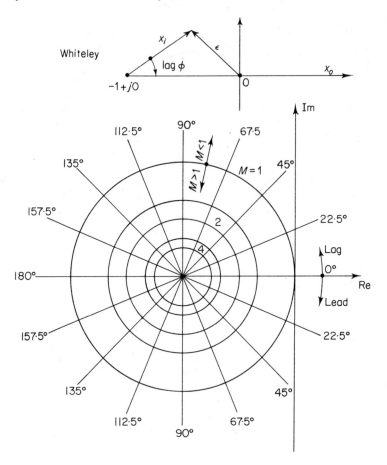

FIGURE 8.42A

We shall see in the next section how M-contours can be used in reshaping the harmonic response locus of an unsatisfactory system. They are also useful as a quick means of determining the resonance curve for a system from its harmonic response locus. The method is to draw on tracing paper to the appropriate scale a set of M-circles. The tracing paper is then superimposed on the harmonic response locus, and so one is able to read off directly the dynamic magnification for any frequency. The same overlay can carry a set of constant ϕ contours, ϕ being the closed-loop phase shift. Figure 8.42 shows sets of M and ϕ contours for both the Nyquist and Whiteley diagrams.

8.16 Closed-loop steady-state performance
We have seen how the general characteristics of the transient response of a closed-loop system may be inferred from its harmonic response locus. The steady-state response to a step input can likewise be deduced from the same

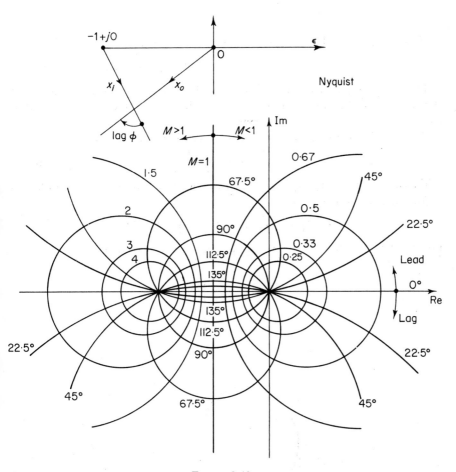

FIGURE 8.42B

locus. The particular points which can be settled at a glance are whether there is a steady-state error in response to a step input, a constant velocity input, a constant acceleration input, etc.

Assuming the general transfer operator to be of the form

$$\frac{x_o}{\varepsilon} = \frac{KQ(\mathrm{D})}{\mathrm{D}^s P(\mathrm{D})}$$

where $P(\mathrm{D})$ and $Q(\mathrm{D})$ are polynomials in D, the associated differential equation is

$$P(\mathrm{D}).\mathrm{D}^s x_o = KQ(\mathrm{D})$$

The differential equation is obtained in terms of the input by substituting $x_i - \varepsilon$ for x_o. We then have

$$[KQ(\mathrm{D}) + \mathrm{D}^s P(\mathrm{D})]\varepsilon = P(\mathrm{D}).\mathrm{D}^s x_i \qquad (8.57)$$

344 Dynamics of Mechanical Systems

or

$$[K(1 + q_1 D + q_2 D^2 + \cdots) + D^s(1 + p_1 D + p_2 D^2 + \cdots)]\varepsilon$$
$$= (1 + p_1 D + p_2 D^2 + \cdots)D^s x_i$$

There can be no variation with time of the steady-state response to a step change in x_i so that all the derivatives must be zero. It follows that $\varepsilon = 0$, if $s \geqslant 1$.

If $s = 0$, the steady-state response to a step change X_i in x_i is

$$\varepsilon = x_i/(1 + K) \tag{8.58}$$

The steady-state response to a constant velocity input $Dx_i = V_i$ is obtained as the particular integral of

$$(KQ(D) + D^s \cdot P(D)) = P(D) \cdot D^{s-1} V_i$$

For the steady-state velocity error to be zero $s \geqslant 2$, and so on.

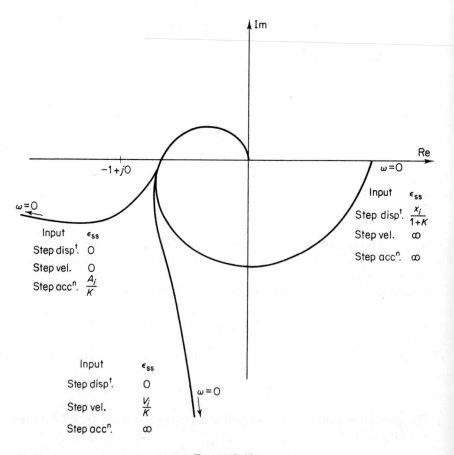

FIGURE 8.43

Inspection of the transfer function

$$\frac{x_o}{\varepsilon} = \frac{KQ(j\omega)}{(j\omega)^s P(j\omega)}$$

shows that the counterpart of setting D equal to zero is to let ω tend to zero. The arguments of $Q(j\omega)$ and $P(j\omega)$ both tend to zero as ω tends to zero so that in the limit

$$\arg(x_o/\varepsilon) \to \arg[1/(j\omega)^s] \text{ as } \omega \to 0$$

If $s = 0$, the Nyquist locus intersects the real axis at a finite point when $\omega = 0$.

If $s = 1$, x_o lags behind ε by 90 degrees as ω tends to zero. As we have seen in section 8.11, this does not imply that the locus tends to coincide with the negative imaginary axis, the implication is that as ω tends to zero the tangent to the locus tends to a direction which is parallel to the negative imaginary axis.

If $s = 2$, the locus tends to a direction parallel to the negative real axis as ω tends to zero, and so on.

These results are summarized in fig. 8.43.

8.17 Improvement of system performance

Early in this chapter we studied a simple position control system and the way in which its performance could be improved. Owing to the simplicity of the system we were able to make the appropriate deductions by algebraic analysis. We will demonstrate now how the same conclusions could have been reached from a consideration of the harmonic response loci.

We have seen that the Nyquist diagram for the simple position control system, which is represented by the block diagram in fig. 8.44 (cf. fig. 8.11), is as given in fig. 8.45 (originally fig. 8.31). The diagram has been drawn for $KT_1 = 2$.

The features of this locus which immediately come to notice are:

1. the gain margin is infinite, so that the system is inherently stable;
2. the phase margins clearly less than the lower limit of 30 degrees suggested above;
3. the locus passes fairly close to $(-1 + j0)$, closer inspection, fig. 8.46, shows that $|x_o/x_i|_{max}$ is 2·24, which is too large; and

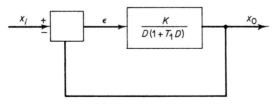

FIGURE 8.44

4 as ω tends to zero arg (x_o/ε) tends to -90 degrees so that the system has a zero steady-state error in response to a step displacement of the input, but must be expected to have a steady-state error of V_i/K in response to a constant velocity input.

As we have observed, the locus passes too close to $(-1 + j0)$ for the transient response to be satisfactory. To improve matters, we must seek means of modifying the Nyquist locus so as to pull it outside the circle corresponding to $M = 1\cdot3$, say. One way of doing this would be simply to reduce the gain K.

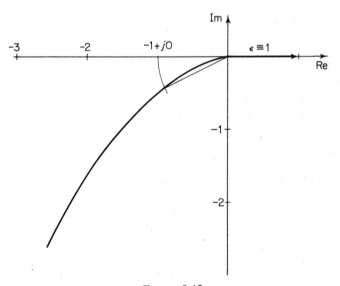

FIGURE 8.45

There would, however, be two adverse effects: first, we would increase the steady-state error V_i/K due to a constant velocity input; second, the transient response would be more sluggish. The speed of response is proportional to the frequency of free vibrations of the system, and, provided that the damping is light, this frequency is approximately equal to the resonance frequency. It can be seen from fig. 8.46 that, if $|x_o/x_i|_{\max}$ is reduced simply by reducing K, the resonance frequency is reduced from ω_1 to ω_2.

The solution to the problem lies in reducing the angle by which x_o lags behind ε in the region $\omega \approx \omega_1$. This is done by introducing into the system a phase advance element. As an example, we have the network given in fig. 8.47 (originally fig. 8.16) which is used to obtain an approximation to proportional plus derivation control. The transfer function for such a network

$$F(j\omega) \equiv \frac{\alpha(1 + j\omega T_d)}{(1 + j\omega\alpha T_d)}$$

is obtained from its transfer operator, eqn. (8.23). As $\alpha < 1$, the introduction of this element without further modification would cause an overall reduction

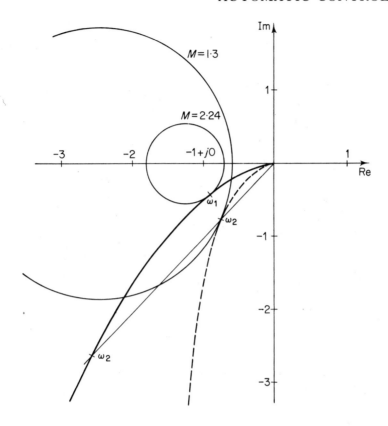

FIGURE 8.46

in the open-loop gain. To prevent this, the gain of the original system is increased from K to K/α. Assuming this to be done, we can take

$$F(j\omega) \equiv \frac{1 + j\omega T_d}{1 + j\omega\alpha T_d} \tag{8.59}$$

The reader is invited to show that the Nyquist locus for this element is a semicircle, as shown in fig. 8.47. It can also be shown that the maximum phase advance, ψ_{max}, is

$$\sin^{-1}\left[(1 - \alpha)/(1 + \alpha)\right]$$

and occurs when $\omega T_d = \alpha^{-\frac{1}{2}}$. The maximum phase advance that is required can be estimated from fig. 8.46, and so α is determined. T_d is determined by the fact that the maximum advance is required at $\omega \approx \omega_1$. The effect of introducing the phase advance network is seen in fig. 8.47.

The steady-state error in response to a constant input velocity will be eliminated if the Nyquist locus is modified so that its argument tends to

348 Dynamics of Mechanical Systems

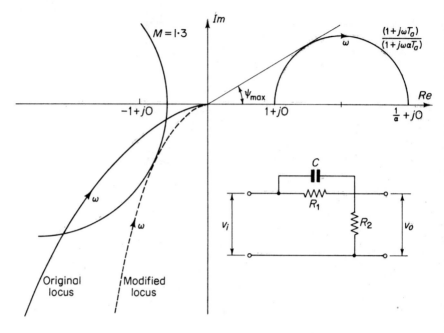

FIGURE 8.47

-180 degrees as ω tends to zero. This can be achieved by introducing an element whose transfer function is of the type

$$F(j\omega) \equiv \frac{1 + j\omega T_i}{j\omega T_i(1 + \beta j\omega T_i)} \qquad (8.60)$$

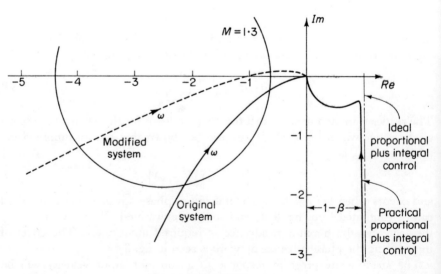

FIGURE 8.48

where $\beta \ll 1$. The hydraulic relay in fig. 8.25 was shown to have this type of transfer function, see p. 318. This transfer function differs in form from that for the phase advance network merely by the additional $j\omega T_i$ in the denominator; its Nyquist locus is as in fig. 8.48, which is drawn for $\beta = 0\cdot 1$. It is of interest to note that an ideal device for giving proportional plus integral control with

$$F(j\omega) \equiv 1 + \frac{1}{j\omega T_i}$$

would be as indicated by the chain-dotted line. The effect on the locus of the original position control system is also shown in fig. 8.48. It is clear that, although the steady-state velocity error has been eliminated, there is a decrease in the overall stability of the system. Indeed, if T_i is made too large the system is unstable. This adverse effect can be corrected by the use of a suitable phase advance network, fig. 8.49.

8.18 Non-unity feedback

In practically all the systems which we have so far considered there has been a direct comparison between the input and response with the feedback being

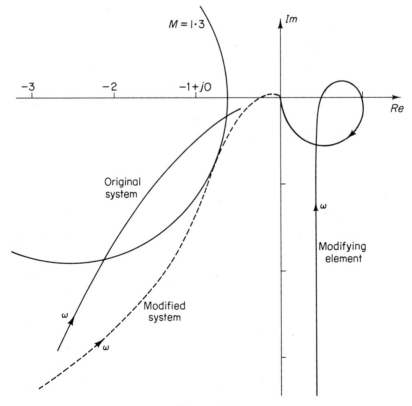

FIGURE 8.49

350 Dynamics of Mechanical Systems

achieved through a single loop. Figure 8.50 shows two systems where the situation is somewhat different. In fig. 8.50(a) there is an element in the feedback limit which modifies the signal so that there is no longer a direct comparison between the input and output signals. In fig. 8.50(b) the direct

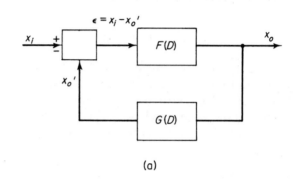

(a)

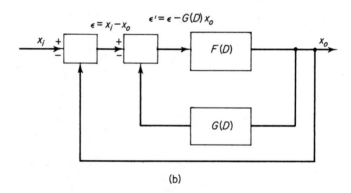

(b)

FIGURE 8.50

comparison is retained, but there is an additional feedback link in parallel with the main one. Both of these types of system cause complications in the Nyquist diagram.

In relation to the system of fig. 8.50(a) we have

$$\frac{x_o}{\varepsilon} = F(D)$$

$$\frac{x_o'}{x_o} = G(D)$$

hence
$$\frac{x_o'}{\varepsilon} = G(D).F(D) \tag{8.61}$$

Also
$$\varepsilon = x_i - x_o' \tag{8.62}$$

AUTOMATIC CONTROL 351

The vector diagram which forms the basis of the Nyquist diagram is therefore as in fig. 8.51(a). As far as the relationship between x_i and ε is concerned, this diagram is of exactly the same type as those which we have classified earlier.

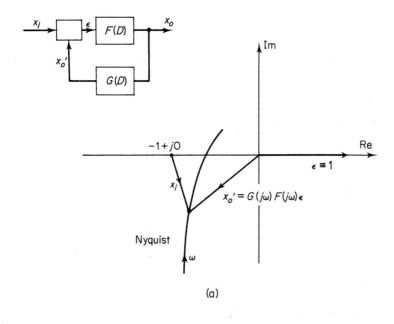

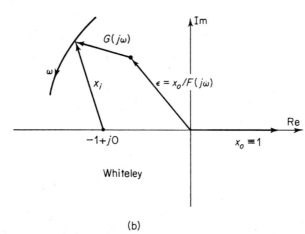

FIGURE 8.51

In particular, the same stability criterion holds. Difficulty occurs, however, if the open-loop response x_o/x_i is required as a function of ω. This cannot be obtained readily because the basic vector relationship involves x'_o not x_o.

A direct comparison of x_o and x_i is still obtainable from the Whiteley diagram, where we have

$$\varepsilon = x_o/F(D) \tag{8.63}$$

and
$$\varepsilon = x_i - x_o'$$
$$= x_i - G(D)x_o \tag{8.64}$$

The open-loop inverse locus is therefore obtained by adding $G(j\omega)$ to $1/F(j\omega)$, as in fig. 8.51(b). The difference between the addition used here, and the multiplication of transfer operators in series, should be noted.

The Whiteley diagram thus obtained is in no way different from previous examples: the stability criterion is unaltered, and M and ϕ contours may be added if required.

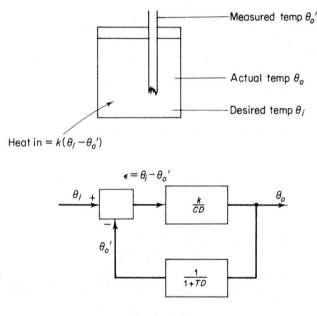

FIGURE 8.52

Figure 8.52 shows a simple system which is of this type. The object of the system is to control the temperature θ_o of the contents of a tank. The temperature is measured by means of a thermometer, and at a given instant records θ_o'. The recorded temperature and the actual temperature are related by the equation

$$(1 + TD)\theta_o' = \theta_o$$

where T is the time constant of the thermometer. Heat is supplied at a rate proportional to the difference between the required temperature θ_i and the recorded temperature. If the thermal capacity of the tank and contents is C we have

$$k(\theta_i - \theta'_o) = C\,D\theta_o$$

The block diagram, fig. 8.52, shows that the system is of the same type as that in fig. 8.50(a).

Finite lag

One type of error signal modification which is sometimes important and may, if unforeseen, cause instability is known as a finite lag. It arises when a finite period of time elapses between the moment at which the output is recorded and the moment when it is compared with the input. As an example of this, consider granular material being discharged from a hopper at a controlled rate on to a conveyor belt. Ideally, the rate of discharge would be measured at the point of discharge. It may, however, be impractical to do this; instead, the quantity of material on the conveyor is weighed at some distance from the point of discharge. In such a system the error signal is always out of date by the time taken by the belt to travel from the hopper to the weighing section. The same type of situation may occur in other continuous-flow situations. The effect is simply to delay the feedback signal by a time T_f without change of amplitude, so that at a frequency ω the feedback signal x'_o, fig. 8.53, lags behind the output

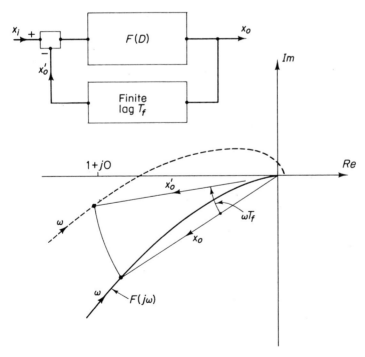

FIGURE 8.53

354 Dynamics of Mechanical Systems

x_o by an angle equal to ωT_f. The Nyquist diagram shows that the overall effect is to make the system less stable. A system is rendered unstable by the introduction of a finite lag $T_f = $ (phase margin)$/\omega'$, where ω' is the frequency at which $|x_o/\varepsilon| = 1$.

Multiple feedback links

For the system shown in fig. 8.50(b) we have

$$x_o = F(D)\varepsilon'$$

and

$$\varepsilon' = \varepsilon - G(D)x_o$$

Hence

$$\frac{x_o}{\varepsilon} = \frac{F(D)}{1 + F(D).G(D)} \tag{8.65}$$

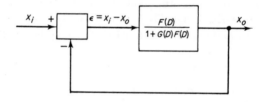

FIGURE 8.54

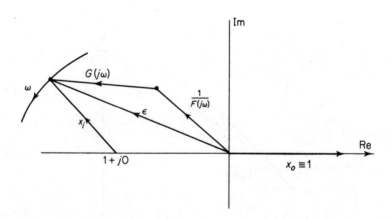

FIGURE 8.55

AUTOMATIC CONTROL 355

The system may therefore be represented as in fig. 8.54, which can be treated in the Nyquist diagram in the usual way, except that there is no simple construction for obtaining the combined locus when $F(j\omega)$ and $G(j\omega)$ are known separately. If the effect of varying one of these components is to be studied this can be done more conveniently in the Whiteley diagram, where we have

$$\frac{\varepsilon}{x_o} = \frac{1}{F(D)} + G(D) \qquad (8.66)$$

fig. 8.55.

Figure 8.56 shows a particular example of the use of a subsidiary feedback link; originally fig. 8.18, it shows the use of velocity feedback as a means of damping a position control system. It can be seen that there is little difficulty in estimating the amount of velocity feedback needed to obtain a required value of M_{max}.

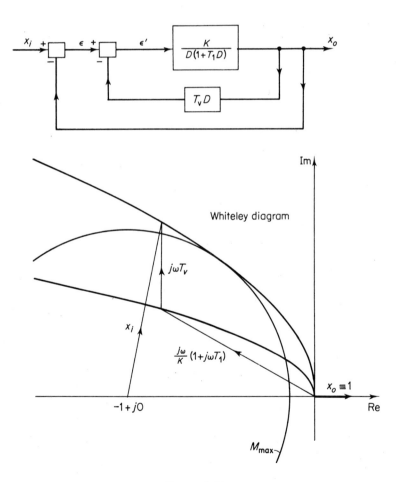

FIGURE 8.56

8.19 Logarithmic plotting — Bode diagrams

Two factors lead one to contemplate the possibility of using logarithmic coordinates for plotting harmonic response data.

1. The amplitude of the transfer function for two elements in series is the product of the amplitudes of the two individual transfer functions, so by taking logarithms we substitute addition for multiplication.
2. The ranges both of the amplitude and frequency often involve factors of 100 or 1000 with interest equally divided over the whole range of operation. By using logarithmic coordinates, quantities in the range 1 to 10 are represented with the same accuracy as quantities in the range 100 to 1000.

Another advantage of logarithmic plotting is revealed in practice. Let us consider the very common transfer operator

$$F(j\omega) \equiv \frac{1}{1 + j\omega T}$$

for which
$$|F(j\omega)| = (1 + \omega^2 T^2)^{-\frac{1}{2}}$$

and
$$\log_{10} |F(j\omega)| = -\tfrac{1}{2} \log_{10} (1 + \omega^2 T^2) \tag{8.67}$$

If $\omega T \ll 1$, $\quad \log_{10} |F(j\omega)| \approx 0$

and if $\omega T \gg 1$, $\quad \log_{10} |F(j\omega)| \approx -\log_{10} \omega T$

These two approximations to $\log |F(j\omega)|$ are plotted against $\log \omega$ together with the exact relationship, in fig. 8.57. Except in the region $\omega T = 1$, eqn. (8.67) is seen to be adequately represented by the two asymptotes. The intersection of the two asymptotes, at the origin in this graph, is known as the *break point*. Since the error at this point is known, together with the very small errors at $\omega T = 0{\cdot}1$ and $\omega T = 10$, it is very easy to sketch in the 'exact' curve. Graphs, such as that in fig. 8.57, are normally drawn on 'semi-log paper', this is graph paper ruled with a logarithmic spacing of the graduations in the ω direction and with linear spacing in the $\log |F(j\omega)|$ direction. Actual frequencies are marked on the frequency scale, e.g., 1, 10, 10^2, 10^3 hertz, each interval being referred to as a decade. It follows that the break point occurs at a frequency equal to T^{-1}. The numbers on the $\log_{10} F(j\omega)$ scale are usually multiplied by 20 and the resulting figures represent the *gain* expressed in *decibels* (db). The slope of the falling branch of the curve in fig. 8.57 is therefore 20 db per decade. It should be noted that this use of the term 'gain' differs from that which we have adopted previously, both usages are common practice, and are readily distinguished by context.

The graph of plane shift, also shown in fig. 8.57, cannot be approximated so easily. As phase angles of elements in series are added directly they are plotted on the vertical linear scale without conversion.

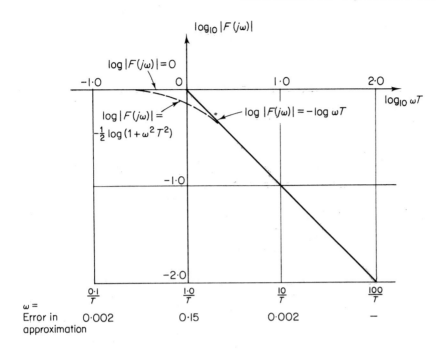
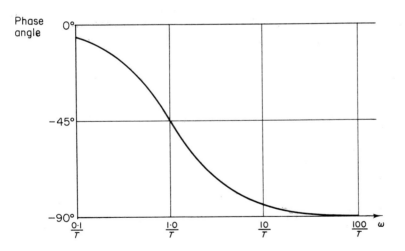

FIGURE 8.57

Figure 8.58 gives the log gain-frequency plots for various other types of element. These graphs may be added to determine the response of elements in series, or to find the response of a single element, e.g., the phase advance network in fig. 8.59. These graphs are known collectively as Bode diagrams.

358 Dynamics of Mechanical Systems

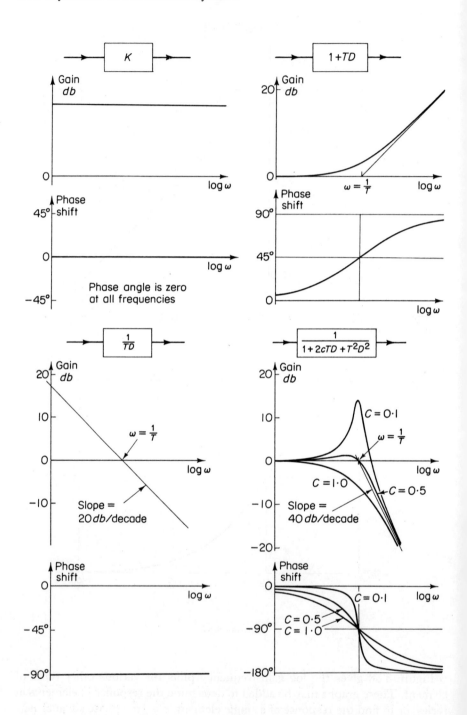

FIGURE 8.58

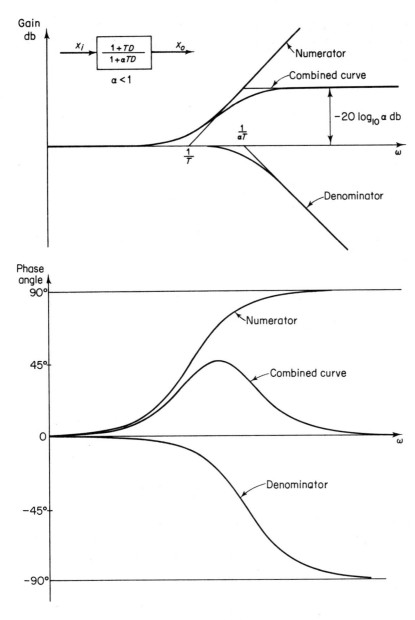

FIGURE 8.59

The gain margin and phase margin, and hence the degree of stability of a closed-loop system, can readily be determined from the Bode diagrams. The gain margin was defined earlier as the reciprocal of the distance from the origin to the point where the Nyquist locus cuts the negative real axis. It could equally well be defined as the reciprocal of the amplitude where the phase shift is

180 degrees. The phase margin is the angle by which the phase shift falls short of 180 degrees when the amplitude is unity. Figure 8.60 shows the gain and phase margins for a particular system.

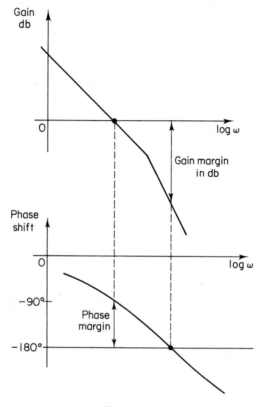

FIGURE 8.60

There is no simple and direct means of deducing the closed-loop response of a system from the Bode diagrams for its open-loop response, so that the Bode diagrams are somewhat less informative than the Nyquist or Whiteley diagrams. Nevertheless, they are very valuable, both for the reasons given at the start of this section and because of the simplicity of construction indicated above.

APPENDIX 8.1

Relationships between harmonic and transient responses

In this chapter we have, at several points, used the argument that the harmonic response of a system must have certain properties in order that the transient response to a step input should have certain other properties, in particular that it should decay fairly rapidly. Such reasoning implies that there must be a formal mathematical relationship between the transient and harmonic responses. We shall not attempt here a rigorous analysis of this relationship, but a brief, if superficial, statement of what is involved is not out of place.

In chapter 6, we saw how the response of a linear system to any input can be deduced if its response to a unit impulse is known. If $u(t)$ is the response to a unit impulse applied to the system at time $t = 0$, the response at time t to a similar impulse applied at time τ is $u(t - \tau)$.

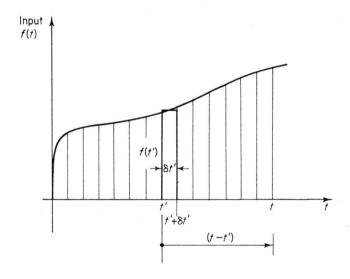

FIGURE 8.61

An input $f(t)$ can be regarded as a series of impulses, fig. 8.61. The input applied over the time interval t' to $(t' + \delta t')$ approximates to an impulse of strength $f(t') \cdot \delta t'$. The response to this impulse at time t is therefore

$$[f(t')\, \delta t']u(t - t')$$

362 Dynamics of Mechanical Systems

The response at time t to the complete input starting at $t = 0$ is

$$\int_0^t f(t')u(t - t')\,dt' \tag{A.8.1}$$

The response to an input $e^{j\omega t}$ starting at $t = 0$ is

$$\int_0^t e^{j\omega t'}u(t - t')\,dt'$$

If τ is substituted for $t - t'$ the integral becomes

$$e^{j\omega t}\int_0^t e^{-j\omega \tau}u(\tau)\,d\tau$$

Assuming that the system is stable, the transient response to this input will decay leaving only the steady-state response to $e^{-j\omega t}$, which is $F(j\omega)\,e^{j\omega t}$, where $F(j\omega)$ is the transfer function for the system. It follows that

$$F(j\omega) = \int_0^\infty e^{-j\omega \tau}u(\tau)\,d\tau \tag{A.8.2}$$

As an example of the use of this integral to determine the harmonic response from the impulse response, consider

$$u(t) = \frac{1}{T}e^{-t/T}$$

which was shown, in chapter 6, to be the impulse response of a simple first order system (exponential lag)

$$F(j\omega) = \frac{1}{T}\int_0^\infty e^{-j\omega \tau - \tau/T}\,d\tau$$

$$= \frac{1}{1 + j\omega T}$$

If the impulse response has been determined experimentally it is not usually possible to determine an explicit expression for $F(j\omega)$. Instead, we write in the form

$$F(j\omega) = \int_0^\infty \cos \omega\tau \cdot u(\tau)\,d\tau - j\int_0^\infty \sin \omega\tau \cdot u(\tau)\,d\tau$$

and evaluate the integrals numerically for a suitable range of ω.

A counterpart to this process enables us to determine the harmonic response from the frequency response. First we consider the steady-state response to a succession of alternately positive and negative unit impulses applied at constant time intervals $T/2$, fig. 8.62. This succession of impulses can be decomposed into a Fourier series and hence the total response is obtained as the sum of the responses due to the individual terms of the Fourier series. The response so

AUTOMATIC CONTROL 363

obtained is in terms of T. If T is now made very large we have the response to a simple impulse applied at time $t = 0$. It can be shown that this response is

$$u(t) = \frac{1}{2\pi} \int_{-\infty}^{\infty} F(j\omega) \, e^{j\omega t} \, d\omega \tag{A.8.3}$$

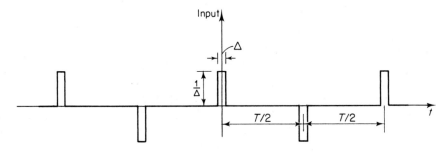

FIGURE 8.62

The evaluation of this integral by elementary methods is often extremely tedious, and is usually unnecessary if one is only interested in obtaining an estimate of the rate at which transient oscillations decay. This can be determined by a simple graphical process on the Nyquist diagram.

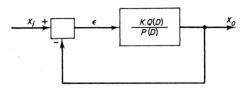

FIGURE 8.63

The governing differential equation of the system shown in fig. 8.63 is

$$[KQ(D) + P(D)]x_o = KQ(D)x_i$$

We have observed that if the open-loop harmonic response locus passes through the point $(-1 + j0)$ then at this point x_i is zero and the system must therefore be capable of sustained oscillations at the corresponding value of ω, ω_o say. This means also that $D = \pm j\omega_o$ must be roots of the equation

$$KQ(D) + P(D) = 0 \tag{A.8.4}$$

Now the Nyquist locus is the locus in the Argand diagram of

$$F(j\omega) \equiv \frac{KQ(j\omega)}{P(j\omega)}$$

We can equally well draw in the same diagram loci for

$$F(\sigma + j\omega) \equiv \frac{KQ(\sigma + j\omega)}{P(\sigma + j\omega)}$$

where:

1 σ is given a series of fixed values and ω is varied; and
2 ω is given a series of fixed values and σ is varied.

If any one of these loci passes through the point $(-1 + j0)$ it will mean that $D = \sigma_1 \pm j\omega_1$, where σ_1 and ω_1 are the corresponding values of σ and ω, are roots of eqn. (A.8.4), and hence that the complementary function contains a term of the form $A\,\mathrm{e}^{\sigma_1 t} \cos(\omega_1 t + \psi)$. Provided that the Nyquist locus ($\omega = 0$) passes sufficiently close to $(-1 + j0)$ it is possible to obtain estimates of σ_1 and ω_1 by sketching in loci (1) and (2) referred to above.

The direction of a tangent to a locus corresponding to constant σ is given by

$$\frac{\partial}{\partial \omega} F(\sigma + j\omega) = jF'(\sigma + j\omega)$$

The direction of a tangent to a locus corresponding to constant ω is given by

$$\frac{\partial}{\partial \sigma} F(\sigma + j\omega) = F'(\sigma + j\omega)$$

Hence
$$\frac{\partial}{\partial \omega} F(\sigma + j\omega) = j\frac{\partial}{\partial \sigma} F(\sigma + j\omega) \qquad \text{(A.8.5)}$$

from which it follows that, at any particular values of σ and ω, the $\sigma = constant$ and $\omega = constant$ loci intersect at right-angles. $\partial F/\partial \omega$ and $\partial F/\partial \sigma$ are positive, respectively, in the directions of ω and σ increasing. Equation (A.8.5) shows us that the directions in which these increases take place are related as shown in fig. 8.64. Furthermore, the rate of increase of ω along the $\omega = constant$ locus is equal to the rate of increase along the $\sigma = constant$ locus. These facts enable us to sketch in ω and σ contours in the Nyquist diagram building out from the Nyquist locus, for which $\sigma = 0$, as in fig. 8.64(b). The grid is a set of curvilinear squares which can be drawn in freehand with surprising accuracy. The actual values of σ and ω which correspond to the point $(-1 + j0)$ are obtained by interpolation.

As a real system cannot oscillate with finite amplitude as ω tends to infinity, a Nyquist locus must approach the origin as ω becomes large. This fact, together with the requirement that for stability σ must be negative, means that for a stable system the locus must pass to the right of $(-1 + j0)$, as shown in fig. 8.64(b). This argument does not, however, constitute a rigorous proof of Nyquist's criterion, because there will be other roots to eqn. (A.8.4) in addition to the complex pair which we have just determined. For each additional complex pair there will be a grid of ω-loci which overlay the grid which we have sketched. Usually, it will be impractical to sketch in the overlying grids because

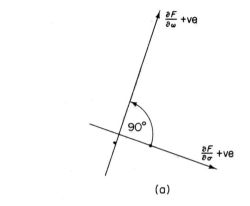

(a)

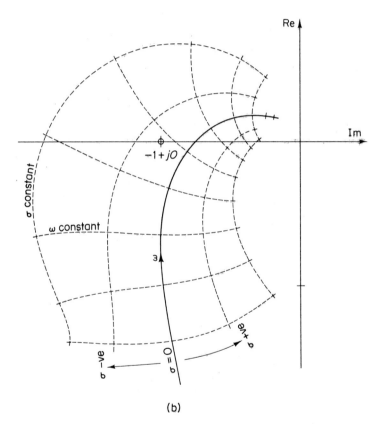

(b)

FIGURE 8.64

of their remoteness from the relevant part of the Nyquist locus. We are therefore unable to draw any general conclusion regarding the stability of such roots.

This construction gives the dominant pair of complex roots: that is to say, that if the roots are $-\sigma_1 \pm j\omega_1$ then we would expect $A\,e^{-\sigma_1 t} \cos(\omega_1 t + \psi)$

to be the dominant component in the complementary function. As a gross simplification, we can regard the system as a second order system with a damping coefficient equal to σ_1/ω_1. Whilst this leads to a good estimate of the rate of decay of transient oscillations, it may give a poor assessment of the amount of overshoot following a step input. To obtain this a more refined treatment is required.[2, 3]

REFERENCES
1 LAWDEN, D. F. *Mathematics of Engineering Systems*, Methuen, 1959.
2 WELBOURN, D. B. *Essentials of Control Theory*, Arnold, 1963.
3 MACMILLAN, R. H. *An Introduction to the Theory of Control in Mechanical Engineering*, C.U.P., 1955.

9

MECHANICAL VIBRATIONS

9.1 Introduction

We have seen in chapter 7 that, with appropriate idealizations, the mathematics for a vibration of a simple spring-mass-damper system is identical with the mathematics for a simple position control system. Nevertheless, in chapter 8, automatic control systems were considered without special reference to mechanical vibrations, and in this chapter we shall concentrate on vibration theory with only limited reference to what has been written in chapter 8. There are two main reasons for this division of attention: first, both fields of study are of immense technological importance so that each merits consideration in its own right; second, the physical circumstances are such, that in order to simplify the mathematics of complex systems we make different types of idealizations which in turn lead to different mathematical treatment. We shall find, for example, that the concept of the transfer operator will play little part in our treatment of mechanical vibrations. At the same time, we shall develop other concepts which have no relevance to control theory. Nevertheless, we believe that although the control engineer, for example, may have only a mild interest in mechanical vibrations, the student engineer who is as yet uncommitted in his choice of career will find that his knowledge of both fields will be enriched by an appreciation of the extent to which they are related to each other.

The importance to human existence and comfort of vibrations is so great that it would be pointless to attempt to list all instances which come to mind. Electrical oscillations, sound waves, waves on the surface of a fluid, physiological phenomena such as the beating of a heart and the movement of lungs, can all be regarded as coming within the scope of vibration theory. We shall, however, concentrate on mechanical vibrations. The range of examples is still virtually endless: the vibrations of an engine on its mounting, torsional vibrations of the crankshaft within an engine, vibrations of an automobile on its suspension, chatter of a lathe tool, and flutter of an aircraft wing, comprise a small sample. Our study will start with further developments of the theory relating to the vibrations of a single spring-mounted mass, which is given in chapter 7.

9.2 Transient vibrations of a single spring-mounted mass

We have seen in chapter 7 that the equation of motion of a mass which is mounted on a spring and viscous damper in parallel, fig. 9.1, is

$$m \frac{d^2x}{dt^2} + \lambda \frac{dx}{dt} + sx = P(t) \qquad (9.1)$$

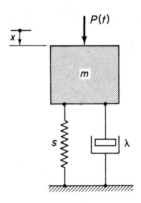

FIGURE 9.1

For the purpose of automatic control theory it is convenient to express this equation in the standard form

$$(T^2 D^2 + 2cT D + 1)x = P(t)/s \qquad (9.2)$$

where $T = \sqrt{(m/s)}$ and $c = \lambda/2\sqrt{ms}$. In vibration theory it is more usual to divide eqn. (9.1) by m rather than s, and so to arrive at the alternative standard form

$$(D^2 + 2c\omega_n D + \omega_n^2)x = \omega_n^2 P(t)/s \qquad (9.3)$$

where $\omega_n = \sqrt{(s/m)} = 1/T$, and $c = \lambda/2\sqrt{ms}$

as before.

The formal solution of eqn. (9.2), when $P(t)$ represents a step change in the force applied at $t = 0$, has been given in chapter 7, and need not be repeated here. It will be remembered that the form of the solution depends on the value of the damping coefficient c, and in particular on whether c is greater than unity, equal to unity, or less than unity, as shown in fig. 7.4. In studying mechanical vibrations we shall be concerned primarily with lightly damped systems for which c is less than unity, and indeed for much of this chapter will concentrate on systems for which the amount of damping is negligible. If $c < 1$, we deduce from eqn. (7.18) that the displacement of mass in fig. 9.1 due to a sudden change P in the force applied to it is

$$x = \frac{P}{s}\left[1 - \frac{e^{-c\omega_n t}}{\sqrt{(1 - c^2)}} \cos(\beta t + \psi)\right] \qquad (9.4)$$

where $\psi = -\tan^{-1} \dfrac{c}{\sqrt{(1-c^2)}}$ and $\beta = \omega_n \sqrt{(1-c^2)}$

If the damping is negligible so that $c = 0$, eqn. (9.4) becomes

$$x = \frac{P}{s}(1 - \cos \omega_n t) \qquad (9.5)$$

In this case, the motion is simple harmonic at a circular frequency $\omega_n = \sqrt{(s/m)}$. This quantity ω_n is referred to as the *natural circular frequency* of the system. If s is expressed numerically in newtons per metre and m is in kilograms, ω_n is expressed in radians per second. The natural frequency is usually stated in hertz and equals $\omega_n/2\pi$. A common laziness, which we will adopt, is to refer to ω_n as the natural frequency, leaving it to be understood that in obtaining numerical values it is necessary to insert the 2π. Equation (9.4) is plotted in fig. 9.2 for various values of c and re-emphasizes the point made in chapter 7

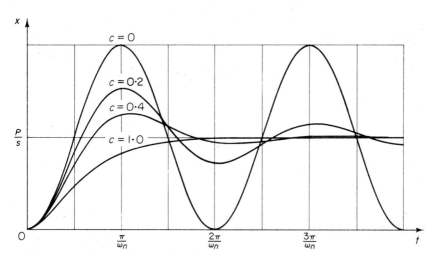

FIGURE 9.2

that, provided c is very much less than unity, it has an insignificant effect on the frequency of vibrations. It is appropriate to recall also that, as we have seen in chapter 7, the amplitude of vibrations due to 0·2 per cent of its initial value after $1/c$ cycles (page 279).

9.3 Harmonic response of a spring-mounted mass — constant excitation

The steady motion of the mass, fig. 9.1, in response to a harmonically varying force of constant amplitude

$$P(t) \equiv P \cos \omega t$$

is the particular integral of

$$m\frac{d^2x}{dt^2} + \lambda\frac{dx}{dt} + sx = P\cos\omega t \tag{9.6}$$

and must be of the form

$$x = X\cos(\omega t - \phi) \tag{9.7}$$

On substituting ω_n^{-1} for T we deduce from eqn. (7.27) that in this case

$$\tan\phi = \frac{2c(\omega/\omega_n)}{1 - (\omega/\omega_n)^2} \tag{9.8}$$

On making the same substitution in eqn. (7.28) and replacing Θ_i by P/S the amplitude is found to be

$$X = \frac{P/S}{\{[1 - (\omega/\omega_n)^2]^2 + 4c^2(\omega/\omega_n)^2\}^{\frac{1}{2}}} \tag{9.9}$$

Equations (9.8) and (9.9) are plotted for various values of c in fig. 9.3. The main feature to be noted is that if the damping is light, $c \ll 1$, the amplitude of vibration is very large when $\omega \approx \omega_n$, and the system is said to be in a state of *resonance*. If the system is undamped, $c = 0$, the amplitude tends to infinity when the forcing frequency coincides with the natural frequency ω_n. In practice, c is never zero and there is an upper limit on the amplitude. It is, however, unlikely that there will be sufficient damping inherent in a system to keep the amplitude within tolerable bounds if it is excited at the natural, or resonant, frequency, and it is usually necessary either to add sufficient damping to limit the amplitude, or to ensure that the natural frequency is outside the usual range of operating frequencies. We will consider later the factors which determine the optimum amount of damping to be used.

It is clear that power is dissipated by damping and that to sustain the motion power must be supplied by the excitation. At any instant the power input is the product of the instantaneous values of the exciting force and the velocity of the mass. If the displacement of the mass is $x = X\cos(\omega t - \phi)$, its velocity is $\dot{x} = -\omega X\sin(\omega t - \phi)$. The instantaneous power input is therefore

$$P\cos\omega t \times [-\omega X\sin(\omega t - \phi)] = -\omega XP\cos\omega t\sin(\omega t - \phi)$$

The work done per cycle is

$$\int_{t=0}^{t=2\pi/\omega} -\omega XP\cos\omega t\sin(\omega t - \phi)\,dt$$

$$= -\tfrac{1}{2}\omega XP\int_0^{2\pi/\omega}[\sin(2\omega t - \phi) + \sin(-\phi)]\,dt$$

$$= (\tfrac{1}{2}\omega XP\sin\phi) \times 2\pi/\omega$$

the first term in the integral being zero.

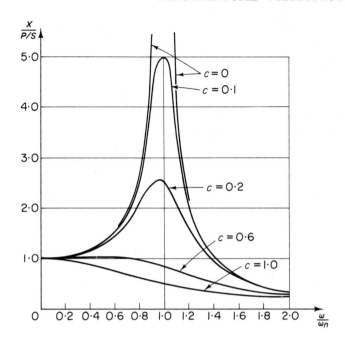

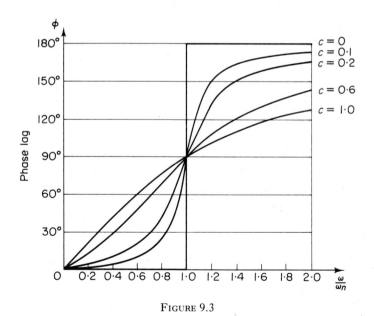

Figure 9.3

The mean power is

$$\tfrac{1}{2}\omega XP \sin \phi \qquad (9.10)$$

We can see from the vector diagram, fig. 9.4, that $P \sin \phi = \lambda \omega X$, so that on eliminating either X or ϕ the above expression becomes

$$\text{mean power} = \tfrac{1}{2}\lambda \omega^2 X^2 \qquad (9.11)$$
$$= P^2 \sin^2 \phi / 2\lambda \qquad (9.12)$$

Equation (9.12) shows that maximum power consumption occurs when $\phi = 90$ degrees, that is when $\omega = \omega_n$. It also shows that an increase in the amount of damping reduces the power consumed.

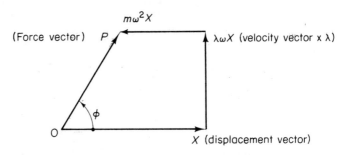

FIGURE 9.4

Equation (9.10) can also be deduced directly from the vector diagram by taking one half the scalar product of the force, P, and velocity, ωX, vectors. This is to be compared with taking the scalar product of the voltage and current vectors in a vector diagram for an a.c. circuit. The factor $\tfrac{1}{2}$ is needed in this case because the vectors in fig. 9.4 represent peak, rather than r.m.s., values of the force and displacement.

It is of interest to consider the power input when the applied force and the vibration are at different frequencies. Assume that the frequency of vibration is α times the exciting frequency, then

$$\text{power input} = (P \cos \omega t)[-\omega X \sin (\alpha \omega t - \phi)]$$
$$= -\tfrac{1}{2}\omega X P\{\sin[(\alpha+1)\omega t - \phi] + \sin[(\alpha-1)\omega t - \phi]\}$$

Taken over a large number of periods, the average values of both terms in this expression tend to zero, unless $\alpha = 1$, when the mean power is $\tfrac{1}{2}\omega XP \sin \phi$ as before. It follows that a harmonically varying force can excite steady motion only at the same frequency.

Equations (9.8) and (9.9) can be used to determine the response to any periodic force. The procedure is to decompose the excitation into a Fourier series, find the response to each component of the series, taking into account as many terms as seems reasonable, and then adding the individual responses to give the overall motion. Figure 9.5 gives some examples. The converse process, whereby an observed motion is analysed into its Fourier components and the excitation needed to produce each is calculated, enables an unknown excitation to be determined.

MECHANICAL VIBRATIONS 373

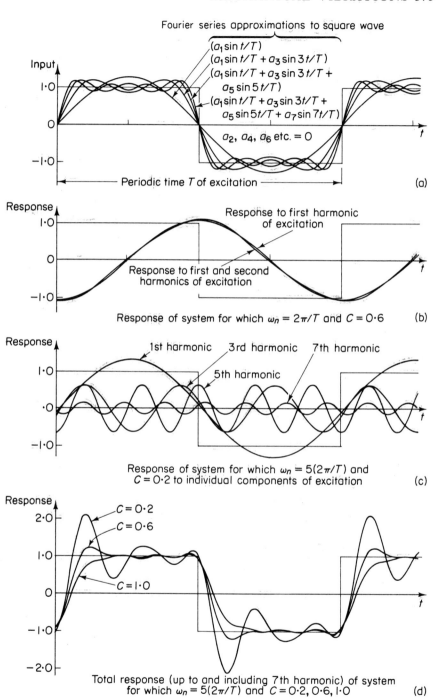

FIGURE 9.5

9.4 Harmonic response of a spring-mounted mass — inertia excitation

Unbalanced moving parts provide a very common source of vibration. At any particular speed the unbalanced inertia forces can be calculated and the phase and amplitude of the resulting motion calculated from eqns. (9.8) and (9.9). However, as the magnitude of the exciting force varies with the running speed, the variations of phase and amplitude differ from those depicted in fig. 9.3. We will consider a system in which the excitation is due to a single unbalanced rotor which turns at a constant angular velocity ω. Let the total vibrating mass be m, and let the out-of-balance be equivalent to a mass m' rotating with eccentricity ε. The system can be represented by the model shown in fig. 9.6.

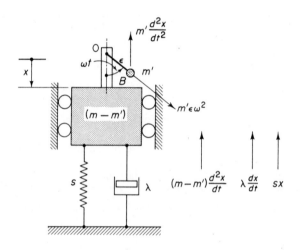

FIGURE 9.6

The acceleration of the eccentric mass m' has two components, the acceleration of O, d^2x/dt^2 vertically, and the acceleration of m' relative to O, $\varepsilon\omega^2$ from m' to O. These two accelerations give rise to the inertia forces $m'\, d^2x/dt^2$ and $m'\varepsilon\omega^2$ as depicted. As the total mass is m we must represent the remainder of the vibrating system by the mass $(m - m')$; the forces which act on this mass are as shown. The equation of motion for the mass $(m - m')$ is obtained by resolving in the x direction all the forces shown in the figure. We have

$$(m - m')\frac{d^2x}{dt} + \lambda \frac{dx}{dt} + sx + m'\frac{d^2x}{dt^2} - m'\varepsilon\omega^2 \cos \omega t = 0$$

or

$$m\frac{d^2x}{dt} + \lambda \frac{dx}{dt} + sx = m'\varepsilon\omega^2 \cos \omega t \tag{9.13}$$

Equation (9.13) is identical with eqn. (9.6), the equation of motion for constant excitation, except that P is replaced by $m'\varepsilon\omega^2$. The phase and amplitude are

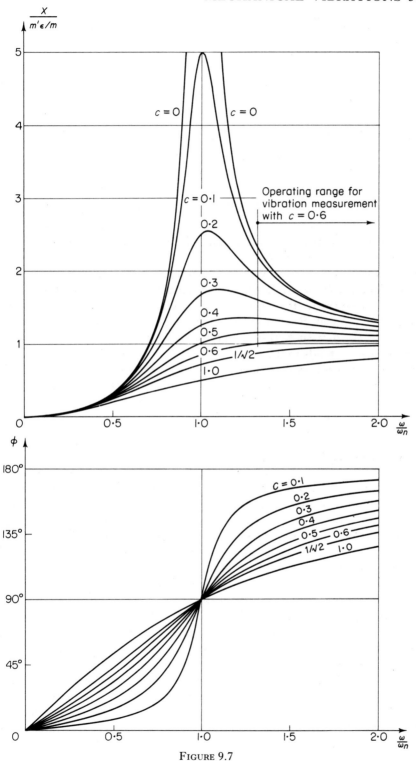

FIGURE 9.7

therefore obtained by replacing P in eqns. (9.8) and (9.9) by $m'\varepsilon\omega^2$. When this is done we have

$$\tan\phi = \frac{2c(\omega/\omega_n)}{1 - (\omega/\omega_n)^2} \tag{9.14}$$

and

$$X = \frac{m'\varepsilon\omega^2/s}{\{[1 - (\omega/\omega_n)^2]^2 + 4c^2(\omega/\omega_n)^2\}^{\frac{1}{2}}}$$

$$= \frac{(\omega/\omega_n)^2}{\{[1 - (\omega/\omega_n)^2]^2 + 4c^2(\omega/\omega_n)^2\}^{\frac{1}{2}}} \cdot \frac{m'\varepsilon}{m} \tag{9.15}$$

The variations in phase and amplitude with frequency computed from eqns. (9.14) and (9.15) are shown in fig. 9.7. Equation (9.14) is identical with eqn. (9.8) so that the curves of phase variation are the same as those shown in fig. 9.3. The graphs of amplitude against frequency are different; first, we see that at low frequencies the motion of the base mass, represented as $(m - m')$ in fig. (9.6), is small. Second, we see that the peak amplitude now occurs for $\omega/\omega_n > 1$. Third, we note that at high frequency the amplitude of the motion of the base mass tends to the value $m'\varepsilon/m$. Remembering that there is a phase shift of 180 degrees at high frequencies, it can be seen that the motion is such as to keep the centre of gravity of the whole stationary, so that the system is self-balancing.

9.5 Vibration isolation

The standard method of preventing unbalanced forces which are generated in machinery from being transmitted to the supporting structure is to interpose a spring mounting between the machine and its foundation. The degree of vibration isolation which is achieved is measured by the ratio of the amplitude of the transmitted force to the amplitude of the exciting force. The transmitted force has two components, the spring force and the damper force, and is represented in the vector diagram, fig. 9.8(a), by the vector R. As the ratio R/P will not be altered by a change of scale in this diagram it is convenient if we divide all the vectors by sX and so deal in the dimensionless quantities shown in fig. 9.8(b). One important conclusion can be drawn by inspection of this diagram; at low frequencies, the diagram having been sketched for $\omega/\omega_n < 1$, the transmitted force exceeds the exciting force. A closer study shows that if the transmitted force is to be less than the exciting force $(\omega/\omega_n)^2$ must be greater than 2. We conclude that for satisfactory operation the exciting frequency must be well in excess of the natural frequency. The precise way in which R/P varies with frequency, for different values of c, is shown in fig. 9.9. The formula from which these curves are plotted is obtained by a two-fold application of Pythagoras' theorem to fig. 9.8(b). This gives

$$\frac{R}{P} = \left\{ \frac{1 + 4c^2(\omega/\omega_n)^2}{[1 - (\omega/\omega_n)^2]^2 + 4c^2(\omega/\omega_n)^2} \right\}^{\frac{1}{2}} \tag{9.16}$$

The phase angle χ by which the transmitted force lags behind the exciting force is given by

$$\chi = \tan^{-1} \frac{2c(\omega/\omega_n)}{1 - (\omega/\omega_n)^2} - \tan^{-1} 2c(\omega/\omega_n)$$

$$= \tan^{-1} \frac{2c(\omega/\omega_n)^3}{1 - (1 - 4c^2)(\omega/\omega_n)^2} \tag{9.17}$$

The graphs of R/P show what we have already noted, that if the mounting is to be of benefit it is necessary to run the machine at a frequency which is well above the natural frequency. Since the running speed of the machine will be determined by its design and purpose it is not generally possible to alter it, but the natural frequency is a matter of choice. It can be reduced either by increasing the mass, or by reducing the stiffness of the supporting springs. The amount of damping to be incorporated is a matter of compromise. Under normal running conditions we see that it is an advantage to have the damping as small as possible, but every machine has to be started up and stopped and this means running through resonance, where a moderate amount of damping may be essential to prevent large amplitude vibrations. The amplitude attained as the system passes through resonance depends upon the rate at which the running speed changes. The amplitudes shown in figs. 9.3, 9.7, and 9.9 are for steady-state conditions and will be attained if the speed change is sufficiently slow. If the driving motor does not have sufficient power to accelerate the machine quickly enough the proper running speed may be unattainable for the state can be reached where all the driving power goes into sustaining vibrations just below the resonant frequency and there is no power available to accelerate the system further. Referring to eqn. (9.12) we see the danger of this happening is reduced by the introduction of more damping.

9.6 Harmonic response of a spring-mounted mass — seismic excitation

In discussing the problem of vibration isolation we have assumed that the object is to prevent vibrations which are generated within a machine from being transmitted to the supporting structure. An equally important problem is that of protecting equipment and personnel from vibrations of their environment; the suspension of a motor-car and the mounting for a camera in a reconnaissance aircraft provide two examples. The idealized system is shown in fig. 9.10 as the basic spring/mass/damper system, but with the excitation now

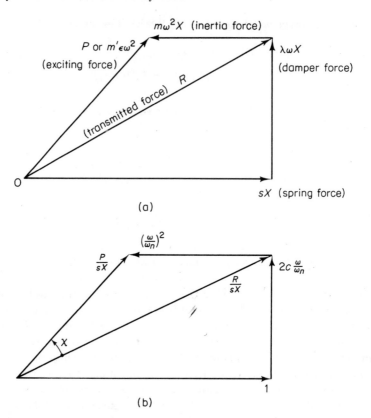

FIGURE 9.8

due to a vibration of the support. If x is the displacement of the mass relative to a fixed datum and y is the displacement of the support the equation of motion is

$$m\frac{d^2x}{dt^2} + \lambda\frac{d}{dt}(x-y) + s(x-y) = 0$$

or
$$\frac{d^2x}{dt^2} + 2c\omega_n\frac{dx}{dt} + \omega_n^2 = 2c\omega_n\frac{dy}{dt} + \omega_n^2 y \tag{9.18}$$

The steady-state amplitude X when $y = Y\cos\omega t$ is obtained from the vector diagram, which follows from eqn. (9.18). The expressions for the amplitude and phase are

$$\frac{X}{Y} = \left\{\frac{1 + 4c^2(\omega/\omega_n)^2}{[1-(\omega/\omega_n)^2]^2 + 4c^2(\omega/\omega_n)^2}\right\}^{\frac{1}{2}} \tag{9.19}$$

and
$$\tan\phi = \frac{2c(\omega/\omega_n)^3}{1-(1-4c^2)(\omega/\omega_n)^2} \tag{9.20}$$

MECHANICAL VIBRATIONS 379

These two expressions are identical with eqns. (9.16) and (9.17) respectively so that the variations of amplitude and phase are as shown in fig. 9.9 with X/Y instead of R/P, and ϕ in place of χ. Again we note that to achieve effective isolation the natural frequency must be small compared with the exciting frequency and that, whilst under normal operating conditions damping to some extent reduces the effectiveness of the mounting, it is essential if there is any danger of prolonged excitation close to resonance.

9.7 Vibration measurement

Equation (9.19) shows that the amplitude of the motion of the mass in fig. 9.10 tends to zero as ω tends to infinity. In practical terms, this means that if ω is large compared with ω_n the mass can be regarded as being stationary in space, so that the motion of the base in space is effectively its motion relative to the mass. We have described in an earlier chapter how this motion can be determined by making the mass a permanent magnet and measuring the current induced in a coil which is attached to the base. Such an instrument is known as a vibration pick-up or vibration transducer.

The displacement of the mass relative to the base is $(x - y)$, fig. 9.10. If we denote this by z and eliminate x from eqn. (9.18) the equation of motion is

$$m \frac{d^2}{dt^2}(z + y) + \lambda \frac{dz}{dt} + sz = 0 \tag{9.21}$$

Assuming $y = Y \cos \omega t$ this equation becomes

$$m \frac{d^2 z}{dt^2} + \lambda \frac{dz}{dt} + sz = m\omega^2 Y \cos \omega t \tag{9.22}$$

Equation (9.22) is of the same form as eqn. (9.13), the equation of motion in response to inertia excitation. The way in which the amplitude Z varies with ω is therefore as given in fig. 9.7, we merely replace $X/(m'\varepsilon/m)$ by Z/Y. With an electromagnetic vibration pick-up damping can readily be obtained by introducing an appropriate resistance across the terminals of the detector coil, and is normally such as to give a damping coefficient of about 0·6. This protects the instrument from damage in the event of its being subjected to vibrations at frequencies which are close to its own natural frequency; it also provides for the maximum range of operating frequencies. This can be seen by reference to fig. 9.7, where the required operating condition of $Z/Y \approx 1$ is obtained at a lower frequency for $c = 0·6$ than it is for any other value of c.

For successful operation of a vibration transducer, the fundamental frequency of the motion being measured must be appreciably higher than the natural frequency of the instrument itself. Given this there is, of course, no difficulty in registering the higher harmonics which may be present in the motion. It follows that the instrument may only be used to record a steady periodic motion which is sustained sufficiently long for the transient response

380 Dynamics of Mechanical Systems

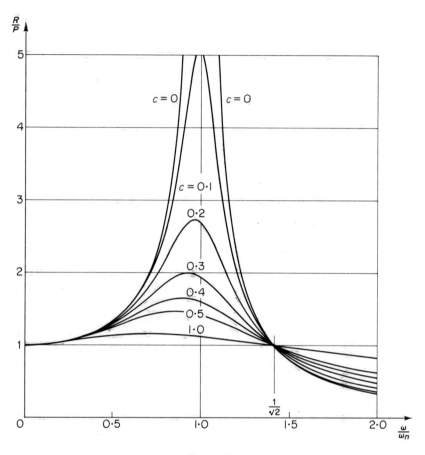

FIGURE 9.9

of the instrument to die out. if it is used to record transient motion the readings may be false. Nevertheless, it is sometimes necessary to measure transient motion, in particular transient accelerations. For this purpose we use an accelerometer. This may also consist of a spring-mounted mass and be idealized by the system shown in fig. 9.10. The same equation of motion holds:

$$m\frac{d^2z}{dt^2} + \lambda\frac{dz}{dt} + sz = -m\frac{d^2y}{dt^2} \tag{9.23}$$

where $z = x - y$ as in eqn. (9.21).

The difference lies in the mode of operation. If the system is in a steady-state condition, with d^2y/dt^2 constant, the acceleration of the mass must also be d^2y/dt^2, and must result from the force in the spring. The compression in the spring is therefore proportional to the acceleration which is to be measured.

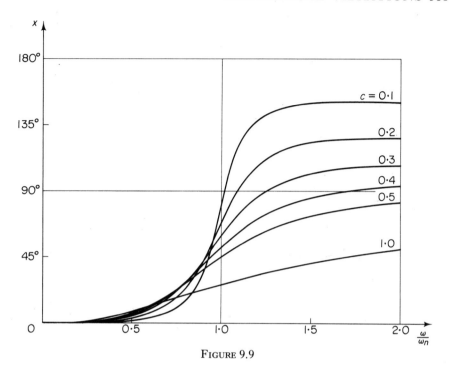

FIGURE 9.9

We can deduce this more formally by evaluating the steady-state response of eqn. (9.23) to a step change in d^2y/dt^2. The steady-state response is

$$z = -\frac{m}{s} \cdot \frac{d^2y}{dt^2}$$

$$= -\frac{d^2y}{dt^2} \bigg/ \omega_n^2$$

For accurate detection of the acceleration d^2y/dt^2 the transient response of the accelerometer must be fast, and must decay rapidly. This means that ω_n must be large and that the damping coefficient should be about 0·6.

9.8 Undamped vibrations

Up to this point our study of mechanical vibrations has been largely concerned with the effect which damping has on the response of a simple spring/mass system to various excitations, visualizing a number of situations in which the introduction of the right amount of damping is an essential feature in the design of a system which is required to have certain vibratory properties. Very often a system vibrates, not because the designer wants it to, but because he is unable to prevent it from doing so. It would be far better if turbine blades, the tubes in boilers, electric transmission lines, bridges, etc., did not vibrate but complete prevention of vibrations may be impractical. In all cases the vibrations are due

382 Dynamics of Mechanical Systems

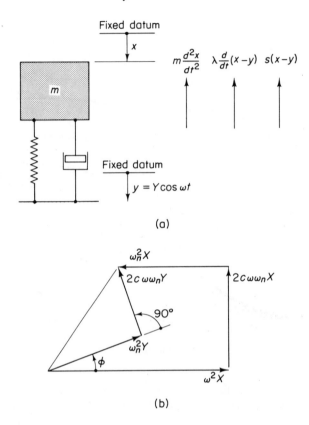

FIGURE 9.10

to the inherent flexibility of the structures, and in all cases the amount of inherent damping is likely to be small. If the damping is small we know that the frequency of free vibrations, assuming that the system has only one degree of freedom, is approximately ω_n, and we know that the system will vibrate with large amplitude if it is subjected to a sustained harmonic excitation at a frequency which is close to ω_n. The design problem is now to ensure that when it is in operation the system is never subjected to excitation at the resonant frequency. The only effect of damping in a lightly damped system is to limit the amplitude of vibration at resonance. If, however, a system is designed so that in practice it never resonates, because if it did the result might be disastrous, the question of exactly how much damping is present becomes irrelevant, and we can ignore it in our analysis. This greatly simplifies the mathematics. Without damping, the equation of motion for a spring/mass system subjected to an exciting force $P(t)$ is

$$m\frac{d^2x}{dt^2} + sx = P(t) \tag{9.24}$$

or
$$\frac{d^2x}{dt^2} + \omega_n^2 x = \omega_n^2 P(t)/s \qquad (9.25)$$

The general solution of eqn. (9.25) consists, as always, of a particular integral and a complementary function. As, however, there is no damping in the system there is no mechanism whereby the complementary function can decay so that 'transient' response is no longer transient but persists for all time. This causes no difficulty if it is the initial motion of the system following, say, an impulse which is of interest. If, on the other hand, it is the 'steady-state' response to a harmonic excitation which is required, we take the line that, although it may be insufficient to affect the steady-state motion, there is always some damping present in a practical system, and that this will in time cause the complementary function to die away. Accordingly, the harmonic response of the system is the particular integral of

$$\frac{d^2x}{dt^2} + \omega_n^2 x = \omega_n^2 (P/s) \cos \omega t \qquad (9.26)$$

To solve this equation we assume the solution

$$x = X \cos \omega t \qquad (9.27)$$

and substitute in the differential equation to obtain

$$(-\omega^2 + \omega_n^2)X = \omega_n^2 (P/s)$$

so that
$$X = \frac{P/s}{1 - (\omega/\omega_n)^2} \qquad (9.28)$$

Equation (9.28) is seen to yield negative values of X when $\omega > \omega_n$, fig. 9.11. This happens because the trial solution, eqn. (9.27), assumes that the phase angle is always zero, whereas it is in fact 180 degrees when $\omega > \omega_n$, see fig. 9.3. We could deduce, independently, that this must be so by noting that

$$-X \cos \omega t = X \cos (\omega t \pm 180°)$$

In drawing the frequency response curve for a single degree of freedom system it is usual to plot the modulus of the amplitude against frequency, fig. 9.11. For systems with more than one degree of freedom it will be seen to be advantageous to plot negative amplitudes and take this as a sign that there is a 180-degree phase shift.

9.9 The phase plane

The general solution of the equation of motion of an undamped spring/mass system, eqn. (9.25), is

$$x = A \sin (\omega_n t + \phi) + \text{particular integral}$$

If the excitation is a constant force P applied at time $t = t_0$ the solution for $t > t_0$ is

$$x = A \sin (\omega_n t + \phi) + P/S \qquad (9.29)$$

384 Dynamics of Mechanical Systems

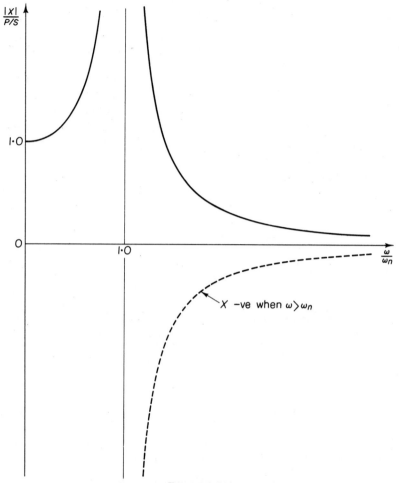

FIGURE 9.11

where the constants A and ϕ depend upon the values of x and dx/dt at time t_o. The obvious way of representing this motion graphically, assuming that A and ϕ have been evaluated, is to plot x against t, as in fig. 9.2, where $t_o = 0$, and x_o and $\dot{x}_o$ are both zero. If the disturbing force changes at time t_1 by an amount ΔP (not necessarily small) we can calculate the subsequent motion in either of two ways:

1 by use of the principle of superposition,
2 by calculating x and $\dot{x}$ at time t_1, and using these values to determine the constants A and ϕ for the next phase of the motion.

Superficially, the second of these methods is much less attractive than the first, but, if instead of plotting x against t we plot x against $\dot{x}$, the second method proves to be simpler in practice.

A graph of displacement against velocity is said to be plotted in the *phase plane*. Let us determine what the shape of this graph is for the motion of a spring-mounted system which is subjected to a constant disturbing force P. As we have already noted, the displacement is given by eqn. (9.29) which may be written as

$$x - P/s = A \sin(\omega_n t + \phi) \qquad (9.29(a))$$

On differentiating with respect to time we have

$$\dot{x} = A\omega_n \cos(\omega_n t + \phi)$$

or
$$\dot{x}/\omega_n = A \cos(\omega_n t + \phi) \qquad (9.30)$$

The relationship between x and $\dot{x}$ is obtained by squaring and adding eqns. (9.29(a)) and (9.30) to give

$$(x - P/s)^2 + (\dot{x}/\omega_n)^2 = A^2 \qquad (9.31)$$

In terms of the variables x and $\dot{x}/\omega_n$, eqn. (9.31) is the equation to a circle of radius A, and with its centre at the point $x = P/s$, $\dot{x}/\omega_n = 0$, fig. 9.12. The radius A of the circle is determined if we can specify any one point through which it must pass. Such a point is $(\dot{x}_0/\omega_n, x_0)$ which is defined by the velocity and displacement at t_0. The graph therefore starts at this point, fig. 9.13, and as

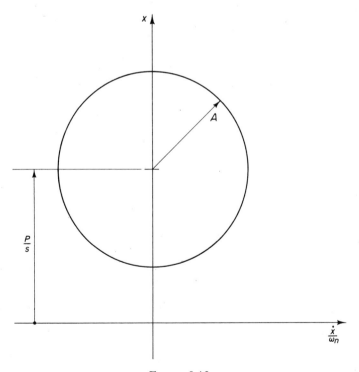

FIGURE 9.12

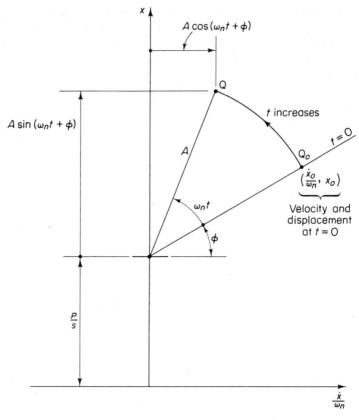

FIGURE 9.13

t increases, the point Q (which defines the velocity and displacement at time t) moves round the circle in an anticlockwise direction with angular velocity ω_n about the centre. The direction must be anticlockwise for if $\dot{x}$ is positive x must increase with time.

If a force P_o is applied at a time when the displacement and velocity are both zero the path, or *trajectory*, in the phase plane is a circle which passes through the origin, as shown in fig. 9.14. It is immediately clear from this diagram that, if the force is maintained, the maximum displacement is $2P_o/s$, and that the velocity varies between $\pm\omega_n P_o/s$. At time t_1 the trajectory will have been traced as far as the point Q_1, in fig. 9.15. There is no instantaneous change in either the displacement or the velocity due to a sudden change in the applied force from P_o to P_1 at time t_1. It follows that for $t > t_1$ the trajectory is a circle passing through Q_1 and centred on point $(0, P_1/s)$, as shown. Subsequent changes in the applied force are dealt with similarly. The response to a continuously changing force is obtained by representing it as a series of steps. In this case, the drawing of the phase-plane diagram is facilitated if the exciting force divided by the spring stiffness is plotted as a function of time alongside it as shown in fig. 9.16.

MECHANICAL VIBRATIONS 387

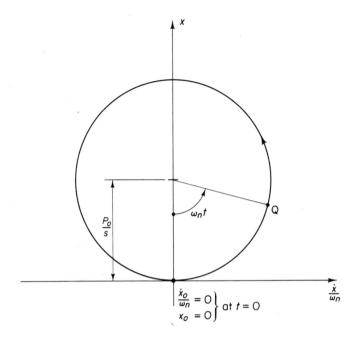

FIGURE 9.14

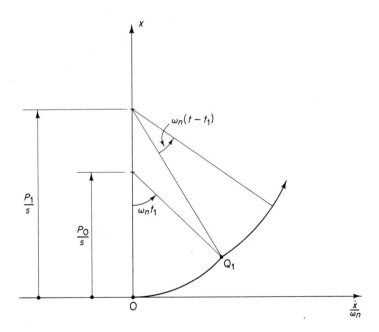

FIGURE 9.15

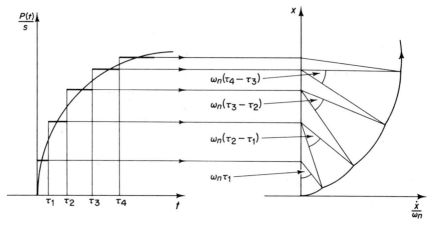

FIGURE 9.16

Precisely the same construction is used when the excitation is due to a movement of the base of the system, seismic forcing, rather than an applied force. The equation of motion for the mass in fig. 9.17 is

$$m\frac{d^2x}{dt^2} + s[x - y(t)] = 0$$

or

$$\frac{d^2x}{dt^2} + \omega_n^2 x = \omega_n^2 y(t) \tag{9.32}$$

This equation differs from eqn. (9.25) only in that $y(t)$ now replaces $P(t)/s$.

A periodic excitation requires a somewhat different treatment. First, we note that if the periodic time of the excitation, or any of its components, happens to coincide with the natural period of the system, the amplitude increases indefinitely and no steady state is reached. Thus the squarewave excitation shown in fig. 9.18, where $T = 2\pi/\omega_n$, causes the trajectory to spiral out continuously as

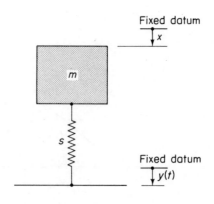

FIGURE 9.17

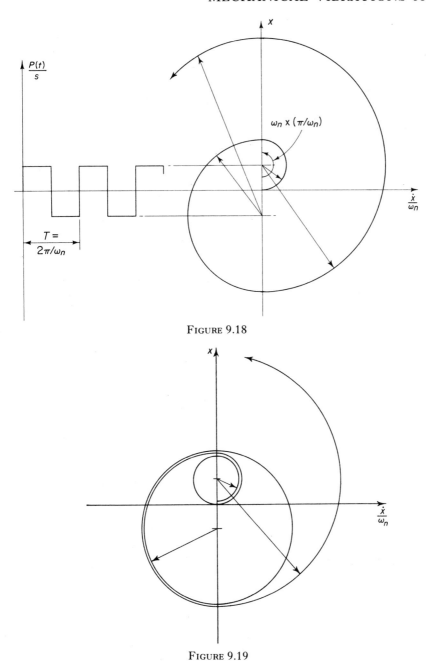

FIGURE 9.18

FIGURE 9.19

shown. 'Resonance' also occurs if T is any odd multiple of $2\pi/\omega_n$; the reader may care to verify that when $T = 6\pi/\omega_n$ the trajectory is as in fig. 9.19. If the periodic time is an even multiple of $2\pi/\omega_n$ the phase-plane trajectory is closed, as in fig. 9.20(a) which shows the response when $T = 4\pi/\omega_n$.

390 Dynamics of Mechanical Systems

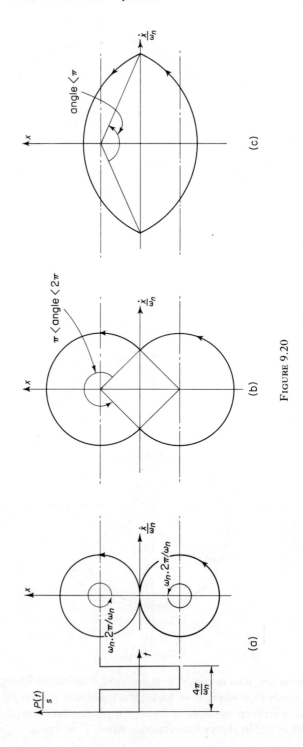

FIGURE 9.20

MECHANICAL VIBRATIONS 391

Let us assume that, whilst the system is vibrating steadily as in fig. 9.20(a), the periodic time is reduced slightly. After transient effects have died down, due to the inevitable small amount of damping, we would expect the phase-plane trajectory to be of roughly the same shape as in fig. 9.20(a), except that the angle subtended by each of the two sectors of the motion must be reduced to something less than 2π. Provided that T is greater than $2\pi/\omega_n$ the steady-state trajectory is as in fig. 9.20(b). It will be noted that the amplitude is greater than it was when $T = 4\pi/\omega_n$. If T is reduced further, the amplitude continues to increase and becomes infinite when $T = 2\pi/\omega_n$. On speeding up further the angle subtended by each of the two sectors becomes less than π, and the trajectory is as shown in fig. 9.20(c). The amplitude is now seen to decrease continuously as T is reduced to zero.

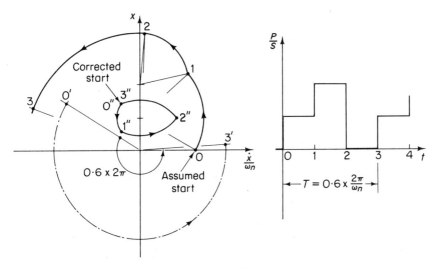

FIGURE 9.21

With waveforms which are more complicated than the one just considered it is less obvious how to select the starting point for the phase-plane trajectory to cause it to be a closed path. The method of dealing with this difficulty is shown in fig. 9.21. A starting point 0 for the trajectory is chosen arbitrarily and the standard construction follows for a complete cycle of the excitation. This takes us from point 0 via points 1 and 2 to point 3 in the phase plane. In the event of point 3 being coincident with point 0 the trajectory is closed and the steady state motion has been determined. It must be recognized, however, that in any case the trajectory drawn represents a possible motion. Other possible motions are obtained by superimposing a free motion such as that represented by the chain-dotted trajectory. The particular free motion chosen subtends at the origin an angle corresponding to the periodic time of the excitation and is of such a radius and so orientated that when the coordinates of point 0′ are added to those of point 0, and the coordinates of 3′ are added to those of point 3, the resulting points 0″ and 3″ are coincident. Taking this point as the corrected start the trajectory is constructed afresh and should close on itself.

One of the features of the phase plane is that it is fairly easy to make allowance for simple non-linearities. Consider, for example, free oscillations of the system shown in fig. 9.22, where two soft springs of stiffness s_1 give effective vibration

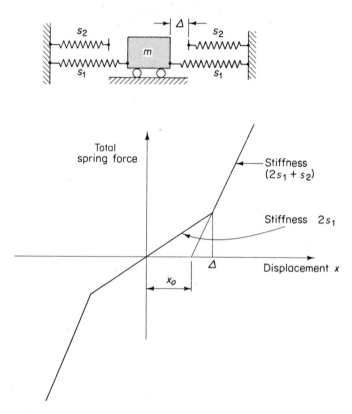

FIGURE 9.22

isolation, and the backing-up springs s_2 prevent excessive movement under transient conditions and at resonance. If the mass oscillates with amplitude less than Δ the two springs s_2 are irrelevant and the phase-plane trajectory is a circle with respect to the axes shown in fig. 9.23 where $\omega_1 = 2s_1/m$. Suppose now that the mass starts from its central position with sufficient velocity to cause it to make contact with one of the springs s_2. The trajectory for this first part of the motion is the arc AB. In order to determine the trajectory for the ensuing motion we must study the equation of motion; as long as contact between the mass and a spring s_2 is maintained the equation is

$$m \frac{d^2x}{dt^2} + (2s_1 + s_2)(x - x_o) = 0$$

or
$$\frac{d^2x}{dt^2} + \omega_2^2 x = \omega_2^2 x_o \qquad (9.33)$$

where $\omega_2 = \sqrt{[(2s_1 + s_2)/m]}$

If we wish to retain the simplicity of circular arc construction in the phase-plane diagram we must, at the instant when eqn. (9.33) starts to hold, change the

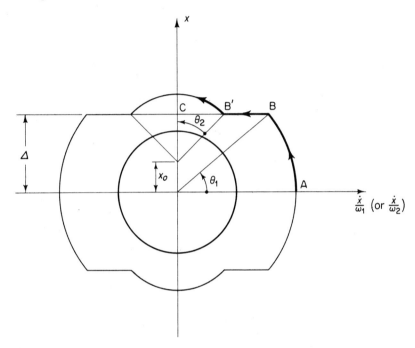

Figure 9.23

scale of the abscissa from $\dot{x}/\omega_1$ to $\dot{x}/\omega_2$. As neither the displacement nor the actual velocity changes this means that the trajectory must suffer a jump from point B to B' such that $CB'/CB = \omega_1/\omega_2$. Thereafter, the trajectory is a circle with its centre at the point $(0, x_o)$. The trajectory for the total motion is as shown in fig. 9.23. The periodic time being $4(\theta_1/\omega_1 + \theta_2/\omega_2)$.

The effect of damping can be included by treating it simply as an additional disturbing force. In order to do this the equation of motion is written in the form

$$\frac{d^2x}{dt^2} + \omega_2^2 x = \frac{\omega_n^2}{s}[P(t) - \text{damping force}] \qquad (9.34)$$

The simplest type of damping, considered in relation to the case of construction of the phase-plane diagram, is Coulomb damping. With Coulomb damping the friction is constant in magnitude and always acts so as to oppose the sliding velocity. The damping term to be included in eqn. (9.34) is therefore $-f$ when

394 Dynamics of Mechanical Systems

dx/dt is positive and $+f$ when dx/dt is negative. The response to a step change in force when the initial displacement and velocity are both zero is as shown in fig. 9.24. Initially, the velocity is positive so that the net disturbing force is

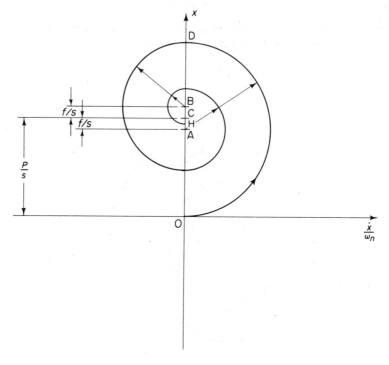

FIGURE 9.24

$(P - f)$, and the centre of the trajectory is at A. When point D is reached the velocity changes sign and the centre of the trajectory moves to B. This process continues, with the trajectory spiralling in towards the equilibrium point C, until it crosses the x-axis at a point such as H between A and B. When this happens, the spring force sx is insufficient to overcome the net force $(P \pm f)$.

With viscous damping or velocity squared damping, such as occurs due to turbulent flow of a liquid in a pipe, the magnitude of the friction force varies with the velocity. This is allowed for by considering the motion in small steps and recomputing the damping force at the end of each step. In this way the damping force is always slightly 'out of date', but the resulting accuracy may nevertheless be quite high. Figure 9.25 shows the trajectories for the step response of underdamped and critically damped systems with viscous damping.

9.10 Vibration of a system with two degrees of freedom

A system is said to have two degrees of freedom when two independent co-ordinates are needed to define its configuration; fig. 9.26 gives a few examples of such systems. Before we go on to study such systems in detail it is convenient

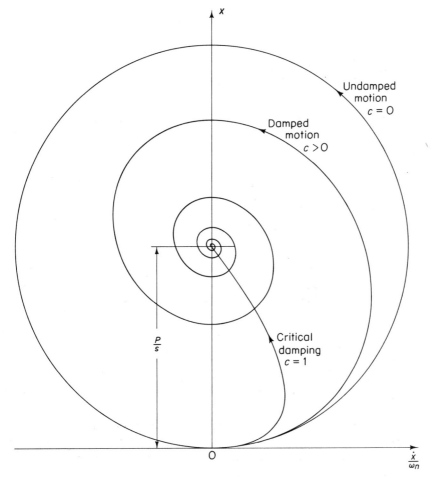

FIGURE 9.25

to recall the main features of the vibrating properties of systems which have one degree of freedom. If it is undamped, a system with one degree of freedom:

1 executes simple harmonic motion when it is disturbed from a stable equilibrium state, the frequency of the vibration being known as the natural frequency,
2 resonates when subjected to harmonic excitation at the natural frequency.

Light damping causes the amplitude force vibrations to decay, and to limit the amplitude of vibration at resonance, but affects the frequency of free vibrations, and the resonance frequency only slightly.

Turning now to systems with two degrees of freedom, we cannot assume that free motion will necessarily be simple harmonic, but it is reasonable to assume that whatever the motion is, light damping will cause it to decay slowly. It is

396 Dynamics of Mechanical Systems

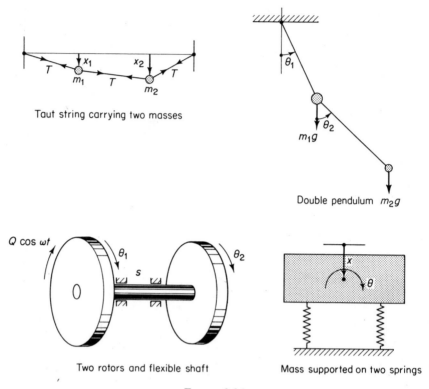

Taut string carrying two masses

Double pendulum $m_2 g$

Two rotors and flexible shaft

Mass supported on two springs

FIGURE 9.26

also reasonable to assume that the system will resonate in response to suitable excitation, and, that the effect of damping will simply be to limit the amplitude of the vibrations. This reasoning suggests that we might well ignore damping when analysing systems with two degrees of freedom. We have only to recall the additional complexity which damping brings to the mathematics for our minds to be made up. The first system which we shall study is given in fig. 9.27.

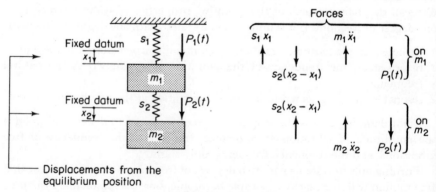

FIGURE 9.27

The equation of motion for each of the masses is obtained by summing the forces which act on it. These forces, the spring and inertia forces, are as shown in fig. 9.27. Care must be taken to ensure that the directions, and hence the signs of the terms are correct. The most troublesome term is that due to the force in the lower spring. It is suggested that the safest approach is to consider the forces induced by x_1 and x_2 separately. A displacement x_1 puts the spring in compression, while x_2 puts it in tension, the net compression is therefore $(x_2 - x_1)$ and the forces applied to the masses are $s_2(x_2 - x_1)$ in the directions shown. The equations of motion are:

$$m_1 \frac{d^2 x_1}{dt^2} + s_1 x_1 - s_2(x_2 - x_1) = P_1(t)$$

and

$$m_2 \frac{d^2 x_2}{dt^2} + s_2(x_2 - x_1) = P_2(t)$$

Or, alternatively

$$\left.\begin{array}{l}(m_1 D^2 + s_1 + s_2)x_1 - s_2 x_2 = P_1(t) \\ (m_2 D^2 + s_2)x_2 - s_2 x_1 = P_2(t)\end{array}\right\} \quad (9.35)$$

and

If x_2 is eliminated from eqns. (9.35) the resulting equation can be written as

$$\left[D^4 + \left(\frac{s_1 + s_2}{m_1} + \frac{s_2}{m_2}\right)D^2 + \frac{s_1 s_2}{m_1 m_2}\right]x_1$$

$$= \left(D^2 + \frac{s_2}{m_2}\right)\frac{P_1(t)}{m_1} + \frac{s_2 P_2(t)}{m_1 m_2} \quad (9.36)$$

Impulse response

If $P_1(t)$ and $P_2(t)$ are impulses applied at $t = 0$ the resulting motion of m_1 is given by the solution of

$$\left[D^4 + \left(\frac{s_1 + s_2}{m_1} + \frac{s_2}{m_2}\right)D^2 + \frac{s_1 s_2}{m_1 m_2}\right]x_1 = 0 \quad (9.37)$$

because for $t > 0$, the applied forces are zero.

We know that a system with one degree of freedom in a state of free vibration moves with simple harmonic motion so let us test the effect of assuming that

$$x_1 = X_1 \cos \omega t \quad (9.38)$$

On substituting this in eqn. (9.38) and carrying out the differentiations we have

$$\left[\omega^4 - \left(\frac{s_1 + s_2}{m_1} + \frac{s_2}{m_2}\right)\omega^2 + \frac{s_1 s_2}{m_1 m_2}\right]X_1 \cos \omega t = 0 \quad (9.39)$$

an equation which must be satisfied if our assumption is to be valid. This means that either $X_1 = 0$, and there is no vibration, or

$$\omega^4 - \left(\frac{s_1 + s_2}{m_1} + \frac{s_2}{m_2}\right)\omega^2 + \frac{s_1 s_2}{m_1 m_2} = 0 \qquad (9.40)$$

Equation (9.40) is known as the *frequency equation* for the system. It is satisfied if

$$\omega^2 = \left(\frac{s_1 + s_2}{m_1} + \frac{s_2}{m_2}\right) \pm \sqrt{\left[\left(\frac{s_1 + s_2}{m_1} + \frac{s_2}{m_2}\right)^2 - 4\frac{s_1 s_2}{m_1 m_2}\right]} \qquad (9.41)$$

Our assumption that mass m_1 moves with simple harmonic motion is therefore valid provided that the frequency has one or other of the values given by eqn. (9.41). The motion of the other mass is obtained by substituting for x_1 in either of eqns. (9.35). The first is the most convenient; remembering that $P_1(t) \equiv 0$ for $t > 0$, it gives

$$x_2 = X_2 \cos \omega t$$

where
$$X_2 = \left(\frac{s_1 + s_2 - m\omega^2}{s_2}\right) X_1 \qquad (9.42)$$

showing that m_2 vibrates at the same frequency as m_1, and is either in phase or 180 degrees out of phase.

When a freely-oscillating system moves so that all its parts execute simple harmonic motion the system is said to be vibrating in a *normal mode* (or *principal mode*); the corresponding frequency is called a *natural frequency* of the system. A system with one degree of freedom has only one normal mode, and, as we have seen, one with two degrees of freedom has two normal modes. In general, a system with n degrees of freedom has n normal modes, and n natural frequencies. Exceptions to this general rule will be considered later in the chapter.

It is helpful and convenient to represent relationship between the amplitudes of the masses, eqn. (9.42), graphically. For the particular system where $m_1 = m_2 = m$ and $s_1 = s_2 = s$, eqn. (9.41) gives

$$\omega^2 = 0.382\sqrt{(s/m)} \quad \text{or} \quad 2.618\sqrt{(s/m)}$$

and eqn. (9.42) gives correspondingly

$$X_2/X_1 = 1.618 \quad \text{or} \quad -0.618$$

A graph which shows the deflected form of the system when the displacements are a maximum, fig. 9.28, is known as a *modal shape*, there being one modal shape for each normal mode.

It must not be assumed from all this that a system automatically vibrates in one of its normal modes when it is disturbed from its equilibrium position. The general solution to eqn. (9.36) is then

$$x_1 = A_1 \cos(\omega_1 t + \phi) + A_2 \cos(\omega_2 t + \phi_2) \qquad (9.43)$$

MECHANICAL VIBRATIONS 399

where ω_1 and ω_2 have the two values given by eqn. (9.41), namely the two natural frequencies. The constants A_1, A_2, ϕ_1, and ϕ_2 are determined by the displacements and velocity of each of the masses at $t = 0$. Unless these initial conditions are such that either A_1 or A_2 is zero, the motion of the masses is not simple harmonic, and indeed unless $\omega_1:\omega_2$ can be expressed as the ratio of two whole numbers, the motion is not even periodic. The system will vibrate

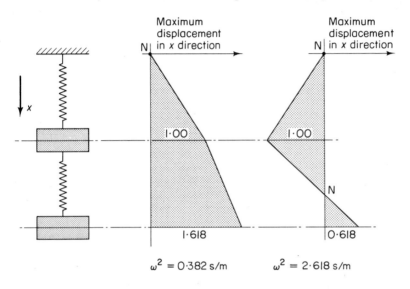

FIGURE 9.28

in one of its normal modes if the strengths of the impulses applied to the masses are in the appropriate ratio. A more practical way of achieving the same effect is to displace the masses so that the system takes up one of its modal shapes, and then to release the masses simultaneously.

Harmonic response
We will now determine the response of the system in fig. 9.27 to a harmonic excitation, in particular when $P_1(t) = P \cos \omega t$ and $P_2(t) \equiv 0$. The equation of motion for m_1 is then, from eqn. (9.36),

$$\left[D^4 + \left(\frac{s_1 + s_2}{m_1} + \frac{s_2}{m_2} \right) D^2 + \frac{s_1 s_2}{m_1 m_2} \right] x_1 = \left(D^2 + \frac{s_2}{m_2} \right) \frac{P}{m_1} \cos \omega t \quad (9.44)$$

The complementary function is given in eqn. (9.43), and although it may persist for many cycles in a lightly damped system it must eventually decay. The steady-state motion is then the particular integral of eqn. (9.44). We know that the response of a linear system to a harmonic excitation must be sinusoidal so let us assume that

$$x_1 = X_1 \cos \omega t$$

Substituting for x_1 in eqn. (9.44), and carrying out the differentiations, we find

$$\left[\omega^4 - \left(\frac{s_1+s_2}{m_1} + \frac{s_2}{m_2}\right)D^2 + \frac{s_1 s_2}{m_1 m_2}\right]X_1 = \left(\frac{s_2}{m_2} - \omega^2\right)\frac{P}{m_1}$$

The left-hand side of this equation is quadratic in ω^2 and can therefore be readily factorized, and we may write

$$(\omega^2 - \omega_1^2)(\omega^2 - \omega_2^2)X_1 = \left(\frac{s_2}{m_2} - \omega^2\right)\frac{P}{m_1} \qquad (9.45)$$

where ω_1^2 and ω_2^2 are the two roots of the equation

$$\omega^4 - \left(\frac{s_1+s_2}{m_1} + \frac{s_2}{m_2}\right)\omega^2 + \frac{s_1 s_2}{m_1 m_2} = 0 \qquad (9.46)$$

Finally, we have

$$x_1 = \frac{(s_1/m_1)[(s_2/m_2) - \omega^2]}{(\omega^2 - \omega_1^2)(\omega^2 - \omega_2^2)}\frac{P}{s_1}\cos \omega t \qquad (9.47)$$

The motion of the second mass is obtained by substituting for x_1 in the second of eqns. (9.35) [remembering that $P_2(t) \equiv 0$]. It is found that

$$x_2 = \frac{s_1 s_2 / m_1 m_2}{(\omega^2 - \omega_1^2)(\omega^2 - \omega_2^2)}\frac{P}{s_1}\cos \omega t \qquad (9.48)$$

It is evident from eqns. (9.47) and (9.48) that the amplitudes X_1 and X_2 tend, simultaneously, to infinity when the frequency of excitation ω equals either ω_1 or ω_2. It can be shown algebraically that $\omega_1^2 < s_2/m_2 < \omega_2^2$ so that X_1 and X_2 vary with ω in the way shown in fig. 9.29. In this diagram a negative value for an amplitude implies that the motion is 180 degrees out of phase with the excitation.

Equation (9.46), from which the resonance frequencies are calculated, is identical to eqn. (9.40), from which the natural frequencies were calculated. It follows that a system with two degrees of freedom will vibrate in resonance when excited at a frequency which equals one or other of its natural frequencies. In general, a system with n degrees of freedom has n resonance frequencies.

The purpose of analysing the vibrating properties of a lightly damped system is generally to ensure that there is no likelihood of it being excited at a resonance frequency, free vibrations are of little interest in themselves. Nevertheless, vibration analysis for lightly damped systems with many degrees of freedom is almost entirely concerned with the determination of natural frequencies and normal modes, the reason being that, once those properties have been determined, the way in which the system will respond to excitation can be predicted.

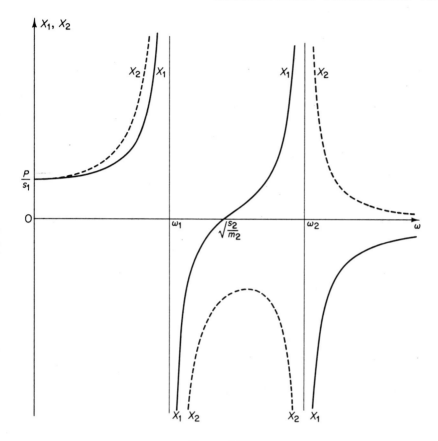

FIGURE 9.29

9.11 Vibration absorber

One of the interesting features of fig. 9.29 is that it shows the amplitude of m_1 to be zero when the exciting frequency is $\sqrt{(s_2/m_2)}$. Except for the fact that its amplitude is not arbitrary, the lower mass can be regarded as behaving as if it were suspended by its spring s_2 from a fixed point and were executing free vibrations at its natural frequency.† The amplitude adjusts itself so that the force in the lower spring counterbalances the excitation $P \cos \omega t$, that is

$$P = s_2 X_2$$

or
$$X = P/s_2$$

This result can also be deduced by putting $x_1 = 0$ in the first of eqns. (9.35).

When the system is vibrating in this way, the lower mass is said to be acting as a *vibration absorber*. The phenomenon itself provides a practical means of eliminating unwanted vibrations. If, for example, a machine mounted on a

† Strictly, the term 'natural frequency' should be used only in relation to free vibrations of the whole system, and it is incorrect to refer to the natural frequency of a part of a system.

flexible structure vibrates excessively, it may be simpler to add a vibration absorber which is tuned to operate at the appropriate frequency than to reduce the nuisance by changing the mass of the machine itself or the stiffness of the supporting structure. Some caution is needed in adopting the vibration absorber because in changing the system from one with a single degree of freedom to one with two degrees of freedom an extra resonance frequency is introduced: in removing one trouble-spot, two have been created elsewhere. This might not matter if the machine operates at a fixed speed, but with varying speed of operation the cure may be worse than the disease. The way out of this difficulty is to introduce viscous damping in parallel with the absorber spring. If the amount of damping is λ_2 the equations of motion are

$$\left. \begin{array}{l} (m_1 D^2 + \lambda_2 D + s_1 + s_2)x_1 - (\lambda_2 D + s_2)x_2 = P\cos\omega t \\ (m_2 D^2 + \lambda_2 D + s_2)x_2 - (\lambda_2 D + s_2)x_1 = 0 \end{array} \right\} \quad (9.49)$$

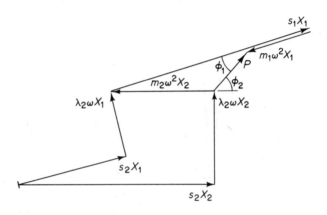

FIGURE 9.30

Equations (9.49) may be represented by the vector diagram in fig. 9.30 from which it is possible to deduce expressions for the amplitudes X_1 and X_2, and the corresponding phase angles ϕ_1 and ϕ_2. The way in which X_1 varies with ω for different amounts of damping in a particular system is shown in fig. 9.31. It can be seem that the effect of light damping is to limit the amplitude at resonance. It also prevents the absorber from being completely effective at $\omega = \sqrt{(s_2/m_2)}$. Very heavy damping has the effect of locking the two masses together so that the resonance curve looks very much like one for a lightly damped system with a single degree of freedom. Between these two extremes there is an optimum amount of damping for which there are no well defined resonance peaks.

9.12 Vibrations of two rotors connected by a flexible shaft

Many practical engineering systems can be considered to consist effectively of two rotors connected by a flexible shaft, for example a motor and electric generator, or the engine and propeller of a ship or an aeroplane. Such a system has two degrees of freedom because, in order to define its configuration, we need two coordinates, one for each rotor. It differs from the one which we have

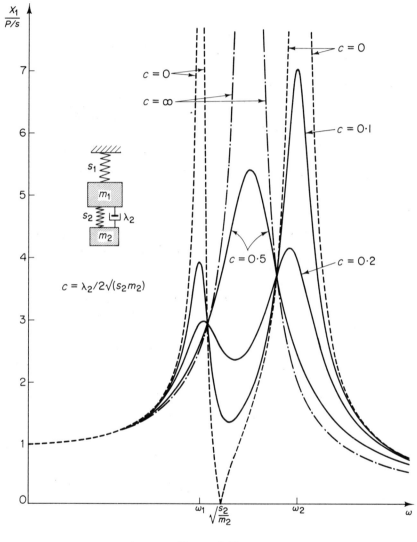

FIGURE 9.31

just studied in that it is not, as it were, tied to the ground. While the displacements of the masses m_1 and m_2 in fig. 9.27 must obviously be limited in any practical system, the rotations θ_1 and θ_2 in fig. 9.32 are without limit. Thinking in terms of a motor and load, the system in fig. 9.32 will execute torsional vibrations if there is a periodic fluctuation in the driving torque or in the load torque.

404 Dynamics of Mechanical Systems

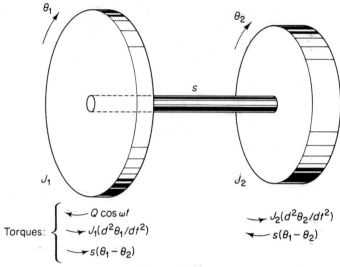

FIGURE 9.32

If we were concerned merely with determining the resonance frequencies it would be sufficient to determine the natural frequencies of the system, but as the way in which the amplitudes at the two discs vary with frequency is also of interest, we will take the excitation into account. If θ_1 and θ_2 are both measured with respect to a position in which the shaft is unstrained, the torques which act on the rotors are as in fig. 9.32, and the equations of motion are:

$$\left. \begin{array}{l} J_1 \dfrac{d^2\theta_1}{dt^2} + s(\theta_1 - \theta_2) = Q \cos \omega t \\[2mm] J_2 \dfrac{d^2\theta_2}{dt^2} - s(\theta_1 - \theta_2) = 0 \end{array} \right\} \quad (9.50)$$

These equations are solved by assuming solutions

$$\theta_1 = \Theta_1 \cos \omega t$$
$$\theta_2 = \Theta_2 \cos \omega t$$

and substituting them in the equations of motion. The amplitudes of motion are then found to be

$$\left. \begin{array}{l} \Theta_1 = \dfrac{\left(\dfrac{s}{J_2} - \omega^2\right)\dfrac{Q}{J_1}}{\omega^2\left(\dfrac{s}{J_1} + \dfrac{s}{J_2} - \omega^2\right)} \\[6mm] \Theta_2 = -\dfrac{\dfrac{s}{J_2} \cdot \dfrac{Q}{J_1}}{\omega^2\left(\dfrac{s}{J_1} + \dfrac{s}{J_2} - \omega^2\right)} \end{array} \right\} \quad (9.51)$$

Figure 9.33 shows the way in which Θ_1 and Θ_2 vary with frequency. Like fig. 9.29, the present graph shows two resonance peaks, but the lower resonance frequency, and hence the lower natural frequency, is now zero. In terms of the free motion of the system, a zero natural frequency means that the system is

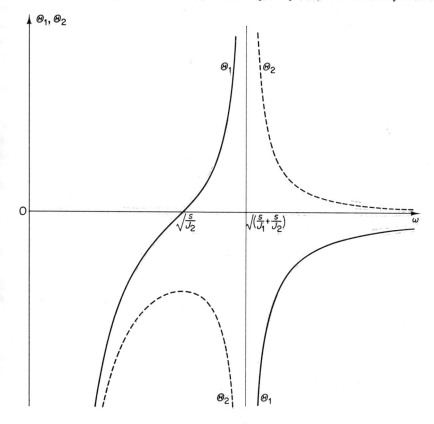

FIGURE 9.33

capable of motion as a rigid body, that is without strain of the flexible members. The equation for free motion of the present system is obtained by omitting the excitation and eliminating either θ_1 or θ_2 from eqns. (9.50). If θ_2 is eliminated the equation is

$$D^2\left(D^2 + \frac{s}{J_1} + \frac{s}{J_2}\right)\theta_1 = 0 \tag{9.52}$$

The general solution of this equation is

$$\theta_1 = A + Bt + C \cos(\omega_2 t + \phi) \tag{9.53}$$

where $\omega_2 = (s/J_1 + s/J_2)^{\frac{1}{2}}$, and A, B, C, and ϕ are the four constants of integration. On substituting this expression for θ_1 in one of eqns. (9.50) (with $Q\cos\omega t$ omitted) θ_2 is found to be

$$\theta_2 = A + Bt - (J_1/J_2)C\cos(\omega_2 t + \phi) \qquad (9.54)$$

The common terms $(A + Bt)$ in eqns. (9.53) and (9.54) represent a steady vibration of the rotors without twisting of the shaft. These terms are associated with the isolated D^2 on the left of eqn. (9.52), which in turn corresponds with the solution $\omega = 0$ of the frequency equation

$$-\omega^2\left(-\omega^2 + \frac{s}{J_1} + \frac{s}{J_2}\right) = 0 \qquad (9.55)$$

A system for which one or more of the natural frequencies is zero is said to be *degenerate*.

Equation (9.55) shows that the non-zero natural frequency for the system is

$$\omega_2 = \left(\frac{s}{J_1} + \frac{s}{J_2}\right)^{\frac{1}{2}} \qquad (9.56)$$

The associated modal slope is determined by substituting

$$\theta_1 = \Theta_1 \cos\omega_2 t \quad \text{and} \quad \theta_2 = \Theta_2 \cos\omega_2 t$$

in either of eqns. (9.50) (with $Q = 0$). It is found that

$$\Theta_2/\Theta_1 = -J_1/J_2 \qquad (9.57)$$

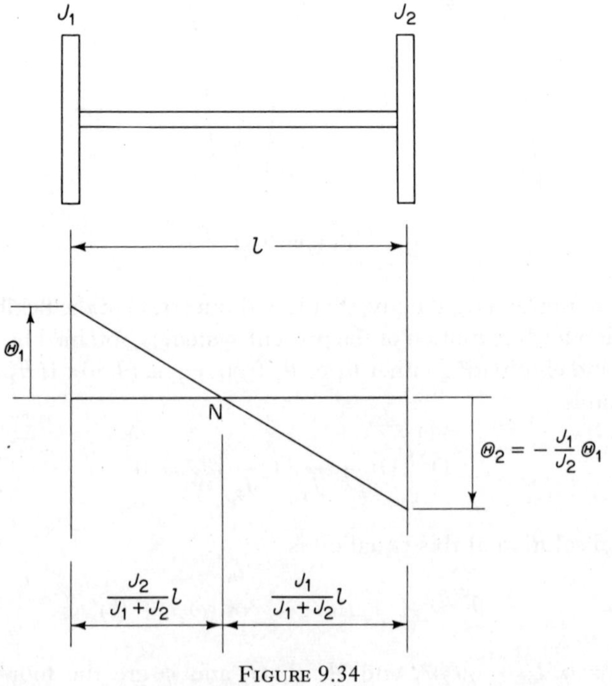

FIGURE 9.34

The amplitude of motion varies linearly along the shaft, as shown in fig. 9.34.

As the natural frequencies of a system are properties of the system and do not depend in any way on the excitation to which the system may be subjected we would expect the system to resonate when excited at the appropriate frequencies, irrespective of the way in which the excitation is applied to the system. Obviously, resonance will occur when $\omega = \omega_2$, whether the excitation is applied to J_1 or to J_2. We will now see how the system responds when the excitation is applied at some point along the length of the shaft, as it might be in a test to determine, experimentally, the natural frequencies of an existing system.

In order to derive the equations of motion it is necessary to consider the equilibrium of the small element of shaft to which the excitation is applied, as well as the equilibrium of the two rotors. If l is the length of the shaft let the excita-

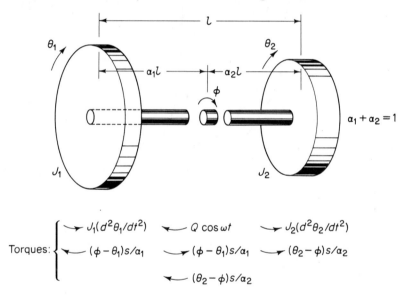

FIGURE 9.35

tion be applied at a section distance $\alpha_1 l$ from one end, fig. 9.35. Let the rotation at this section be ϕ. The twist in the length of shaft $\alpha_1 l$ is $(\phi - \theta_1)$, and its stiffness is s/α_1; the torque in it is therefore $(\phi - \theta_1)s/\alpha_1$. The torque in the other part of the shaft is similarly $(\theta_2 - \phi)s/\alpha_2$. The complete system of torques is as shown in fig. 9.35, and the equations of motion are:

$$(J_1 D^2 + s/\alpha_1)\theta_1 - \phi s/\alpha_1 = 0$$
$$(s/\alpha_1 + s/\alpha_2)\phi - \theta_1 s/\alpha_1 - \theta_2 s/\alpha_2 = Q \cos \omega t$$

and

$$(J_2 D^2 + s/\alpha_2)\theta_2 - \phi s/\alpha_2 = 0$$

Substituting Ω_1^2 for $s/(\alpha_1 J_1)$ and Ω_2^2 for $s/(\alpha_2 J_2)$, and eliminating θ_1 and θ_2

from these equations we find that the amplitude of motion at the section where the excitation is applied is

$$\Phi = \frac{(\Omega_1{}^2 - \omega^2)(\Omega_2{}^2 - \omega^2)Q\alpha_1\alpha_2/s}{\omega^2(\omega_2{}^2 - \omega^2)} \tag{9.58}$$

where ω is the frequency of excitation and ω_2 is the non-zero natural frequency. The graph of Φ against ω shows, in general, two resonance peaks, one at $\omega = 0$ and the other at $\omega = \omega_2 = \sqrt{(s/J_1 + s/J_2)}$, see the full-line curve in fig. 9.36.

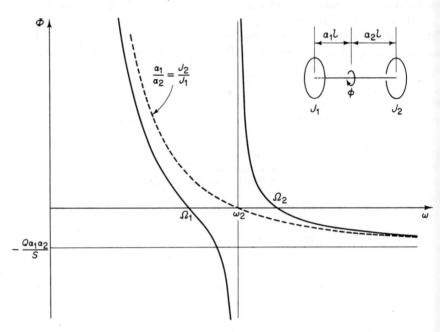

FIGURE 9.36

It can be seen that Φ is zero at two frequencies, Ω_1 when the left-hand portion of the system is acting as a vibration absorber, and Ω_2 when the right-hand portion is acting as a vibration absorber. The values of Ω_1 and Ω_2 change if the torque is applied at a different section along the shaft, because α_1 and α_2 change. A particularly interesting situation occurs when $\alpha_1 = J_2/(J_1 + J_2)$ and $\alpha_2 = J_1/(J_1 + J_2)$, for then $\Omega_1{}^2 = \Omega_2{}^2 = (s/J_1 + s/J_2) = \omega_2{}^2$, and eqn (9.58) reduces to

$$\Phi = \frac{Q\alpha_1\alpha_2}{s\omega^2}(\omega_2{}^2 - \omega^2) \tag{9.58(a)}$$

so that the variation of Φ with ω now follows the broken curve in fig. 9.36. The curve has only one resonance, that is when $\omega = 0$, and the peak at $\omega = \omega_2$ has been completely suppressed. This is because when $\alpha_1/\alpha_2 = J_2/J_1$, the section at which the excitation is applied coincides with the position of the node for vibrations in the second normal mode. Now, in order to excite a particular mode of vibration the excitation must supply energy to the system, and it cannot do this if the point at which the excitation is applied is stationary, hence there is no vibration in the second mode even though the frequency coincides with the second natural frequency.

Some attention must be paid to the possibility of a resonance being suppressed in this way when tests are made to determine, experimentally, the modes of vibration and natural frequencies of a system. If the shaker which applies the exciting force to the system is attached at random to the system there is a chance that it may coincide with a node for one of the modes of vibration, and if this happens the existence of the mode in question may be overlooked.

9.13 Geared systems

A practical example of the two rotor systems which has been studied in section 9.12 is a motor and generator set. Very often the two rotors are not mounted on a single shaft, as we have envisaged, but are on separate shafts which are connected by gearing, fig. 9.37. If the gears and shafting have negligible polar moments of inertia, the configuration of the system at any instant is completely determined if θ_1 and θ_2 and are specified, so that the system has two degrees of freedom. The reasoning and analysis is, however, much simplified if we introduce an additional variable θ_g, the rotation of the gear wheel which is attached to shaft 1. The equations of motion for free vibrations of the two rotors are then

$$J_1 \frac{d^2\theta_1}{dt^2} + s_1(\theta_1 - \theta_g) = 0 \\ J_2 \frac{d^2\theta_2}{dt^2} + s_2(\theta_2 - \theta_g/n) = 0 \quad \quad (9.59)$$

The equilibrium equation for the gears is

$$n \times s_1(\theta_1 - \theta_g) = -s_2(\theta_2 - \theta_g/n) \quad \quad (9.60)$$

If θ_g is eliminated from eqns. (9.59) and (9.60) we find that the resulting pair of equations can be written as

$$J_1 \frac{d^2\theta_1}{dt^2} + s_1^*[\theta_1 - (\theta_2/n)] = 0 \\ \frac{J_2}{n^2} \frac{d^2(\theta_2/n)}{dt^2} - s_1^*[\theta_1 - (\theta_2/n)] = 0 \quad \quad (9.61)$$

where s_1^* is given by the equation

$$\frac{1}{s_1^*} = \frac{1}{s_1} + \frac{1}{(s_2/n^2)} \quad \quad (9.62)$$

If eqns. (9.61) are compared with eqns. (9.50) (ignoring the forcing term $Q \cos \omega t$ in the latter) it can be seen that the geared system is equivalent to a simple ungeared two rotor system having rotors with moments of inertia of J_1 and J_2/n^2 respectively and a shaft of stiffness s_1^*, except that the rotation of rotor 2 is defined by the variable θ_2/n instead of θ_2, fig. 9.37(a). Equation (9.62) shows that the stiffness s_1^* of the equivalent shaft is the same as the overall stiffness of a shaft of stiffness s_1 joined end to end to a shaft of stiffness s_2/n^2.

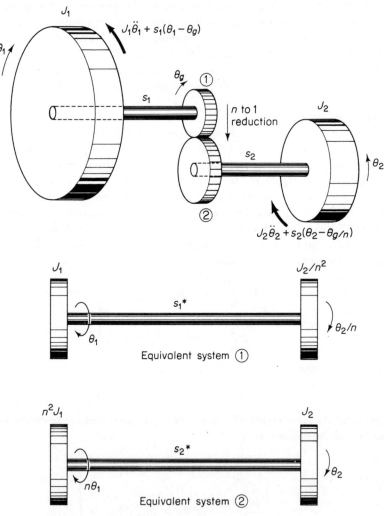

FIGURE 9.37

There is no need to solve eqns. (9.61) formally, as we can deduce immediately from eqn. (9.56) that the non-zero natural frequency for the system is

$$\omega_2 = \left(\frac{s_1^*}{J_1} + \frac{s_1^*}{J_2/n^2}\right)^{\frac{1}{2}} \tag{9.63}$$

The system in fig. 9.37(a) is the equivalent system referred to shaft 1, that is to say we have taken the rotation θ_1 of the shaft 1 as reference and replaced θ_2 by θ_2/n. We could equally well have taken shaft 2 as reference and in this case the equivalent system would have been as in fig. 9.37(b), where s_2^* is given by the equation

$$\frac{1}{s_2^*} = \frac{1}{n^2 s_1} + \frac{1}{s_2} \tag{9.64}$$

The non-zero natural frequency is now given by

$$\omega_2 = \left(\frac{s_2^*}{n^2 J_1} + \frac{s_2^*}{J_2}\right)^{\frac{1}{2}} \tag{9.65}$$

The frequency given by eqn. (9.65) is, of course, identical with that obtained from eqn. (9.63), because $s_2^* = n^2 s_1^*$.

The concept of an equivalent system can be extended to cover several reductions, and to take into account intermediate rotors, and the moments of inertia of the gears. The general rule is:

Select any shaft as reference and leave its stiffness and any moments of inertia carried by it unaltered. If the speeds of the other shafts are p_1, p_2, p_3, etc., times the speed of the reference shaft then the stiffness of each must be multiplied by p^2 and the moments of inertia carried by it must be multiplied by p^2.

The equivalent gearless system thus formed has the same natural frequencies as the original geared system.

9.14 Principal coordinates — coupling

In analysing the systems which we have just considered, there has been no problem about the choice of coordinates with which to define the configuration of the system. The displacement of the masses from their equilibrium position in one case, fig. 9.27, and the rotation of the discs in the other, fig. 9.32, seem to be the obvious coordinates to use. Now consider the system in fig. 9.38: it consists essentially of a single body supported on two springs. Assuming that there is a mechanism, not shown, which prevents sideways displacement of its

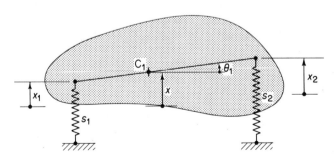

FIGURE 9.38

centre of gravity, the body has two degrees of freedom, and two coordinates are necessary to specify its position. The displacement of the body is fully defined by the extensions x_1 and x_2 of the supporting springs. It is equally well defined by the displacement x of the centre of gravity and the rotation θ. The equations of motion will be different for these two sets of coordinates, but the natural frequencies and normal modes, which are properties of the system, are not going to be affected by the way in which we choose to look at it. The choice of coordinate system is decided simply as a matter of convenience. The practical example of a system of this type which we shall have in mind in making the analysis is the motor-car suspension (somewhat simplified) where it is natural to think of the motion as a combination of bouncing and pitching. Accordingly, we shall choose as our coordinate system the displacement x of the centre of gravity and rotation θ.

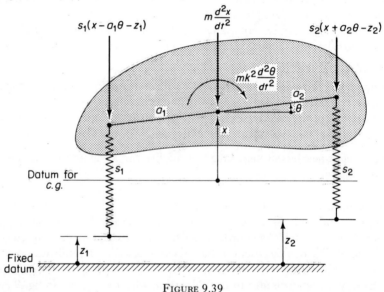

FIGURE 9.39

In the case of a motor-car, vibrations are excited as a result of unevenness in the road surface so that we will take as excitation displacements z_1 and z_2 as shown in fig. 9.39. The equations of motion obtained by equating the forces which act vertically on the body, and the moments about the centre of gravity are:

$$m\frac{d^2x}{dt^2} + (s_1 + s_2)x = (a_1 s_1 - a_2 s_2)\theta + (s_1 z_1 + s_2 z_2)$$

and

$$mk^2\frac{d^2\theta}{dt^2} + (a_1^2 s_1 + a_2^2 s_2)\theta = (a_1 s_1 - a_2 s_2)x - (a_1 s_1 z_1 - a_2 s_2 z_2)$$

respectively.

(9.66)

MECHANICAL VIBRATIONS 413

The interesting thing about these two equations is that if $a_1 s_1 = a_2 s_2$ the terms in θ and x on the right-hand sides disappear, so that, instead of a pair of simultaneous equations involving x and θ as unknowns, there are two independent equations:

$$\left.\begin{array}{l} m \dfrac{d^2 x}{dt^2} + (s_1 + s_2) x = s_1 z_1 + s_2 z_2 \\[6pt] mk^2 \dfrac{d^2 \theta}{dt^2} + (a_1{}^2 s_1 + a_2{}^2 s_2) \theta = -a_1 s_1 z_1 + a_2 s_2 z_2 \end{array}\right\} \quad (9.67)$$

This independence is emphasized if we consider free motion (i.e., z_1 and z_2 both zero) when the equations are

$$\left.\begin{array}{l} m \dfrac{d^2 x}{dt^2} + (s_1 + s_2) x = 0 \\[6pt] mk^2 \dfrac{d^2 \theta}{dt^2} + (a_1{}^2 s_1 + a_2{}^2 s_2) \theta = 0 \end{array}\right\} \quad (9.68)$$

When the equations of motion are independent of each other the coordinates x and θ are said to be *uncoupled*. They may also be called *principal coordinates* because the principal (normal) modes are defined by

$$\left.\begin{array}{l} x = X \cos \omega_1 t \\ \theta = \Theta \cos \omega_2 t \end{array}\right\} \quad (9.69)$$

where $\quad \omega_1{}^2 = \dfrac{s_1 + s_2}{m} \quad$ and $\quad \omega_2{}^2 = \dfrac{a_1{}^2 s_1 + a_2{}^2 s_2}{mk^2} \quad (9.70)$

It must be emphasized that the modes defined by eqns. (9.69) occur as a result of the special condition $a_1 s_1 = a_2 a_2$, and that although the pitching (θ) and heaving (x) motions have visual significance, and consequently it seems natural to describe the motion as being one or the other, it is wrong to do so unless the condition is fulfilled.

The results of making $a_1 s_1 = a_2 s_2$ are of interest and importance. Consider, in particular, the second of eqns. (9.67) which becomes

$$mk^2 \dfrac{d^2 \theta}{dt^2} + (a_1{}^2 s_1 + a_2{}^2 s_2) \theta = a_1 s_1 (z_2 - z_1) \quad (9.71)$$

If $z_1 = z_2$, the right-hand side of eqn. (9.71) is zero so that there is no excitation of the pitching mode of vibration. The passage of a motor-car over a bump in the road can be regarded as a step change in z_1 followed by a step change in z_2 of the same amplitude. If the car is moving at a high speed the time interval between the inputs is small compared with the periodic times for the two normal modes so that the change in z_1 and z_2 can be considered to be simultaneous. The resulting oscillations of the vehicle are therefore almost entirely in the heaving mode with very little undesirable pitching motion.

14*

414 Dynamics of Mechanical Systems

The absence of pitching response when $z_1 = z_2$ (given that $a_1 s_1 = a_2 s_2$) holds whatever the way in which z_1 and z_2 vary with time. This means that if the system is supported on a solid horizontal platform which moves vertically so that $z_1 = z_2 = Z \cos \omega t$ there will be no resonance when $\omega = \omega_2$, even though ω_2 is one of the natural frequencies of the system. The resonance at $\omega = \omega_1$ is similarly suppressed if the excitation is a rocking motion such that $s_1 z_1 = -s_2 z_2$. This is a further example of the principle which we discussed in section 9.12 in relation to a simple two rotor system: it is always possible to suppress a particular resonance by an appropriate choice of the method of excitation.

9.15 Equal natural frequencies

We have described a system for which one of the natural frequencies is zero as being degenerate. Another form of degeneracy occurs when the natural frequencies are equal. It is not always possible for this to happen; inspection of eqn. (9.40) shows that the two natural frequencies for the two-mass system in fig. 9.27 must be different. In the case of the system in fig. 9.38, however, if in addition to making $a_1 s_1 = a_2 s_2$ we make $a_1 a_2 = k^2$ then the two natural frequencies, eqn. (9.70), both equal $\sqrt{[(s_1 + s_2)/m]}$. This means that there is only one resonance frequency, notwithstanding the two degrees of freedom. There are still two distinct normal modes, the uncoupled pitching and heaving modes, but since they are at the same frequency they can be combined in any proportions and the resultant motion will still be simple harmonic.

9.16 Two degrees of freedom — concluding comments

Our study of vibrations of lightly damped systems with two degrees of freedom has revealed that there are important differences in behaviour between them and systems with one degree of freedom. The latter have one natural frequency and their response to a transient disturbance is a simple harmonic vibration at that frequency. When the system is subjected to a harmonic excitation, resonance occurs when the frequency of excitation coincides with the natural frequency, irrespective of the way in which the system is excited (e.g., constant excitation, seismic excitation, etc.). Systems with two degrees of freedom have two natural frequencies and two normal modes of vibration. This means that although it is possible for such systems to execute free simple harmonic vibrations they will, in general, do so only for appropriate sets of initial conditions. For other sets of initial conditions the motion is made up of two simple harmonic motions and is usually non-periodic. An exception to this occurs when the two natural frequencies are equal, and in this case the free motion is simple harmonic.

Harmonic excitation of a system with two degrees of freedom usually reveals two resonance frequencies, one corresponding to each natural frequency. It is, however, possible to choose the excitation in such a way that one of the resonances is suppressed, so that the system may appear to have only one resonance.

It is natural to speculate at this point as to whether there are such fundamental differences between the behaviour of systems of two and three degrees of freedom, as there are between those of one and two degrees of freedom. In fact there are not, and the study of higher order systems does not lead to conclusions which are in any way inconsistent to those which we have already reached. It is evident that the amount of algebra involved is going to increase appreciably as more degrees of freedom are introduced, and that the methods of analysis which we have used so far might become unmanageable. For this reason, the study of vibration theory in relation to linear higher order systems is concerned mainly with the development of practical methods for calculating natural frequencies and normal modes. Before reviewing some aspects of these methods we will consider some particular higher order systems for which no special computational techniques are necessary.

9.17 Torsional oscillations of three rotors on a flexible shaft

The three rotor system depicted in fig. 9.40 has three degrees of freedom. Normally, the frequency equation for a system with three degrees of freedom is a cubic equation in ω^2, and it is not possible to deduce explicit expressions for the three natural frequencies. With the system shown, however, the possibility of a rigid body rotation of the whole system means that the lowest natural frequency is zero, and so only two non-zero natural frequencies are left to be determined. The equations of motion of the three rotors are

$$\left. \begin{array}{l} J_1 \dfrac{d^2\theta_1}{dt^2} + s_1(\theta_1 - \theta_2) = 0 \\[6pt] J_2 \dfrac{d^2\theta_2}{dt^2} - s_1(\theta_1 - \theta_2) + s_2(\theta_2 - \theta_3) = 0 \\[6pt] J_3 \dfrac{d^2\theta_3}{dt^2} - s_2(\theta_2 - \theta_3) = 0 \end{array} \right\} \quad (9.72)$$

Assuming
$$\theta_1 = \Theta_1 \cos \omega t$$
$$\theta_2 = \Theta_2 \cos \omega t$$
$$\theta_3 = \Theta_3 \cos \omega t$$

we find from the first and third of eqns. (9.72), respectively,

$$\Theta_1 = \frac{s_1 \Theta_2}{s_1 - J_1 \omega^2}; \qquad \Theta_3 = \frac{s_2 \Theta_2}{\Theta_2 - J_2 \omega^2} \qquad (9.73)$$

On substituting for Θ_1 and Θ_3 in the second of eqns. (9.72) we find that the resulting equation can be arranged thus:

$$\omega^2 \left[\omega^4 - \left(\frac{s_1}{J_1} + \frac{s_1 + s_2}{J_2} + \frac{s_2}{J_3} \right) \omega^2 + s_1 s_2 \left(\frac{1}{J_1 J_2} + \frac{1}{J_2 J_3} + \frac{1}{J_3 J_1} \right) \right] = 0$$
$$(9.74)$$

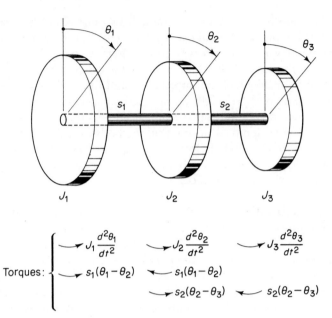

FIGURE 9.40

This equation is satisfied if $\omega = 0$ (rigid body rotation) or if the quadratic expression for ω^2 within the square brackets is zero. The general solution of the quadratic equation

$$\omega^4 - \left(\frac{s_1}{J_1} + \frac{s_1 + s_2}{J_2} + \frac{s_2}{J_3}\right)\omega^2 + s_1 s_2 \left(\frac{1}{J_1 J_2} + \frac{1}{J_2 J_3} + \frac{1}{J_3 J_1}\right) = 0 \quad (9.75)$$

in terms of the given parameters must clearly be very cumbersome, but there is no difficulty in evaluating the two frequencies when numerical values are assigned to the shaft stiffness and the moments of inertia of the rotors. The relative amplitudes of the rotors are found by substituting for ω^2 in eqns. (9.73).

The system shown in fig. 9.41 has two degrees of freedom and two natural

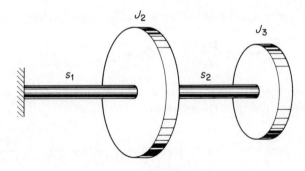

FIGURE 9.41

MECHANICAL VIBRATIONS 417

frequencies. The two frequencies are found from eqn. (9.75) by putting $J_1 = \infty$. The frequency equation is then

$$\omega^4 - \left(\frac{s_1 + s_2}{J_2} + \frac{s_2}{J_3}\right)\omega^2 + \frac{s_1 s_2}{J_2 J_3} = 0 \qquad (9.76)$$

9.18 Symmetrical systems
If the system in fig. 9.40 is symmetrical about the central rotor so that $s_1 = s_2 = s$ (say) and $J_3 = J_1$ eqn. (9.75) becomes

$$\omega^4 - 2s\omega^2\left(\frac{1}{J_1} + \frac{1}{J_2}\right) + \frac{s}{J_1}\left(\frac{2}{J_2} + \frac{1}{J_1}\right) = 0$$

The roots of this equation are

$$\omega^2 = s/J_1 \quad \text{and} \quad s(1/J_1 + 2/J_2)$$

On substituting $\omega^2 = s/J_1$ in eqns. (9.73) we find that $\Theta_2 = 0$. If the central rotor does not move, the torques applied to it by the two shafts must at all times be equal and opposite to each other, and as the system is symmetrical this means that $\Theta_3 = -\Theta_1$.

When $\omega^2 = s(1/J_1 + 2/J_2)$ we have from eqns. (9.73)

$$\frac{\Theta_1}{\Theta_2} = \frac{\Theta_3}{\Theta_2} = -\frac{J_2/2}{J_1}$$

A modal shape for a symmetrical system, fig. 9.42, is therefore either symmetrical or antisymmetrical. This conclusion must obviously be true for any symmetrical system and is important. It is important because advantage can be taken of it to simplify the calculation of normal modes. If, for example, we know that there is a node at the centre of a system we can regard the system as being anchored at that point. If it is anchored there it will make no difference to the motion of one half of the system if the other half is discarded. Accordingly, for antisymmetrical motion (i.e., a node at the centre) the system shown in fig. 9.43(a) must vibrate with the same frequency as the half-system shown in fig. 9.43(b) for which the natural frequency is known to be $\sqrt{(s/J_1)}$.

When the motion is symmetrical the torque transmitted across the central plane of the system must be zero — if it is not zero, what is its direction? If there is no torque transmitted between two halves of a system the system can be separated at the centre and neither half will know the difference. For symmetrical motion, the system in fig. 9.43(a) vibrates in the same way as the half-system shown in fig. 9.43(c), for which the non-zero natural frequency is known from eqn. (9.56) to be $\sqrt{[s(1/J_1 + 2/J_2)]}$. This example illustrates very well the good computational maxim that it is generally easier to do two simple jobs than one hard one. The point is perhaps more forcibly made in relation to the symmetrical five rotor system shown in fig. 9.44. Determination of the four non-zero natural frequencies of the system by way of the equations of

418 Dynamics of Mechanical Systems

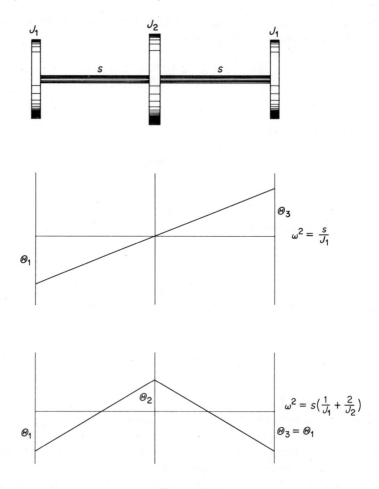

FIGURE 9.42

motion and the frequency equation would be a very heavy task. It is very much easier to use eqn. (9.76) to determine the two natural frequencies for antisymmetrical motion, and eqn. (9.75) (with J_1 replaced by $J_1/2$) for the non-zero frequencies for symmetrical motion.

9.19 Holzer's method

The method of solution which we have used in previous sections for the determination of natural frequencies and normal modes has three phases. First, the equations of motion are set up, then the frequency equation is derived; finally, the frequency equation is solved. In general terms, the amount of work involved in setting up the equations of motion is roughly proportional to the number of degrees of freedom involved. The other two tasks, however, become relatively much more onerous. The work involved in setting up and solving the

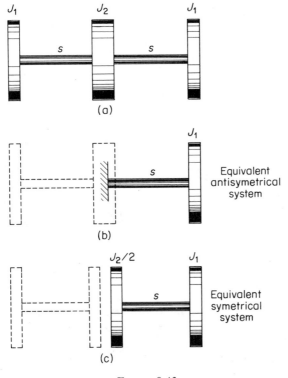

FIGURE 9.43

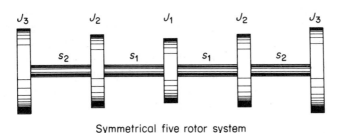

Symmetrical five rotor system

FIGURE 9.44

frequency equation can be avoided by guessing the way in which a system vibrates and substituting the guess in the equation of motion to see if it is correct. It is not necessary to guess both the frequency and the modal shape, one implies the other. Obviously, the checking in the original equations of motion must be done in an organized way if the method is to be practicable. We shall introduce and elaborate on one such method — the Holzer method. Holzer's method applies specifically to the determination of natural frequencies and normal modes of torsional vibrations of shaft/rotor systems such as we have just studied. It was developed with particular reference to the torsional vibrations of engine crankshafts.

The basis of the Holzer method is that it takes an estimated value of a natural frequency and tests to see if the equations of motion are satisfied. The way in

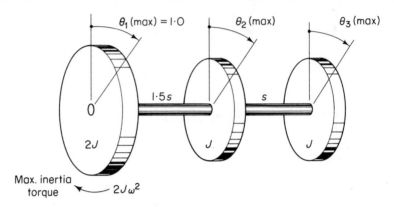

FIGURE 9.45

which this is done will be described with reference to the particular system shown in fig. 9.45. In general terms, the method is as follows:

The left-hand disc is assumed to oscillate with simple harmonic motion of unit amplitude at a particular (estimated or guessed) frequency. The maximum inertia torque ($2J\omega^2$ in the same direction as θ) is calculated and is the maximum torque carried by the adjacent sections of shaft. The maximum angle of twist in the shaft can therefore be calculated ($2J\omega^2/1\cdot5s$) and subtracted from the unit amplitude assumed at the first disc to give the amplitude at the central disc. The inertia torque at the central rotor follows and, after proper allowance for the torque in the left-hand shaft, the torque — and hence the angle of twist of the right-hand shaft — is calculated, and so the calculation moves on to the third disc. If the frequency has been estimated correctly, the torque in the right-hand section of shaft will exactly counterbalance the inertia torque of the third disc. Any lack of equilibrium means that the system is not capable of executing free simple harmonic vibrations at the assumed frequency.

Normally, the numerical values of the moments of inertia of the discs and the torsional stiffnesses of the shafts are known, and the computations are entirely numerical. However, in the hope that the steps in the calculation will be more readily followed we shall work in terms of the parameters J and s given in fig. 9.45. This means that our initial estimate for the natural frequency must be expressed in terms of J and s, and we are brought to the problem of making this estimate. Our thoughts might run on these lines: when the system is vibrating at its lower non-zero natural frequency there will be one node. As the system is not symmetrical the single node will not be located at the central disc; even so we would not expect the central disc to move very much so that perhaps we can, without too much error, ignore it. The system can then be regarded as consisting of two discs connected by a shaft of stiffness s', where s' is found from the equation

$$\frac{1}{s'} = \frac{1}{1 \cdot 5s} + \frac{1}{s}$$

to be $0 \cdot 65s$. The natural frequency of our supposed equivalent system is deduced from eqn. (9.56):

$$\omega = \left[0 \cdot 6s \left(\frac{1}{2J} + \frac{1}{J} \right) \right]^{\frac{1}{2}}$$

so that
$$\omega^2 = 0 \cdot 9s/J$$

The calculations which must now be made to check this estimate are most conveniently set out in tabular form:

Disc No.	(1) Moment of inertia I	(2) $I\omega^2$	(3) Amplitude Θ	(4) Inertia torque $I\omega^2\Theta$	(5) $\Sigma I\omega^2\Theta = Q$	(6) Shaft stiffness s	(7) $\delta\Theta = \dfrac{Q}{s}$
1	$2J$	$1 \cdot 8s$	$1 \cdot 0$			$1 \cdot 5s$	
2	J	$0 \cdot 9s$				s	
3	J	$0 \cdot 9s$			*	—	—

Columns (1) and (6) show the known values of the moments of inertia and shaft stiffnesses respectively and have been completed for the given example. Column (2) contains the corresponding entries in column (1) multiplied by the estimated value of ω^2. The first entry in column (3) shows that we intend to start with an assumed amplitude of unity at the first disc. The last entry in column (5), indicated by an asterisk, will give the residual torque which is zero if the natural frequency has been estimated correctly.

The table is now completed line by line. The inertia torque at disc No. 1 is $1 \cdot 8s \times 1 \cdot 0$ and is the total torque carried by the first length of shaft. The angle of twist in that shaft is therefore $1 \cdot 8s/1 \cdot 5s = 1 \cdot 2$ radians. The maximum angular displacement of disc No. 2 is therefore $(1 \cdot 0 - 1 \cdot 2) = -0 \cdot 2$ radians and the inertia torque is $-0 \cdot 18s$. The total torque carried by the second length of shaft is $1 \cdot 8s - 0 \cdot 18s = 1 \cdot 62s$, and so the angle of twist is $1 \cdot 62$ radians. The third line is completed similarly and we find that the residual torque is $-0 \cdot 02s$.

Disc No.	(1) I	(2) $I\omega^2$	(3) Θ	(4) $I\omega^2\Theta$	(5) $\Sigma I\omega^2\Theta$	(6) s	(7) $\delta\Theta$
1	$2J$	$1\cdot 8s$	$1\cdot 0$	$1\cdot 8s$	$1\cdot 8s$	$1\cdot 5s$	$1\cdot 2$
2	J	$0\cdot 9s$	$-0\cdot 20$	$-0\cdot 18s$	$1\cdot 62s$	s	$1\cdot 62$
3	J	$0\cdot 9s$	$-1\cdot 82$	$-1\cdot 64s$	$-0\cdot 02s$	—	—

As the residual torque is small compared with the inertia torques, due to the individual discs, we can deduce that the error in the estimated value of ω is small, and we may be satisfied with the result as it stands. If the residual torque is too large to be accepted it is necessary to repeat the calculation for a different value of ω, preferably one which causes the residual torque to change sign so that the exact value of ω can then be obtained by interpolation. The modal shape associated with the natural frequency which we have just verified is defined by the entries in column (3) of the table. The initial assumption that the motion of the central disc is small is shown to be reasonable.

It can readily be seen that no change in principle is involved in applying this method to systems which have more degrees of freedom; each extra disc means an extra line in the Holzer table, but the basic method is the same. The argument which was used to arrive at an initial estimate of the natural frequency obviously does not apply if there are more discs. A somewhat different method of arriving at an estimated frequency is given in section 9.22.

Except for the problem of making the initial estimate, the method of calculation is exactly the same for other natural frequencies.

The residual torque is the resultant inertia torque acting on the system and is equal and opposite to the torque which must be applied to the last disc to cause the system to execute forced vibrations at the assumed frequency with unit amplitude at the first disc. Returning to the system in fig. 9.45, the way in which Θ_3 varies with frequency is shown in fig. 9.46(a), the three points at which the graph crosses the frequency axis being the three natural frequencies of the system. Figure 9.46(b) shows the variation of Θ_3 with frequency. The two values of ω for which the amplitude at disc 3 is zero are the two natural frequencies when the right-hand disc is rigidly held. This suggests the way in which the Holzer analysis is applied to shaft and disc systems where one end of the shaft is fixed. However, the immediate purpose of drawing fig. 9.46(b) is to lead up to the resonance curves in fig. 9.46(c). This curve shows the way in which the amplitude of motion induced at disc 3 by a unit exciting torque applied at disc 3 varies with frequency. It is obtained by dividing the ordinates of fig. 9.46(b) by the corresponding ordinates of the residual torque curve with the sign reversed. The change of sign is necessary because the driving torque must be in equilibrium with the inertia torques.

Although it is somewhat indirect this method of calculating a resonance curve has much to commend it, in that it circumvents the need to derive and

MECHANICAL VIBRATIONS 423

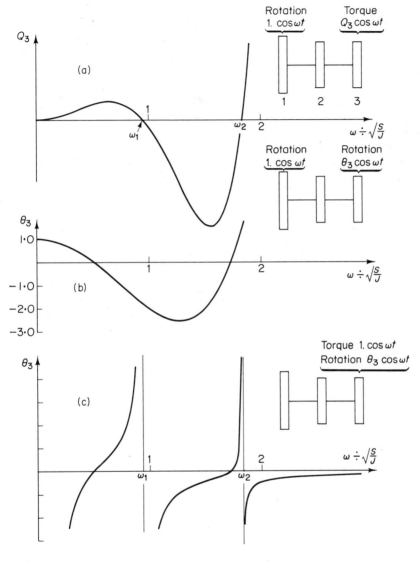

FIGURE 9.46

solve the equations of motion. If it is done by hand, the computation is of course lengthy, but if a computer is available the programme is very short and simple.

9.20 Composite systems
We have seen that the analysis of automatic control systems is very much concerned with the prediction of the overall behaviour of a system from a knowledge of the way in which the component parts behave. The basic theorem used

424 Dynamics of Mechanical Systems

in this prediction is that the transfer operator for a string of non-interacting elements is the product of the individual transfer operators. The restriction to non-interacting elements was crucial to the subsequent development of the argument and it was pointed out particularly that this restriction precluded the possibility of applying the subsequent analysis to mechanical vibrations. Nevertheless, the feeling persists that if modifications are made to a system whose vibrational properties have already been analysed, it ought to be possible to use that analysis in predicting the behaviour of the modified system, and that the work involved ought to be less than that involved in repeating the whole calculation from scratch.

As an example, we will determine the natural frequencies of the system shown in fig. 9.47, which can be considered as consisting of two subsystems

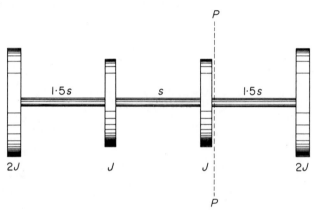

FIGURE 9.47

which are connected together at the junction P–P. The system to the left is identical with that shown in fig. 9.45 and is one for which the response curves shown in fig. 9.46 have already been determined. It is fig. 9.46(c) which will be of particular interest. This curve may be represented by the equation

$$\Theta_3 = Q_3 f(\omega)$$

or
$$\Theta_3/Q_3 = f(\omega) \qquad (9.77)$$

It would be legitimate to refer to $f(\omega)$ as the transfer function of the subsection, but in view of the different context it is preferable to use the term *receptance*.

Because the subsystem to the right of the junction P–P is simple, it is convenient to express its receptance in algebraic form. It can readily be shown from the equations of motion that

$$\Phi_3/T_3 = \left(\omega^2 - \frac{1\cdot 5s}{2J}\right)\bigg/ 1\cdot 5s\omega^2$$

$$= g(\omega) \qquad (9.78)$$

When the system is vibrating freely there is no external torque applied at the junction P–P so that

$$Q_3 + T_3 = 0 \tag{9.79}$$

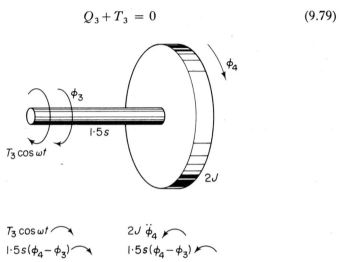

FIGURE 9.48

Furthermore, the displacements of the two subsystems must be identical at this point so that

$$\Theta_3 = \Phi_3 \tag{9.80}$$

Equations (9.77) to (9.80) enable us to deduce that for free vibrations of the whole system

$$f(\omega) = -g(\omega) \tag{9.81}$$

The frequencies at which eqn. (9.81) is satisfied are determined by plotting $f(\omega)$ and $-g(\omega)$ with respect to the same coordinate axes and reading off the values of ω at which the curves cross each other. From fig. 9.49, the curves intersect at infinity when $\omega = 0$, and as indicated when $\omega = \omega_1$, ω_2, or ω_3. The modal shapes for each of these frequencies is obtained by drawing up a Holzer table for the whole system. It is suggested that the reader does this as an exercise, and verifies that the modal shapes do assume the symmetrical and antisymmetrical forms which would be expected for a symmetrical system.

This example is intended to do no more than introduce the notion of receptance and to indicate how the vibrational properties of a composite system can be determined from the properties of the components. The particular practical application which we have in mind is an investigation of the way in which the natural frequencies and normal modes of, say, an aircraft engine and propeller system alter when different propellers are used. For further developments of the method itself the reader is referred elsewhere.

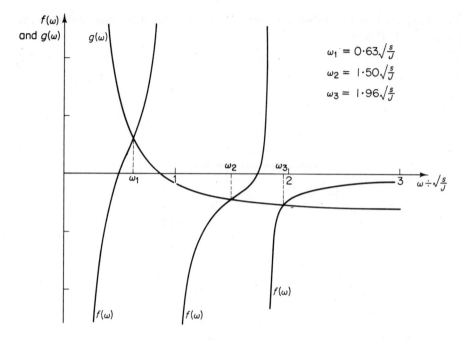

FIGURE 9.49

9.21 Vibrations of beams

The analysis of the vibration of beams is fundamentally more complicated than the problem of torsional vibrations in shafts. The reason for this is that at any cross-section of the shaft of a torsional system only two variables are significant, the torque and the angular displacement; with a beam there are four variables to be taken into account, the shear force, the bending moment, the slope, and the deflection. We shall consequently review the vibration of beams in the light of the study which we have already made of torsional vibrations but will carry the study to somewhat less depth. As with our studies of torsional vibrations we shall, for the time being, concentrate on those systems for which it is reasonable to ignore the mass of the beam itself in relation to the masses which it supports. The simplest type of structure which we can consider in this category is a single concentrated mass supported on a uniform beam or cantilever, fig. 9.50.

In each of the examples given the beam applies a restoring force to the mass when it is deflected from its equilibrium position. The force is proportional to the displacement x and may be denoted by sx where s is the stiffness of the structure measured at the mass. In standard cases, such as those shown in fig. 9.50, the stiffness in a particular case is readily deduced from standard formulae given in textbooks on theory of structures. The deflection produced at the end of a cantilever of length l, second moment of area of cross-section I and Young's modulus E, by a transverse force F, is $\delta = Fl^3/3EI$. The stiffness, which is the force required to cause unit deflection, is $F/\delta = 3EI/l^3$. Other expressions are given in fig. 9.50. In non-standard cases, as for example, when the beam is of varying cross-section along its length, it is necessary to solve the differential equation for the deflection of the beam in order to determine the

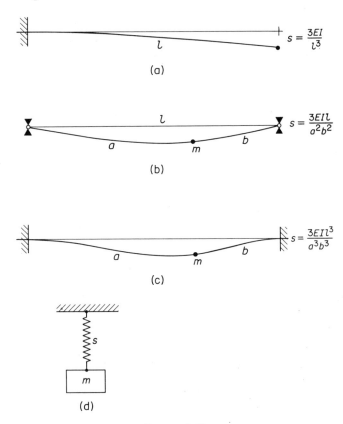

FIGURE 9.50

stiffness. If the beam actually exists the stiffnesses can be determined experimentally by measuring the deflection caused by a known force.

As far as the motion of the mass is concerned the systems in fig. 9.50(a), (b), and (c) are each equivalent to the simple spring/mass system of fig. 9.50(d), for which the natural frequency is $\omega = \sqrt{(s/m)}$. The problem of determining the natural frequency of a simple concentrated mass supported on a light elastic beam is therefore relatively trivial once the appropriate stiffness has been calculated.

If a light beam carries more than one concentrated mass, as in the case of the cantilever in fig. 9.51(a), the ideal of stiffness must be generalized. If the beam is simply supported at mass 2 and subjected to a static transverse load F_1 at mass 1, as in fig. 9.51(b), the displacement x_1 at mass 1 and the reaction R_2

at mass 2 are both proportional to F_1. Phrasing this somewhat differently, the load F_1 and the reaction R_2 are both proportional to the displacement x_1 which has been imposed on the system. Let

$$F_1 = s_{11}x_1 \quad \text{and} \quad R_2 = s_{21}x_1$$

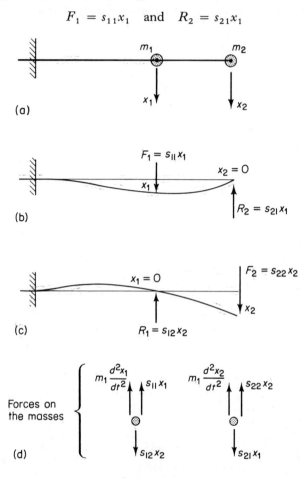

FIGURE 9.51

where s_{11} and s_{21} are constants which depend on the dimensions of the beam and its Young's modulus, and are known as *stiffness coefficients*. If the beam is now supported at mass 1 and the force is applied at mass 2 we have

$$F_2 = s_{22}x_2 \quad \text{and} \quad R_1 = s_{12}x_2$$

By virtue of the principle of superposition the two situations depicted in fig. 9.51(b) and (c) can be combined in any proportions. The forces which must be applied to the beam to cause the displacements x_1 and x_2 simultaneously are $(F_1 - R_1)$ and $(F_2 - R_2)$. The forces applied to the masses by the beam when the system is in motion are $-(F_1 - R_1)$ and $-(F_2 - R_2)$, and together with inertia effects make up the force system shown in fig. 9.51(d), so that for free vibrations of the system we have:

$$\left.\begin{array}{l} m_1 \dfrac{d^2 x_1}{dt^2} + s_{11} x_1 - s_{12} x_2 = 0 \\ \\ m_2 \dfrac{d^2 x_2}{dt^2} - s_{21} x_1 + s_{22} x_2 = 0 \end{array}\right\} \quad (9.82)$$

These equations yield

$$\omega^4 - \left(\dfrac{s_{11}}{m_1} + \dfrac{s_{22}}{m_2}\right)\omega^2 + \dfrac{s_{11} s_{22} - s_{12} s_{21}}{m_1 m_2} = 0 \quad (9.83)$$

as the frequency equation.

The stiffness coefficients may be calculated by the standard methods of analysing the bending of beams which are to be found in any elementary textbook on the theory of structures. Details of the various methods used need not concern us here, it can be expected that, except when standard formulae can be used, the calculations would be fairly heavy. It can also be expected that the elementary approach which has just been given becomes impracticable when a beam carries several concentrated masses. Various practical methods exist for calculating the natural frequencies and normal modes of multi-mass systems, but most of these are beyond the scope of what it would be reasonable to present here. This study of the vibration of multi-degree of freedom systems will be concluded with a method which utilizes an approach which is entirely different from the method of setting up and solving the equations of motion which have hitherto been studied.

9.22 Energy method and Rayleigh's principle

The equation of motion for an undamped single spring/mass system executing small free vibrations about its equilibrium position is

$$m \dfrac{d^2 x}{dt^2} + sx = 0 \quad (9.84)$$

Now

$$\dfrac{d^2 x}{dt^2} = \dfrac{d}{dt}\left(\dfrac{dx}{dt}\right) = \dfrac{dx}{dt}\dfrac{d}{dx}\left(\dfrac{dx}{dt}\right) = \dfrac{1}{2}\dfrac{d}{dx}\left(\dfrac{dx}{dt}\right)^2$$

so that eqn. (9.84) becomes

$$\dfrac{d}{dx}\left[\tfrac{1}{2} m \left(\dfrac{dx}{dt}\right)^2\right] + sx = 0$$

and is in a form which can be integrated directly with respect to x to give

$$\tfrac{1}{2} m \left(\dfrac{dx}{dt}\right)^2 + \tfrac{1}{2} s x^2 = \text{constant} \quad (9.85)$$

The first term in the left-hand side of this equation is the kinetic energy of the mass, and the second represents the change in potential energy of the system due to its displacement from the equilibrium position. It is tempting to refer to the second term as a strain energy term but to do so can cause confusion and is often incorrect. The strain energy stored in a spring is $\frac{1}{2}se^2$ where e is the extension, or compression, of the spring from its unstrained state; x, on the other hand, is the change in length from the equilibrium state. Unless the spring happens to be unstrained when the system is in its equilibrium state, x and e are different quantities. Let us consider two examples: in fig. 9.52(a) the

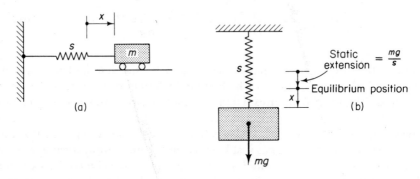

FIGURE 9.52

mass is supported on horizontal frictionless bearings and when it is in equilibrium the spring is unstrained, and in this case it is correct to refer to $\frac{1}{2}sx^2$ as the strain energy of the system. In fig. 9.52(b) the extension of the spring in the equilibrium position is mg/s, and when the mass is displaced a further amount x the strain energy is

$$\tfrac{1}{2}s(x + mg/s)^2$$
$$= \tfrac{1}{2}sx^2 + mgx + \tfrac{1}{2}mg/s$$

This, however, is only part of the potential energy, we must also take into account the potential energy of the mass due to its position in the gravitational field. The displacement x causes this component to be diminished by the amount mgx. Thus the net change in potential energy is $\frac{1}{2}sx^2$. The last term $\frac{1}{2}mg/s$ in the expression for the strain energy is a constant and can be regarded simply as part of the right-hand side of eqn. (9.85). It is to be noted that the change in potential energy due to the displacement x must be positive, due to x being squared. As the datum from which the gravitational potential energy is measured is quite arbitrary, it is convenient to define the total potential energy as being zero when it has its minimum value, that is when x is zero. Equation (9.85) tells us that the sum of the kinetic and potential energies is constant so that the kinetic energy has its maximum value when the potential energy is a minimum, i.e., zero. Conversely, the potential energy is a maximum when the kinetic energy is a minimum. When the potential energy is a maximum, x is a maximum, so that dx/dt is zero. It follows that:

$$\text{maximum kinetic energy} = \text{maximum potential energy} \quad (9.86)$$

This equation is the basis of the energy method for determining natural frequencies. Assuming that the motion of the mass of either of the systems shown in fig. 9.52 is

$$x = X \cos \omega_n t$$

the maximum velocity is $\omega_n X$. Equation (9.86) therefore gives

$$\tfrac{1}{2} m(\omega_n X)^2 = \tfrac{1}{2} s X^2$$

so that
$$\omega_n^2 = s/m$$

The analysis is similar for other systems with one degree of freedom. In the case of the torsional system in fig. 9.53, for a displacement $\theta = \Theta \cos \omega t$, the

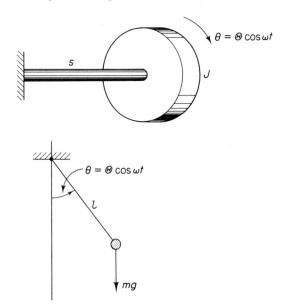

FIGURE 9.53

maximum kinetic energy is $\tfrac{1}{2} J(\omega_n \Theta)^2$ and the maximum potential energy is $\tfrac{1}{2} s \Theta^2$ giving $\omega_n^2 = s/J$. With the simple pendulum the potential energy is due to gravity alone. Its maximum value is $mgl(1 - \cos \Theta)$, which for small values of Θ is approximately $\tfrac{1}{2} mgl \Theta^2$. The maximum kinetic energy $\tfrac{1}{2} m(\omega_n l \Theta)^2$, so that $\omega_n^2 = g/l$.

These few examples of systems each with one degree illustrate the use of the energy method, but do not in themselves provide very convincing evidence of its value: equating the maximum kinetic and potential energies is really no

432 Dynamics of Mechanical Systems

simpler than substituting the assumed motion in the governing differential equation. It is when we come to consider systems with several degrees of freedom that the virtues of the energy method become apparent, the system in

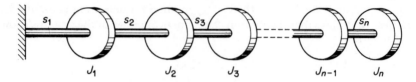

FIGURE 9.54

fig. 9.54 will be our first example. If the motions of $\theta_1 = \Theta_1 \cos \omega t$, $\theta_2 = \Theta_2 \cos \omega t$, etc., the maximum kinetic energy of the system is

$$\tfrac{1}{2}J_1(\omega\Theta_1)^2 + \tfrac{1}{2}J_2(\omega\Theta_2)^2 + \cdots + \tfrac{1}{2}J_n(\omega\Theta_n)^2$$
$$= \tfrac{1}{2}\omega^2 \sum_{i=1}^{i=n} J_i\Theta_i^2$$

The maximum potential energy is the sum of $\tfrac{1}{2}$(stiffness) × (maximum angle of twist)² for all the sections of shaft and is

$$\tfrac{1}{2}s_1\Theta_1^2 + \tfrac{1}{2}s_2(\Theta_2 - \Theta_1)^2 + \tfrac{1}{2}s_3(\Theta_3 - \Theta_2)^2 + \cdots + s_n(\Theta_n - \Theta_{n-1})^2$$
$$= \frac{1}{2}\sum_{j=1}^{j=n} s_j(\Delta\Theta_j)^2$$

where $\Delta\Theta_j$ = angle of twist in the jth shaft = $\Theta_j - \Theta_{j-1}$. On equating these two energy expressions we have

$$\omega^2 = \frac{\sum_{j=1}^{j=n} s_j(\Delta\Theta_j)^2}{\sum_{i=1}^{i=n} J_i\Theta_i^2} \tag{9.87}$$

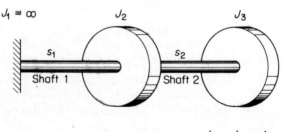

$J_2 = J_3 = J$
$s_1 = s_2 = s$

FIGURE 9.55

and are immediately faced with the problem that, whereas when dealing with systems with one degree of freedom the amplitude cancelled out of the final expression for the frequency, eqn. (9.87) involves the amplitudes which can be eliminated only if their relative values are known. Fortunately, although we do not know these relative values it is generally possible to make a good guess. Let us now study the particular system in fig. 9.55. This is a system for which we can readily obtain exact values of ω with which to compare values given by eqn. (9.87).

We know from previous experience that at the lowest natural frequency the two discs oscillate in phase with each other, that is to say the two inertia torques act in the same direction. This means that the torque in shaft 1 must be greater than the torque in shaft 2, and, as $s_1 = s_2$, that the angle of twist in shaft 1 must be greater than that in shaft 2. So that we have

$$\Theta_3 - \Theta_2 \leqslant \Theta_2$$

i.e.,
$$0.5\Theta_3 \leqslant \Theta_2$$

If $\Theta_3 = 2\Theta_2$, the limiting case, the two shafts carry the same torque and there is nothing to balance the inertia torque due to J_1. On the other hand

$$\Theta_2 \leqslant \Theta_3$$

because if Θ_3 were less than Θ_2 the torque in the two shafts would be in opposite directions. Combining these inequalities we have

$$0.5\Theta_3 \leqslant \Theta_2 \leqslant \Theta_3$$

If Θ_2 is expressed in terms of Θ_3 in eqn. (9.87) it is found that Θ_3 cancels out, so that it is convenient to assume, arbitrarily, that $\Theta_3 = 1$. The above inequality suggests that a reasonable guess for Θ_2 would be 0.75. When these values are substituted in eqn. (9.87) we have

$$\omega^2 = \frac{0.75^2 s + 0.25^2 s}{0.75^2 J + 1.00^2 J}$$

$$= 0.400 s/J$$

Substitution of appropriate values for the shaft stiffnesses and disc inertias in eqn. (9.76) (section 9.17) gives

$$\omega^4 - \frac{3s}{J}\omega^2 + \frac{s^2}{J^2} = 0$$

as the frequency equation. On solving this we find $\omega^2 = \frac{1}{2}(3 \pm \sqrt{5})s/J$, that is to say $\omega^2 = 0.382s/J$ or $2.618s/J$.

The amplitude ratio corresponding with $\omega^2 = 0.382s/J$ is found from eqn. (9.73) to be $\Theta_2/\Theta_3 = 0.618$. The interesting feature which emerges from a comparison of the approximate and exact calculations is that although the energy calculation is based on an amplitude ratio which is 21 per cent in error (0.75 assumed against 0.618 exact) the resulting error in ω^2 is only 4.7 per cent

434 Dynamics of Mechanical Systems

($0.400s/J$ approximate against $0.382s/J$ exact); the error in ω is of course approximately one half the error in ω^2. Thus the error in the frequency is very much less than the error in the initial estimate on which the calculation is based. This result is an illustration of *Rayleigh's principle* which can be stated thus: provided that a reasonable approximation to the modal shape is assumed, the error in the corresponding natural frequency, as calculated by the energy method, will be much smaller than the error in the modal shape.

When the system has more than two degrees of freedom it is not so easy to specify the error in the assumed modal shape: if a third disc (J_4) is added to the system in fig. 9.55 we should have to guess Θ_4/Θ_2 as well as Θ_3/Θ_2 and the errors in these two guesses are likely to be different, nevertheless, Rayleigh's principle still holds good. It applies to all the natural frequencies of the system, but it is not so easy to guess the modal shape for the second and higher natural frequencies as it is for the lowest natural frequency.

Turning again to our particular example, we note that the natural frequency calculated by the energy method is greater than the true natural frequency. If the reader should repeat the calculation for different values of Θ_2/Θ_3 he will find that this is always the case. It can be proved that the energy method always overestimates the lowest natural frequency, and always underestimates the highest natural frequency of a system. With intermediate frequencies the estimate may be high or low. The inherent overestimate of the lowest natural frequency is a very useful property because it means that if the assumed modal shape is varied in a systematic way the lowest value obtained for the frequency

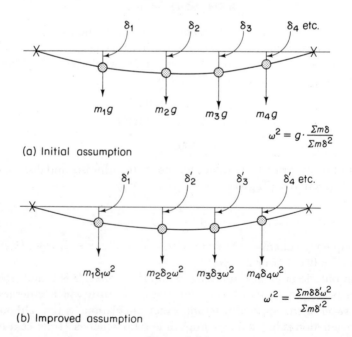

FIGURE 9.56

is the best estimate. The converse applies to estimates of the highest natural frequency but this is generally of less practical interest. With torsional vibrations the accuracy of a frequency estimate can, of course, always be verified by the use of Holzer's method, indeed the energy method is probably the best means of obtaining an initial estimate prior to computing a Holzer table.

When it is applied to the calculation of the natural frequencies of transverse vibrations of a number of concentrated masses attached to a light beam the energy method is usually developed by way of an argument which differs somewhat from that which we have used in relation to torsional vibrations. The reason for the difference lies in the fact that, whereas it is a comparatively simple task to compute the torsional strain energy of a shaft given the angular displacements at a number of sections along its length, heavy calculations are usually needed to compute the strain energy stored in a beam which has been subjected to specified transverse displacements. The calculations are very much simpler if, in effect, we base them on an assumed set of inertia forces rather than an assumed set of deflections. As a convenient starting point we assume that the maximum inertia forces to a series of concentrated masses m_1, m_2, m_3, etc., fig. 9.56(a) are respectively $m_1 g$, $m_2 g$, $m_3 g$, etc. These are the transverse forces to which the beam would be subjected due to the weights of the mass if it were held horizontally. Integration of the beam bending equation

$$EI \frac{d^2 y}{dx^2} = \text{bending moment}$$

either numerically or graphically, gives the 'static' deflection δ at each of the masses, from which we calculated the maximum change in potential energy

$$\tfrac{1}{2} m_1 g\, \delta_1 + \tfrac{1}{2} m_2 g\, \delta_2 + \tfrac{1}{2} m_3 g\, \delta_3 + \cdots = \tfrac{1}{2} g \sum m\, \delta$$

and the maximum kinetic energy,

$$\tfrac{1}{2} m_1 (\omega\, \delta_1)^2 + \tfrac{1}{2} m_2 (\omega\, \delta_2)^2 + \tfrac{1}{2} m_3 (\omega\, \delta_3)^2 + \cdots = \tfrac{1}{2} \omega^2 \sum m\, \delta^2$$

Equating these two energies we have

$$\omega^2 = g\, \frac{\sum m\, \delta}{\sum m\, \delta^2} \qquad (9.88)$$

Experience shows that eqn. (9.88) usually gives a very good approximation to the lowest (non-zero) natural frequency of the system. If a confirmation of the accuracy is required the deflected form should be recalculated assuming that the maximum inertia loads are $m_1\, \delta_1 \omega^2$, $m_2\, \delta_2 \omega^2$, etc., as in fig. 9.56(b). It will probably be found that the new set of deflections δ_1', δ_2', etc. differ substantially from the original set. Nevertheless, if the new change in potential energy $\sum \tfrac{1}{2}(m\, \delta \omega^2)\, \delta'$ is equated to the new maximum kinetic energy $\sum \tfrac{1}{2} m (\delta' \omega)^2$, the modified value of ω^2 given by

$$\omega'^2 = \frac{\sum m\, \delta \delta'\, \omega^2}{\sum m\, \delta'^2} \qquad (9.89)$$

436 Dynamics of Mechanical Systems

is not likely to differ appreciably from that originally calculated. If the difference is too great the inertia forces are taken to be $m_1 \delta_1' \omega'^2$, etc. and the procedure is repeated.

It can be proved that this process is convergent and that it must lead eventually (normally after only one or two iterations) to the correct value for the lowest natural frequency. With the higher natural frequencies a modified procedure must be adopted.

9.23 Continuous systems

In all our calculations of natural frequencies we have so far assumed that only the inertias of the rigid masses need to be taken into account, and that the inertia forces due to the masses of the springs are negligible in comparison: the springs are said to be 'light'. Such systems are referred to in the jargon of the applied mathematician as *lumped parameter* or *discrete* systems. Although it is reasonable to treat many practical systems in this way, real springs and shafts do have masses, and sooner or later we are faced with the problem of taking this mass into account. As an example we will determine the way in which a uniform circular shaft fixed at one end responds to a torque $Q \cos \omega t$ applied to its free end, fig. 9.57(a).

The equation of motion of the shaft is derived by considering the state of equilibrium of a typical element of thickness δx distance x from the fixed end. If the second polar moment of area of the cross-section is J and ρ is the density of the shaft material the inertia torque is

$$\rho J \, \delta x \, \frac{\partial^2 \theta}{\partial t^2}$$

The torques applied to the element by the rest of the shaft are q and $(q + (\partial q/\partial x) \, \delta x)$ as indicated. The equilibrium equation for the element is therefore

$$q + \rho J \frac{\partial^2 \theta}{\partial t^2} \delta x = q + \frac{\partial q}{\partial x} \delta x$$

or

$$\rho J \frac{\partial^2 \theta}{\partial t^2} = \frac{\partial q}{\partial x}$$

The torque q, which is transmitted across any section, of the shaft depends upon the rate of change of the angle of twist and is given by the equation

$$q = CJ \frac{\partial \theta}{\partial x}$$

where C is the modulus of rigidity of the material of the shaft. On eliminating q between these two equations the equation of motion for the shaft is found to be the partial differential equation

$$\frac{\partial^2 \theta}{\partial x^2} = \frac{\rho}{C} \frac{\partial^2 \theta}{\partial t^2}$$

MECHANICAL VIBRATIONS

In terms of the overall stiffness, $s = CJ/l$ as measured at the free end of the shaft, and the total moment of inertia $J_s = \rho lJ$; the equation of motion is

$$\frac{\partial^2 \theta}{\partial x^2} = \frac{J_s}{l^2 s} \frac{\partial^2 \theta}{\partial t^2} \tag{9.90}$$

There is no reason to suppose that the response of this system to a simple harmonic excitation will differ in principle from the response of the systems which we have hitherto considered, so that it would be reasonable to assume

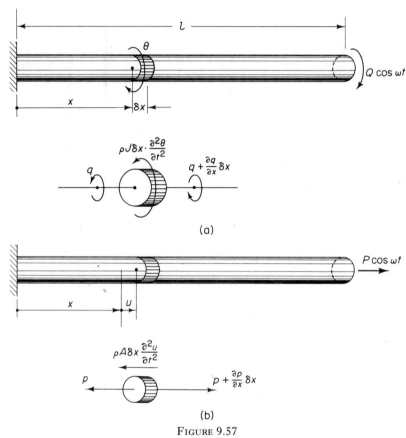

FIGURE 9.57

that in the steady state all the parts of the system vibrate in phase (or 180 degrees out of phase) with the excitation. The amplitude of vibration varies with position along the shaft so that we will assume the solution

$$\theta = f(x) \cos \omega t \tag{9.91}$$

Substituting this assumed solution in eqn. (9.90) we find

$$\frac{d^2 f(x)}{dx^2} \cos \omega t = -\frac{J_s}{l^2 s} f(x) \omega^2 \cos \omega t$$

15

438 Dynamics of Mechanical Systems

The time-varying terms $\cos \omega t$ cancel out to show that the assumed solution is valid if $f(x)$ can be found to satisfy

$$(D^2 + \omega^2 J_s/l^2 s)f(x) = 0$$

Putting $\omega^2 J_s/s = \alpha^2$ the general solution of this equation is

$$f(x) = A \cos \alpha x/l + B \sin \alpha x/l$$

The constants A and B are determined by the known end conditions. When $x = 0$ the amplitude is zero so that $A = 0$, and

$$f(x) = B \sin \alpha x/l$$

The torque in the shaft is

$$q = CJ \frac{\partial \theta}{\partial x}$$

$$= ls \frac{\partial \theta}{\partial x}$$

$$= \alpha s B \cos (\alpha x/l) \cos \omega t \tag{9.92}$$

At $x = l$, $q = Q \cos \omega t$ so that $B = (Q/\alpha s) \cos \alpha$. The complete solution of eqn. (9.90) for this problem is therefore

$$\theta = \frac{Q}{\alpha s \cos \alpha} \sin \frac{\alpha x}{l} \cos \omega t \tag{9.93}$$

At the free end ($x = l$) the amplitude of motion is

$$\Theta = \frac{Q}{s} \frac{\tan \alpha}{\alpha} \tag{9.94}$$

There are an infinite number of values of ω for which the amplitude is infinite, and hence an infinite number of resonance frequencies. These occur when

$$\alpha = \pi/2, \ 3\pi/2, \ 5\pi/2, \ \text{etc.}$$

and correspond to

$$\omega = \frac{\pi}{2}\sqrt{\frac{s}{J_s}}, \ \frac{3\pi}{2}\sqrt{\frac{s}{J_s}}, \ \frac{5\pi}{2}\sqrt{\frac{s}{J_s}}, \ \text{etc.}$$

In general,

$$\omega_n = (2n - 1)\frac{\pi}{2}\sqrt{\frac{s}{J_s}}, \quad n = 1, 2, 3, \ldots, \infty$$

When free vibrations are studied precisely the same reasoning is used as far as eqn. (9.92). If we now put $q = 0$ at $x = l$ eqn. (9.92) gives

$$\alpha s B \cos \alpha = 0$$

This equation may be interpreted in two ways: either $B = 0$ and there is no vibration, or

$$\alpha \cos \alpha = 0$$

It follows that the bar can execute free torsional vibrations only if

$$\alpha = \pi/2,\ 3\pi/2,\ 5\pi/2,\ \text{etc.}$$

that is, if the frequency of vibration is given by

$$\omega_n = (2n - 1)\frac{\pi}{2}\frac{s}{J_s}, \qquad n = 1, 2, \ldots$$

These frequencies are, as expected, identical to the resonance frequencies.

It will be recalled that in dealing with discrete systems it was concluded that an undamped system has as many resonance frequencies, or natural frequencies, as it has degrees of freedom. Reversing this statement, it is deduced that the shaft whose motion we have just analysed has an infinite number of degrees of freedom. This is indeed the case because a system has as many degrees of freedom as the number of coordinates required to define its configuration, i.e., the relative positions of its masses. In this example, the mass is continuously distributed throughout the system so that to define the configuration completely we must specify the rotation at every point along the shaft, and there are an infinity of such points. Systems of this type are described as *continuous* systems.

The equation of motion for longitudinal vibrations in a uniform bar is identical in form to eqn. (9.90). The equilibrium equation for a typical element, fig. 9.57(b), gives

$$\rho A \frac{\partial^2 u}{\partial t^2} = \frac{\partial p}{\partial x}$$

where A is the area of cross-section, u is the displacement of the element from its mean position, and p is the tensile force on the element. The tensile strain in the bar is given by the rate of change of longitudinal displacement so that

$$p = AE\frac{\partial u}{\partial x}$$

On eliminating p, and putting $s = AE/l$ and $m = \rho l A$, these equations yield

$$\frac{\partial^2 u}{\partial x^2} = \frac{m}{l^2 s}\frac{\partial^2 u}{\partial t^2} \tag{9.95}$$

The solution of this equation for an excitation $P \cos \omega t$ at the free end of the bar follows precisely the same steps as the solution of eqn. (9.90) and yields

$$u = \frac{P}{\alpha s \cos \alpha}\sin\frac{\alpha x}{l}\cos \omega t$$

where

$$\alpha = \omega \sqrt{(m/s)}$$

440 Dynamics of Mechanical Systems

Transverse vibrations of a uniform beam present a more difficult problem. The forces which act on a typical element, the inertia force, the shear forces, and the bending moments shown in fig. 9.58. The deflection of the beam has

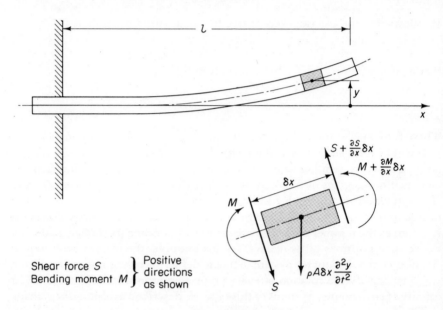

FIGURE 9.58

been much exaggerated in the diagram and on equating forces in the y direction it is permissible to ignore the slight differences in direction due to the slope so that we have

$$S + \rho A \, \delta x \, \frac{\partial^2 y}{\partial t^2} = S + \frac{\partial S}{\partial x} \, \delta x$$

This reduces to

$$\frac{\partial S}{\partial x} = \rho A \frac{\partial^2 y}{\partial t^2}$$

By taking moments about the centre of gravity of the element we obtain the relationship

$$S + \frac{\partial M}{\partial x} = 0 \qquad (9.96)$$

as the relationship between the shear force and the bending moment. Elimination of S between these two equations gives

$$\rho A \frac{\partial^2 y}{\partial t^2} = -\frac{\partial^2 M}{\partial x^2}$$

Finally, taking account of the relationship

$$M = EI \frac{\partial^2 y}{\partial x^2} \qquad (9.96(a))$$

we obtain as the equation of motion:

$$EI \frac{\partial^4 y}{\partial x^4} + \rho A \frac{\partial^2 y}{\partial t^2} = 0 \qquad (9.97)$$

Assuming that the beam vibrates according to the equation

$$y = f(x) \cos \omega t$$

and substituting this solution in eqn. (9.97), we find that the term $\cos \omega t$ cancels out yielding

$$EI \frac{d^4}{dx^4} f(x) - \rho A \omega^2 f(x) = 0$$

or

$$\left(D^4 - \frac{\rho A}{EI} \omega^2 \right) f(x) = 0 \qquad (9.98)$$

as the differential equation to the deflected form of the beam.

An alternative way of looking at this is to regard the problem as one of determining the static deflection $f(x)$ of the beam due to a distributed load of intensity $\rho A \omega^2 f(x)$, this being the inertia loading when the displacement is a maximum. The resulting differential equation is the same.†

The general solution of this equation is

$$f(x) = A \cos \alpha x/l + B \sin \alpha x/l + C \cosh \alpha x/l + D \sinh \alpha x/l \qquad (9.99)$$

where

$$\alpha = l \sqrt[4]{\frac{\rho A \omega^2}{EI}}$$

The four constants of integration A, B, C, and D are determined by four independent end conditions. For free vibrations of the cantilever shown in fig. 9.58 these conditions are $y = 0$ and $dy/dx = 0$ at $x = 0$, and $M = EI \, d^2y/dx^2 = 0$ and $S = -EI \, d^3y/dx^3 = 0$ at $x = l$. Substitution of these four conditions in eqn. (9.99) gives respectively

$$\left. \begin{array}{l} A + C = 0 \\ B + D = 0 \\ -A \cos \alpha - B \sin \alpha + C \cosh \alpha + D \sinh \alpha = 0 \\ A \sin \alpha - B \cos \alpha + C \sinh \alpha + D \cosh \alpha = 0 \end{array} \right\} \qquad (9.100)$$

† The displacement y of a uniform beam due to a distributed load w, which may vary with x, is the solution of the differential equation, $EI \, d^4y/dx^4 = w$.

On elimination of C and D these equations reduce to

$$\left. \begin{array}{l} A\,(\cos \alpha + \cosh \alpha) = -B\,(\sin \alpha + \sinh \alpha) \\ A\,(\sin \alpha - \sinh \alpha) = B\,(\cos \alpha + \cosh \alpha) \end{array} \right\} \quad (9.101)$$

If A and B are both zero, one possible solution, there is no vibration. Assuming that A and B are not zero they are eliminated from these two equations by cross-multiplication of the two left- and right-hand sides. This gives

$$(\cos \alpha + \cosh \alpha)^2 = -(\sin^2 \alpha - \sinh^2 \alpha)$$

and simplifies to

$$\operatorname{sech} \alpha = -\cos \alpha$$

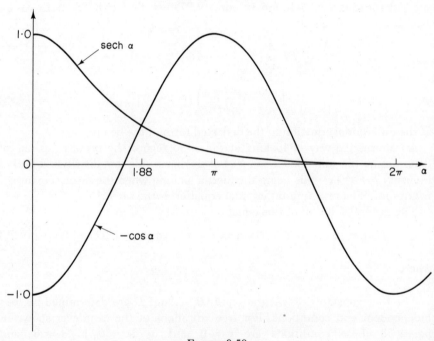

FIGURE 9.59

The solution of this equation is best found graphically, fig 9.59, which shows that possible values for α are

$$\alpha \approx 1{\cdot}88,\ 3\pi/2,\ 5\pi/2, \ldots$$

so that corresponding values of ω are

$$\omega = 3{\cdot}57\sqrt{\frac{EI}{\rho A l^4}},\quad 22{\cdot}2\sqrt{\frac{EI}{\rho A l^4}},\quad 61{\cdot}5\sqrt{\frac{EI}{\rho A l^4}}\ \ldots$$

The associated values of A/B are found from eqns. (9.101) to be

$$B/A = 0{\cdot}736,\ 1{\cdot}018,\ 1{\cdot}00, \ldots$$

C/A and D/A follow respectively from the first two of eqns. (9.100), and hence the modal shapes in fig. 9.60.

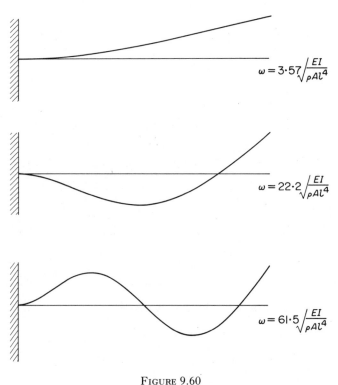

FIGURE 9.60

Calculations similar to those just described enable the natural frequencies for transverse vibrations of a uniform beam to be calculated for any combination of end conditions. The frequency equation for each of a number of different types of support, including the one just studied, are given in fig. 9.61.

9.24 Energy method applied to continuous systems

Determination of the natural frequencies of continuous systems by formal solution of the differential equations is practicable for only a very limited class of problems: a single concentrated mass attached to a uniform beam introduces such complexity into the solution that it is as well to turn one's attention immediately to alternative methods. One possibility is to regard the distributed mass as if it were concentrated into a number of strategically placed discrete masses, and then to adopt the methods of analysis which we have already discussed. Another possibility, and the one which we will now consider, is to make a direct appeal to the energy principle rephrased in forms suitable to the consideration of continuous systems. Torsional oscillations of a uniform circular bar will provide the first example, fig. 9.62.

444 Dynamics of Mechanical Systems

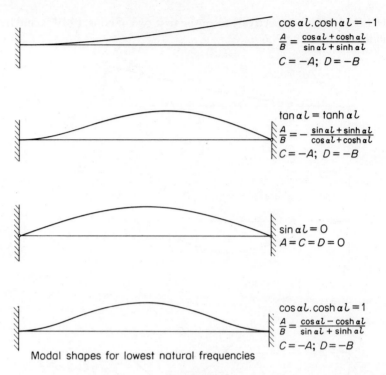

$\cos \alpha l \cdot \cosh \alpha l = -1$
$\dfrac{A}{B} = \dfrac{\cos \alpha l + \cosh \alpha l}{\sin \alpha l + \sinh \alpha l}$
$C = -A; \; D = -B$

$\tan \alpha l = \tanh \alpha l$
$\dfrac{A}{B} = -\dfrac{\sin \alpha l + \sinh \alpha l}{\cos \alpha l + \cosh \alpha l}$
$C = -A; \; D = -B$

$\sin \alpha l = 0$
$A = C = D = 0$

$\cos \alpha l \cdot \cosh \alpha l = 1$
$\dfrac{A}{B} = \dfrac{\cos \alpha l - \cosh \alpha l}{\sin \alpha l + \sinh \alpha l}$
$C = -A; \; D = -B$

Modal shapes for lowest natural frequencies

FIGURE 9.61

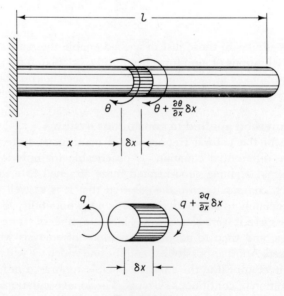

FIGURE 9.62

Torsional vibrations

The potential energy of the system is the torsional strain energy stored in the shaft. For an element of the shaft of length δx the strain energy δV stored is one half the product of the torque in the element and the relative rotation of its two ends, i.e.,

$$\delta V = \tfrac{1}{2} q \frac{\partial \theta}{\partial x} \delta x$$

For the whole shaft, the energy is

$$V = \frac{1}{2} \int_0^l q \frac{\partial \theta}{\partial x} dx$$

The torque and the angle of twist are related by the equation

$$q = CJ \frac{\partial \theta}{\partial x} \tag{9.102}$$

so that our final expression for the potential (or strain) energy is

$$V = \frac{1}{2} \int_0^l CJ \left(\frac{\partial \theta}{\partial x} \right)^2 dx$$

The kinetic energy of the shaft is

$$T = \frac{1}{2} \int_0^l \rho A \left(\frac{\partial \theta}{\partial t} \right)^2 dx$$

If the shaft is executing simple harmonic motion,

$$\theta = \theta_0 \cos \omega t$$

so that on equating the maximum potential energy to the maximum kinetic energy we have

$$\frac{1}{2} \int_0^l CJ \left(\frac{d\theta_0}{dx} \right)^2 dx = \frac{\omega^2}{2} \int_0^l \rho A \theta_0^2 \, dx$$

where the change from a partial to a total differential coefficient is made because θ_0 is a function of x only, i.e., unlike θ it does not vary with t. This equation gives

$$\omega^2 = \frac{\int_0^l CJ \left(\frac{d\theta_0}{dx} \right)^2 dx}{\int_0^l \rho A \theta_0^2 \, dx} \tag{9.103}$$

15*

446 Dynamics of Mechanical Systems

If the cross-section of the beam varies along its length, A and J must be expressed as functions of x. With a uniform bar these parameters are constants and eqn. (9.102) may be written in the form

$$\omega^2 = \frac{CJ}{\rho A} \frac{\int_0^l \left(\frac{d\theta_0}{dx}\right)^2 dx}{\int_0^l \theta_0^2 \, dx} \tag{9.104}$$

In accordance with Rayleigh's principle, a reasonable estimate of the modal shape for a natural frequency should give a good estimate of the value of the frequency; furthermore, the calculation will overestimate the lowest natural frequency. In the case of a uniform bar fixed at one end, as in fig. 9.62, the simple assumption which we can make for the modal shape is

$$\theta_0 = \alpha x \tag{9.105}$$

that is, θ_0 increases uniformly along the length of the bar. This is not a very good guess because, as eqn. (9.102) shows, it amounts to an assumption that the torque is constant along the bar, and this is obviously not true. Nevertheless, we will evaluate the corresponding value of ω^2. Substitution from eqn. (9.105) in eqn. (9.104) gives

$$\omega^2 = \frac{CJ}{\rho A} \frac{\int_0^l \alpha^2 \, dx}{\int_0^l \alpha^2 x^2 \, dx} = 3 \frac{CJ}{\rho A l^2}$$

Substitution of s, the overall stiffness, for CJ/l, and J_s, the total moment of inertia, for $\rho A l$ gives

$$\omega = 1 \cdot 732 \sqrt{(s/J_s)}$$

This result does not compare very well with the exact value $\frac{1}{2}\pi s/J_s$ obtained by formal solution of the governing differential equation, and demonstrates that although we are working from a guessed modal shape, we must not be too casual in the way that we make the guess. We shall do much better if we take account of the fact that the torque is zero at the free end of the shaft by choosing a curve from which $\partial \theta/\partial x$ is zero at the free end. To achieve this, let us assume that

$$\theta_0 = x + \beta x^2$$

By differentiating this equation with respect to x, and setting $d\theta_0/dx$ equal to zero for $x = l$, we find $\beta = -1/2l$ so that

$$\theta_0 = x - x^2/2l$$

We now have

$$\omega^2 = \frac{CJ}{\rho A} \frac{\int_0^l (1-x/l)^2 \, dx}{\int_0^l (x - x^2/2l)^2 \, dx} = 2.5 \frac{CJ}{\rho A l^2}$$

giving
$$\omega = 1.581\sqrt{(s/J_s)}$$

which is less than 1 per cent in excess of the exact value of $1.571\sqrt{(s/J_s)}$. The true modal shape $\theta_0 = A \sin \pi x/2l$ gives, of course, the exact value of ω.

The effect of a disc at the free end of the shaft can readily be included in the calculations. Assuming that the disc is rigid it has no direct effect on the expression for the maximum strain energy in the shaft which is still

$$\frac{CJ}{2} \int_0^l \left(\frac{d\theta_0}{dx}\right)^2 dx$$

The maximum kinetic energy is now

$$\underbrace{\tfrac{1}{2}J_d\omega^2(\theta_0)^2_{x=l}}_{\text{due to the rotor}} + \underbrace{\tfrac{1}{2}\rho A\omega^2 \int_0^l \theta_0^2 \, dx}_{\text{due to the shaft}}$$

where J_d is the moment of inertia of the disc.

It would be wrong now to choose $\theta_0 = f(x)$ such that $d\theta_0/dx = 0$ at $x = l$, for the torque in the shaft is no longer zero at this section. Indeed if $J_d \gg J_s$ the major part of the torque in the shaft is attributable to the inertia torque due to the rotor so that it may be reasonable to revert to our original assumption

$$\theta_0 = \alpha x$$

which implies that the torque in the shaft is constant. On equating the maximum potential and kinetic energies we have

$$\tfrac{1}{2}CJ \int_0^l \alpha^2 \, dx = \tfrac{1}{2}J_d\omega^2(\alpha l)^2 + \tfrac{1}{2}\rho A\omega^2 \int_0^l (\alpha x)^2 \, dx$$

which, on evaluation of the integrals gives

$$\omega^2 = \sqrt{(s/J)}$$

where
$$J = J_d + \tfrac{1}{3}J_s$$

Longitudinal vibrations
Longitudinal vibrations in a uniform bar are governed by a differential equation which is identical in form to that for torsional vibrations [compare eqns. (9.90)

448 Dynamics of Mechanical Systems

and (9.94)]. The relationship corresponding to eqn. (9.103) for a displacement

$$u = u_0 \cos \omega t$$

is
$$\omega^2 = \frac{E}{\rho} \frac{\int_0^l \left(\frac{du_0}{dx}\right)^2 dx}{\int_0^l u_0^2 \, dx} \tag{9.106}$$

Transverse vibrations

The strain energy δV stored in a small element of a beam under flexure, fig. 9.63, is one half the product of the bending moment applied to the element and the relative angle of rotation of the two ends so that

$$\delta V = \tfrac{1}{2} M \frac{\partial \phi}{\partial x} \delta x$$

Now $\phi = \dfrac{\partial y}{\partial x}$ so that $\dfrac{\partial \phi}{\partial x} = \dfrac{\partial^2 y}{\partial x^2}$.

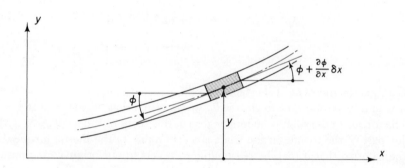

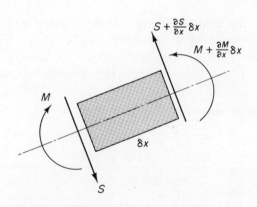

FIGURE 9.63

Furthermore, $M = EI\, \partial^2 y/\partial x^2$, so that the total potential energy for the beam may be expressed as

$$V = \frac{1}{2}\int_0^l EI\left(\frac{\partial^2 y}{\partial x^2}\right)^2 dx$$

The maximum kinetic energy is

$$T = \frac{\omega^2}{2}\int_0^l \rho A y_o^2\, dx$$

and on equating this to the maximum potential energy we have

$$\omega^2 = \frac{\int_0^l EI\left(\dfrac{d^2 y_o}{dx^2}\right) dx}{\int_0^l \rho A y_o^2\, dx} \qquad (9.107)$$

If the beam is uniform

$$\omega^2 = \frac{EI}{\rho A}\,\frac{\int_0^l \left(\dfrac{d^2 y_o}{dx^2}\right)^2 dx}{\int_0^l y_o^2\, dx} \qquad (9.108)$$

As an example of the use of this equation we will determine the lowest natural frequency for a uniform cantilever of length l. In accordance with Rayleigh's principle eqn. (9.108) leads to quite accurate assessments of the natural frequencies of a beam provided that the assumed modal shape $y_o = f(x)$ is reasonably accurate. Even so, as has been pointed out in relation to the use of eqns. (9.103) and (9.104) and the determination of natural frequencies of torsional vibrations, it pays to take some care in guessing the modal shape. For example, it might be thought that a parabolic deflection curve $y_o = \alpha x^2$ should be a reasonable approximation to the modal shape for our uniform cantilever, fig. 9.64(a). In fact, this leads to a rather poor estimate of the natural frequency, for although the shape looks right, and the equation satisfies the end conditions $y_o = 0$, $dy_o/dx = 0$ at $x = 0$, the implied distributions of bending moment (constant along the beam), and shear force (everywhere zero), are quite incorrect. The estimated lowest natural frequency for this assumption is given by

$$\omega^2 = \frac{EI}{\rho A}\,\frac{\int_0^l (2\alpha)^2\, dx}{\int_0^l (\alpha x^2)^2\, dx} = 20\,\frac{EI}{\rho A l^4}$$

or

$$\omega = 4\cdot 47\sqrt{\frac{EI}{\rho A l^4}}$$

This result has to be compared with the $3.53\sqrt{(EI/\rho Al^4)}$ which we obtained previously by exact solution of the governing differential equation.

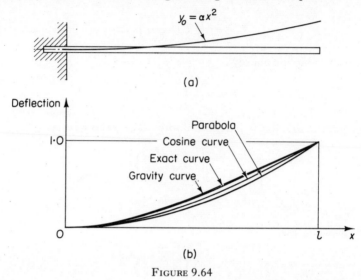

FIGURE 9.64

A better estimate is obtained if the modal shape is assumed to be given by $y_o = (1 - \cos \pi x/2l)$. This satisfies the requirements of $y_o = 0$, $dy_o/dx = 0$ at $x = 0$, and in addition satisfies the condition of zero bending moment at $x = l$. The implied shear force is zero at $x = 0$ and a maximum at $x = l$, and is obviously wrong. Nevertheless, the estimated frequency is found to be $3.66\sqrt{(EI/\rho Al^4)}$, which is only 3.7 per cent in excess of the true value.

An excellent estimation of the frequency is obtained if the modal shape is assumed to be identical with the deflected form of the beam due to its weight when it is held horizontally. The deflection along the beam is given by the equation

$$y_o = \tfrac{1}{24}(x^4 - 4lx^3 + 6l^2x^2)\rho Ag/EI$$

This equation must automatically satisfy all four boundary conditions and might therefore be expected to lead to a good estimate of the natural frequency. Ignoring the factor $\rho Ag/24EI$, which will in any case cancel out, we have

$$\frac{d^2y_o}{dx^2} = 12x^2 - 24lx + 12l^2$$

so that

$$\omega^2 = \frac{EI}{\rho A} \frac{\int_0^l 144(x - l)^4 \, dx}{\int_0^l (x^4 - 4lx^3 + 6l^2x^2)^2 \, dx}$$

$$= 12.4 EI/\rho Al^4$$

and

$$\omega = 3.54\sqrt{(EI/\rho Al^4)}$$

MECHANICAL VIBRATIONS 451

The accuracy of this result is paid for by the relatively large amount of labour incurred in the evaluation of the integrals in the expression for ω^2. As a matter of interest, the various deflected forms which we have considered in this calculation are compared in fig. 9.64(b). In each case, it is assumed that the deflection at the tip is unity.

9.25 Critical speeds of shafts

There is no obvious reason why a straight shaft, properly balanced and with its axis vertical, should not run steadily at any speed. In practice, it is found that

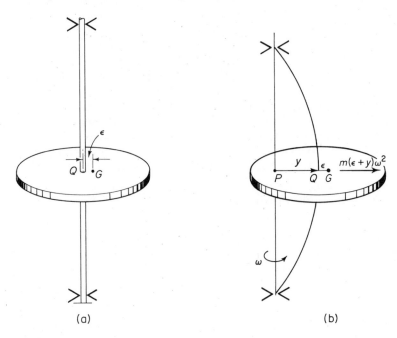

FIGURE 9.65

at certain speeds the shaft takes up a deflected form and the shaft 'whirls'. If the speed is maintained at one of the *critical speeds* the deflections may become sufficiently large to cause failure of the shaft. The critical speeds are well defined and there is usually no evidence of any whirling at any other speeds. Indeed, if the shaft is initially slightly bent it will run more steadily above the lowest critical speed than it does below it. This feature can be explained and in practice rotors are often run above the first critical speed to take advantage of it. The explanation is most easily given in relation to a light vertical shaft which carries at its centre an eccentrically mounted disc as shown in fig. 9.65. If the shaft is turned at low speed we would expect the shaft to bend and take up a form in which the inertia force due to the eccentricity of the disc is just counteracted by the restoring force due to the elasticity of the shaft. If the stiffness of the shaft is measured by the transverse force required to cause unit deflection at its centre is s and the elastic deflection is y we have

$$sy = m(y + \varepsilon)\omega^2$$

where ω is the angular velocity of the shaft, and m is the mass of the disc. On rearranging the equation we have

$$\varepsilon\omega^2 = (s/m - \omega^2)y$$

Now s/m is ω_n^2, where ω_n is the natural frequency for transverse vibrations of the disc, so that

$$y = \frac{\omega^2 \varepsilon}{\omega_n^2 - \omega^2} \qquad (9.109)$$

Equation (9.109) shows that the deflection of the shaft might be expected to become very large as the speed of rotation ω approaches the natural frequency ω_n. This calculation presupposes, however, that the shaft is allowed to reach an equilibrium state. If the shaft speeded up sufficiently quickly it is possible to pass through the critical speed before the displacement has time to build up. When this is done the shaft can then settle down to a steady state in which y has a finite negative value. The negative value indicates that the deflection of the shaft is in the opposite direction to what it was at speeds less than ω_n. The total eccentricity of the disc is

$$y + \varepsilon = \frac{\omega_n^2 \varepsilon}{\omega_n^2 - \omega^2}$$

$$= \frac{\varepsilon}{1 - (\omega/\omega_n)^2}$$

and tends to zero as ω becomes very much greater than ω_n. This tendency for the disc to rotate about its centre of gravity at high speeds accounts for the steadiness which is observed when the speed exceeds the critical speed. In fact, provided that $\omega > \omega_n\sqrt{2}$, the total eccentricity is less than ε, whereas for $\omega < \omega_n$ it is always greater than ε.

The stability of the motion, whether above or below the critical speed, can be accepted as an observed fact. Analysis of the stability, even for a perfectly balanced rotor, is moderately complicated and will not be given in detail here, but there is one point worth comment. In order to test the stability of a mechanical system we can disturb it from its equilibrium state and observe whether the subsequent motion tends to die out leaving the system once more in equilibrium, or whether it tends to grow and to take the system further from its equilibrium state. If the motion is to decay there must be damping in the system. Obviously there will be external damping due partly to the motion of the shaft and rotor relative to the atmosphere, and partly to the bearings. Such damping might reasonably be regarded as equivalent to a viscous damping force acting at the centre of the disc, the magnitude of the force being proportional to the velocity of the centre of the disc. Analysis shows that the effect of this type of damping is to promote stability at all speeds. There will, however, be additional damping due to the deformation of the shaft causing hysteretic damping. The effect of this is to reinforce stability as long as the running speed is less than the critical speed, but when the speed is greater than the critical its effect is unstabilizing. Usually, the external damping dominates and the shaft is stable at all speeds. Trouble may occur if the disc is not firmly attached to the shaft because friction due to relative motion at the surfaces is internal to the system.

Equation (9.109) shows that the critical speed of the shaft coincides with the natural frequency for transverse vibrations, both being expressed in radians per second. This is easily explained by considering the rotating out-of-balance force $m(\varepsilon + y)\omega^2$ as consisting of two forces, $m(\varepsilon + y)\omega^2 \cos \omega t$ and $m(\varepsilon + y)\omega^2 \sin \omega t$, in fixed directions at right-angles to each other. Each force excites resonance when the frequency ω coincides with the natural frequency for transverse vibrations. This particular system has been considered as consisting of a light shaft with a central massive disc and so it has but one natural frequency. By adding more masses to the shaft we increase the number of natural frequencies and correspondingly the number of critical speeds. Provided that gyroscopic effects are small, the critical speeds coincide with the natural frequencies.

We have so far contemplated the shaft being vertical to avoid any confusion of the argument by considerations of the effects of gravity. Gravity has no effect if the shaft is of circular cross-section. The shaft deflects under its own weight and whirling takes place about the deflected centre lines so that in determining the critical speeds of a uniform cantilever, such as that shown in fig. 9.66, we can ignore the static deflection due to the weight of the shaft. Assuming that the shaft is whirling the inertia load on a small element of the shaft is $\rho A\, \delta x y \omega^2$. This is counterbalanced by the shear forces at each end of the element so that

$$S + \frac{dS}{dx}\delta x + \rho A\, \delta x y \omega^2 = S$$

or

$$\frac{dS}{dx} + \rho A y \omega^2 = 0$$

On substituting for S from eqn. (9.96) this equation becomes

$$\frac{d^2 M}{dx^2} - \rho A y \omega^2 = 0$$

and on substituting for M from eqn. (9.96(a)) we have finally

$$EI \frac{d^4 y}{dx^4} - \rho A y \omega^2 = 0$$

or

$$\frac{d^4 y}{dx^4} - \frac{\rho A \omega^2}{EI} y = 0 \qquad (9.110)$$

Equation (9.110) is precisely the same as eqn. (9.98), which determines the modal shape for transverse vibrations of the cantilever, and precisely the same mathematical argument is used to solve it. It follows that for whirling to be possible the speed of vibration must coincide with a natural frequency of transverse vibrations of the shaft.

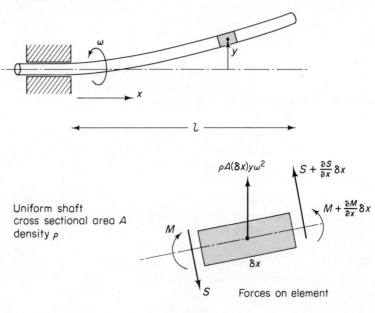

FIGURE 9.66

The conclusion, that whirling occurs when the angular velocity of a shaft coincides with a natural frequency for transverse vibrations, is valid provided that gyroscopic effects are insignificant. In the case of the example which was used to introduce the topic of whirling, the disc (fig. 9.65) was considered to be carried at the centre part of the shaft so that there was no tendency for the disc to be deflected out of its own plane and consequently no gyroscopic couple to be considered. The effect of the gyroscopic couple is to cause whirl to occur at speeds which do not in general coincide with the natural frequencies for transverse vibrations of the stationary shaft. Other features which may cause whirl at speeds other than the expected speeds are variations in the stiffness of the shaft and bearings about axes round to the axis of rotation, and hydrodynamic effects in journal bearings.

9.26 Gyroscopic vibrations

Mechanical vibrations generally involve an interchange between the kinetic and potential energies of a system, and indeed, as we have seen, calculations based on this interchange enable us to calculate natural frequencies. By way of conclusion we will now study briefly a vibrating system in which all the energy is stored as kinetic energy: the system in question is a simple gyroscope, fig. 9.67. It consists of a rotor which spins at high speed about a horizontal axis in bearings which are carried in a frame, the inner frame. The inner frame is itself supported in bearings whose axis is horizontal and perpendicular to the spin

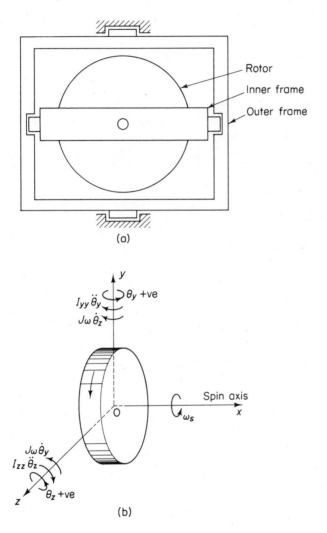

FIGURE 9.67

axis of the rotor, the bearings themselves are carried in a second frame, the outer frame. This is free to rotate about the vertical axis, which intersects the bearing axes at the centre of gravity of the rotor. In accordance with the theory given in chapter 5, a constant torque Q applied about the axis of the inner frame causes the system to rotate (precess) about the vertical axis with constant angular velocity. As we have seen, the behaviour can be explained by considering the angular momentum of the system, and we have derived the relationship between the velocity of precession ω_p and the applied torque Q to be

$$Q = J\omega_s\omega_p$$

where J is the polar moment of inertia of the rotor, and ω_s is its spin velocity. This equation relates to steady-state conditions. The way in which these steady-state conditions appear to be achieved in practice is even more surprising than the steady-state behaviour itself, for they appear to be reached instantaneously. The explanation of this clearly requires a fuller description of the motion than is provided by the steady-state equation, and must lie in the complete solution of the equations of motion. We will set up these equations in relation to Cartesian axes $Oxyz$, which are fixed in space, assuming that during the motion the plane of the rotor does not deflect very far from the Oyz plane. Given this assumption, we may split up the precessional velocity into two components $d\theta_y/dt$ and $d\theta_z/dt$, and the angular accelerations into $d^2\theta_y/dt^2$ and $d^2\theta_z/dt^2$. The resulting inertia torques are as shown in fig. 9.67(b). Assuming that the motion is caused by a constant torque Q_z applied about the O_z axis, the equations of motion are

$$I_{zz} D^2\theta_z - J\omega_s D\theta_y = Q_z \tag{9.111}$$

and
$$I_{yy} D^2\theta_y + J\omega_s D\theta_z = 0 \tag{9.112}$$

On eliminating θ_z these equations reduce to

$$(I_{yy}I_{zz} D^2 + J^2\omega_s^2) D\theta_y = -J\omega_s Q_z \tag{9.113}$$

The particular integral of this equation is

$$D\theta_y = -Q/J\omega_s \tag{9.114}$$

Equation (9.112) shows that the corresponding particular integral for θ_z is $D\theta_z = 0$. Consequently, eqn. (9.114) gives the steady-state precessional velocity ω_p and shows that

$$J\omega_s\omega_p = -Q$$

as we have already deduced by other means; the negative sign in this instance results from the choice of axes.

The complementary function of eqn. (9.113) is the solution of

$$(I_{yy}I_{zz} D^2 + J^2\omega_s^2) D\theta_y = 0$$

and is
$$D\theta_y = A \cos \omega_n t + B \sin \omega_n t$$

where $\omega_n = J\omega_s(I_{yy}I_{zz})^{-\frac{1}{2}}$, and A and B are constants of integration. The complete solution of eqn. (9.113) is therefore

$$D\theta_y = -Q/J\omega_s + A \cos \omega_n t + B \sin \omega_n t$$

From eqn. (9.112)

$$D\theta_z = \frac{I_{yy}\omega_n}{J\omega_s}(A \sin \omega_n t - B \cos \omega_n t)$$

At $t = 0$, $D\theta_z = 0$; therefore $B = 0$.
Also, at $t = 0$, $D\theta_y = 0$; therefore $A = Q/J\omega_s$.
On substituting for A and B we have

$$D\theta_y = -\frac{Q}{J\omega_s}(1 - \cos \omega_n t) \qquad (9.115)$$

Now ω_n is of the same order as ω_s which would not normally represent less than about 120 hertz, which is too fast for the eye to perceive, so that it appears that the response is simply

$$D\theta_y = -Q/J\omega_s$$

Integration of eqn. (9.115) gives

$$\theta_y = -\frac{Q}{J\omega_s\omega_n}(\omega_n t - \sin \omega_n t)$$

The corresponding equation for θ_z is

$$\theta_z = \frac{Q}{J\omega_s\omega_n}\left(\frac{I_{yy}}{I_{zz}}\right)^{\frac{1}{2}}(1 - \cos \omega_n t)$$

Unless Q is very large, the amplitude of this motion is, in general, very small and is not seen by eye; consequently, there is apparently no movement about the axis of the torque.

EXERCISES

Chapter 1

1.1 Determine the number of degrees of freedom possessed by each of the plane mechanisms shown in the figure.

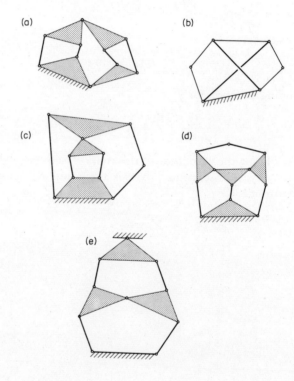

Exercise 1.1

1.2 Determine the possible basic forms of a single degree of freedom plane hinged mechanism with eight members.

1.3 Determine the possible basic forms of a plane rigid pin-jointed framework ($F = 0$) with nine members.

1.4 Show that the mechanisms in the figure both have one degree of freedom.

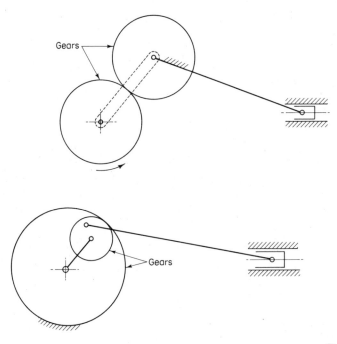

Exercise 1.4

1.5 A body is to be supported by a number of binary links which connect the body itself directly to the ground. What is the minimum number of such links if the body is to be held rigidly in space and if all connections to the body and the ground are by means of ball and socket joints?

1.6 How many links are required to solve the problem stated in exercise 1.5 if hinged joints (one degree of freedom) are used as well as ball and socket joints? How many of each type of joint are needed?

Chapter 2

2.1 The transfer relationship for a reciprocating cam and roller follower is

$$y = a(1 - \cos 2\pi x/l)$$

y being the displacement of the follower, x the displacement of the cam from its initial position, and l the total travel of the cam. Show that the maximum radius which the roller can have, if there is to be no interference, is given by the expression $l^2/4\pi^2 a$.

2.2 To reduce undesirable dynamic effects a cam and roller follower with a continuous rise and fall transfer relationship (no dwells) is designed so that it has zero acceleration at minimum and maximum lift. If the angle of rotation is θ, and the cam profile is symmetrical with respect to the $\theta = \pm \pi/2$ diameter, a simple acceleration relationship which does what is required for $-\pi/2 < \theta < \pi/2$ is

$$\frac{d^2 r}{d\theta^2} = a\theta\left(\frac{\pi^2}{4} - \theta^2\right)$$

460 Dynamics of Mechanical Systems

Show that the total lift is $2\cdot54a$. Determine what limitations, if any, are imposed on the radius R of the base circle, taking $r = R$ when $\theta = 0$,
(a) by a requirement that the profile should at all points be convex outwards,
(b) by interference, if the roller radius is $1\cdot5\,a$.

2.3 Investigate whether interference imposes any restriction on the size of a cam designed with the same transfer relationship as in exercise 2.2, but using a flat-footed follower.

2.4 Two parallel shafts carry circular discs, evenly graduated as shown in the figure.

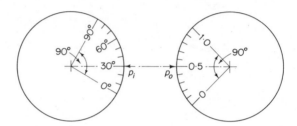

Exercise 2.4

The drive between the two shafts on which the discs are mounted is to be by means of a cam and follower which are so designed that their relative motion is pure rolling, and the readings p_o and p_i are related by the equation

$$p_o = \sin p_i$$

Draw the profiles of the cam and follower and verify their correct operation by means of a cardboard model.

If practical considerations limit the velocity ratio to the range $0\cdot2$ to $5\cdot0$, over what range is the mechanism usable?

2.5 An involute gear has 30 teeth, an addendum circle radius of 80 mm and a base circle radius of $71\cdot5$ mm. The plane of the tips of the teeth of the 20-degree rack with which it meshes is 71 mm from its axis.

Find the smallest radius of curvature of a gear tooth flank at a line of contact (a) at any time, and (b) when only one pair of teeth are in contact.

2.6 Find the number of teeth on the smallest wheel that can mesh without interference with a wheel having 60 teeth. The pressure angle is 20 degrees and the addendum for each wheel is equal to the module.

2.7 Two involute gears, having 30 and 40 teeth respectively, mesh together without interference, their module being $2\cdot5$ mm. If the pressure angle is 20 degrees and the addendum is $2\cdot5$ mm, calculate the base-circle pitch and the length of the path of contact.

2.8 Two parallel shafts, with their axes 75 mm apart, are to be connected by a pair of involute gears so that they shall rotate in the same direction with a velocity ratio $3:2$. A standard pinion is used, having a pressure angle 20 degrees, addendum 5 mm, and module 5 mm. If there are always to be two pairs of teeth in contact, determine the addendum of the teeth on the annulus.

2.9.1 Determine a suitable gear train to give a gear ratio which approximates to $e = 2\cdot7128$.

2.10 The figure shows an epicyclic gear in which the wheel P, having 45 teeth, is geared with Q through the intermediate wheel R at the end of the arm A. When P is rotating

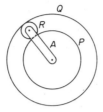

Exercise 2.10

at 63 rev/min in a clockwise direction and A is rotating at 9 rev/min, also in a clockwise direction, Q is required to rotate at 21 rev/min in an anticlockwise direction. Find the necessary numbers of teeth in Q and R.

2.11 Design a compound epicyclic gear train having two annuli such that if one is fixed the gear gives a reduction of 2:1 and if the other is fixed the gear gives a reduction of 3:1. The direction of rotation is to be unchanged.

2.12 In the epicyclic gear shown diagrammatically in the figure, the sun wheel B is attached to the input shaft A, the sun wheel E and the cage carrying the planet wheel

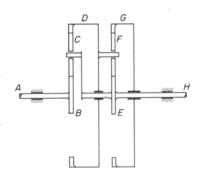

Exercise 2.12

C are both attached to the output shaft H, the annulus D forms the cage for the planet wheel F, and both the annuli D and G are free on the shaft H. The two sun wheels each have 28 teeth and the two annuli each 80 teeth.

Determine the velocity ratio of A to H when G is held fixed.

2.13 The rocker CD of a four-bar crank and rocker mechanism ABCD is at either end of its movement when the crank AB is colinear with the coupler BC, AD being the fixed link. Show, that if the times for the outward and return swings of the crank are equal, the crank AB turning with constant angular velocity,

$$DA^2 + AB^2 = BC^2 + CD^2$$

and that the angle θ between the extreme positions of the rocker is $2\sin^{-1}(AB/CD)$.

2.14 By considering the inversion of the mechanism in exercise 2.12, whereby BC becomes the fixed link and AD becomes the coupler show that the angle BCD between the coupler and rocker of the original mechanism is a maximum or a minimum when crank AB is colinear with the fixed link AD. If $\angle BCD = \phi$ show that

$$BC \cos \phi_{min} = AD \sin \theta/2$$

Determine the lengths of the crank, coupler, and rocker of such a mechanism if $\theta = 60°$, $\phi_{min} = 30°$ and the fixed link is of unit length.

2.15 Design a slider-crank mechanism so that a 90-degree rotation of the crank from position 1 to position 2 causes a 0·1 m displacement of the slider, as shown in the figure.

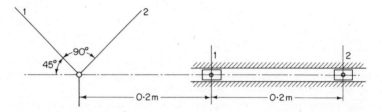

Exercise 2.15

[*Note:* It is advised that this be taken as an exercise in the use of inversion even though the solution may be obvious.]

2.16 Find a point, such as E, attached to the crank AB in the figure which is equidistant from the slider in each of the three positions indicated. This done the original crank

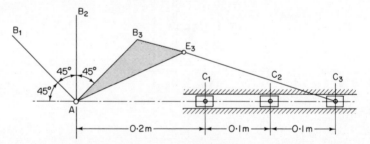

Exercise 2.16

AB can be replaced by AE and exact correlation has been obtained between two rotations of the crank and two displacements of the slider; this is to be compared with the single rotation and displacement in exercise 2.15. (See also exercise 2.20.)

2.17 Design a four-bar mechanism so that in the figure:

(a) a rotation $\phi = 0$ to $\phi = 90°$ of AB causes a rotation from $\psi = 60°$ to $\psi = 120°$ of CD

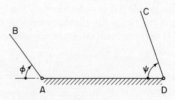

Exercise 2.17

(b) a rotation $\phi = 30°$ to $\phi = 150°$ of AB causes a rotation from $\psi = 60°$ to $\psi = 120°$
(c) crank positions $\phi = 0, 55°, 90°$ of AB correspond with positions $\psi = 60°, 90°, 120°$ respectively of CD.

Take AD to be of unit length and check your solutions by making a cardboard model.

2.18 Design a four-bar linkage to generate $y = x^2$ over the range $0 \leqslant y \leqslant 1$. Use three precision points at $x = 0, 0·6$, and $1·0$.
The cranks are to swing through the angles shown in the diagram.

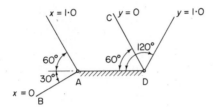

Exercise 2.18

2.19 A link BC is to be guided from position 1 to position 2, as indicated in the figure by cranks AB and CD, A and D being fixed hinges on the line of $B_2 C_1$ produced.

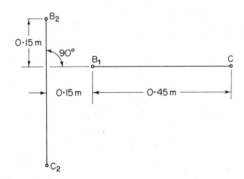

Exercise 2.19

Determine the positions of A and D and verify that BC is able to move continuously from one position to the other.

2.20 A point G is situated 0·1 m to the right of the crank hinge A in the figure for exercise 2.16, and on the centre-line of the slider. Determine the point P on the coupler which is equidistant from G in the three given positions. If a crank GP is now introduced and the slider guides are removed we have a four-bar mechanism which, for specified rotations of the crank AE, guides a point C on the coupler EP through three specified locations. Draw the locus of C.

2.21 A four-bar mechanism ABCD has links of arbitrary length; AD is the fixed link and BC is the coupler. If J_B and J_C are the points at which the inflection circle cuts AB and CD respectively, show that, in the configuration where BC is parallel to AD (assuming such a configuration to exist), $J_B J_C$ is also parallel to AD.

Design a four-bar mechanism such that the centre point E of the coupler moves for a limited range along an approximately straight line parallel to the fixed link and centrally placed with respect to it. The distance between the 'straight' part of the coupler curve and the fixed link is equal to the length of the fixed link.

2.22 In the plane four-bar mechanism in the figure, (a) AB = BC = EB and CD, AD, and EC are of arbitrary lengths. By using Roberts' theorem, sketch in the cognate

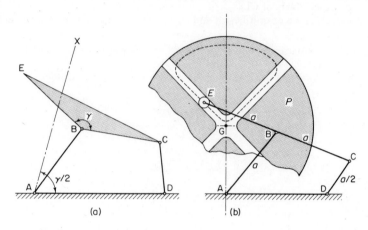

Exercise 2.22

mechanisms, and hence show that the coupler curve traced by the point E is symmetrical about AX positioned as shown with $\angle XAD = \frac{1}{2} \angle EBC$.

A particular application which has been suggested for this class of mechanism is shown in (b). The roller E which engages a cruciform slot on the plate P, is guided along the chain dotted path by a four-bar mechanism ABCD, in which AB = BC = EB = a = 2DC.

Show that, for the configuration in which $\angle BCD$ is zero, point E lies on the inflection circle for the coupler BC. Hence show that, if the two 'straight' parts of the coupler curve are to be at 90 degrees to each other, AD = $a(1 + \sqrt{7})/2\sqrt{2}$, and that the distance AG should be $a(5 - \sqrt{7})/2\sqrt{2}$.

(The plate P will now rotate through 90 degrees for each complete revolution of the crank CD with a dwell, whilst the roller E traverses the 'straight' portions of the coupler path.)

Chapter 3

3.1 A mechanism consists of four links AE, BE, CE, and D, as shown. Determine for the given configuration, (a) by statics, (b) by virtual work, the force P which must be applied at A as shown to overcome a resistance W at D. Neglect friction but see exercise 3.9.

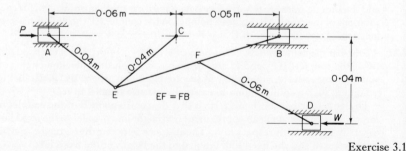

Exercise 3.1

3.2 The tongs shown in the figure are made of rigid bars, such as ab and cd, hinged together at their centres and hinged to adjacent bars at their ends. If there are n pairs of bars, plus two half-bars at the left, determine the reaction R_x and R_y in terms of P, n, and α, neglecting friction.

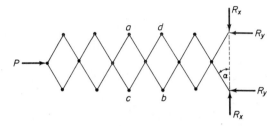

Exercise 3.2

3.3 The figure shows the mechanism of a platform weighing-machine in which the links are so proportioned that the weight W can be placed anywhere on the platform without altering the value of P, the balancing force. If the links are balanced among themselves and $P = W/3$, find the length of the arm AB.

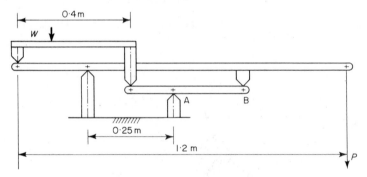

Exercise 3.3

3.4 The figure shows a light frictionless four bar mechanism ABCD and a fixed point

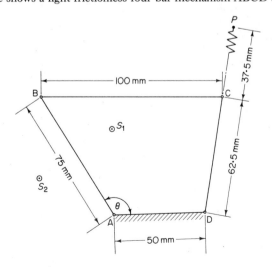

Exercise 3.4

P in its plane. P and C are joined by a light compression spring with freely pivoted ends, so that in the position shown, where θ is 120 degrees and C lies in the straight line PD, the mechanism is in unstable equilibrium: if disturbed slightly it will move until AB comes up against one of two fixed stops S_1, S_2 which limit θ to the range 90 to 150 degrees.

A torque of 0·5 N m in the appropriate sense must be applied to AB to remove it from either stop. Hence find the force in the spring in the positions $\theta = 90$ degrees and $\theta = 150$ degrees, the stiffness of the spring and the load in it in the position $\theta = 120$ degrees.

3.5 A bar AB lies on a flat surface and a concentrated load W acts on it. The bar is very

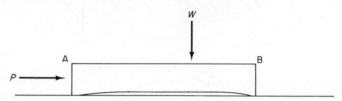

Exercise 3.5

slightly curved so that contact with the surface occurs at A and B only. If the law of friction between bar and surface is

$$F = KN^\gamma$$

where F is the limiting friction, N the normal force, and K and γ are constants find an expression for the force P to move the rod when the load W divides the bar in the ratio of $r : (1 - r)$.

Deduce that the magnitude of P is independent of the position of W only if γ is equal to unity.

3.6 A light square peg A is a loose fit in a hole in a fixed plate B, as shown. The coefficient of friction between the peg and the plate is μ.

If a force P is applied to the peg as shown, prove that it will not slip provided that

$$\tan\theta > \frac{a}{\mu(c + \mu b)}$$

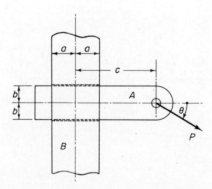

Exercise 3.6

3.7 A rectangular box of weight W rests against an incline along its edge A, which is horizontal, and is supported with edge AC horizontal as shown in the figure by a force P applied to a lever BCD which is hinged to the ground at D. The angle of friction for the surfaces in contact at A and C is 30 degrees.

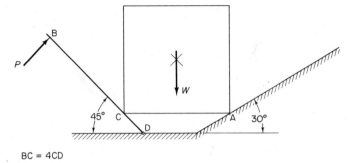

Exercise 3.7

Show that slipping will start at A or C according as the block is raised or lowered from this position, and determine the corresponding values of P.

3.8 If, for the tongs of exercise 3.2, a friction couple Q opposes the relative motion at each joint, find the magnitude of the pair of equal and opposite forces R_x which will just make them operate with $P = 0$, the length of each bar being l.

3.9 Determine the instantaneous efficiency of the mechanism of exercise 3.1 if the slides at A and D and all hinges are frictionless, but at B there is a coefficient of friction of 0·5 between the slider and its guides.

3.10 The mechanism shown in the figure works in a vertical plane. All the members are light, but a load of 60 kN acts vertically downwards through G. OA is a fixed horizontal guide, and the coefficient of friction between this guide and the slider at A is 0·2. The pivot at A is frictionless, but those at O and B have both a constant resistance to turning of 0·5 kN m regardless of load. What is the force at P required to move A to the right?

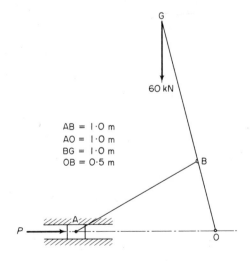

Exercise 3.10

468 Dynamics of Mechanical Systems

3.11 Four equal uniform thin rods, each of length a, are freely jointed together to form a rhombus. They are then hung with one diagonal vertical on a rough circular peg of radius r ($r < a/10$). If equilibrium is possible in the configuration in which the length of the horizontal diagonal is one-half that of the vertical diagonal, what is the minimum coefficient of friction at the peg?

3.12 Two identical involute gears each with N teeth are mated to form a gear drive, and are so proportioned that there is always exactly one pair of teeth in contact. The output torque from the drive is constant. Show that, if the coefficient of friction μ at the point of contact between the teeth is small, the average efficiency of the drive for a continuous rotation of the input and output shafts is approximately $(1 + \pi\mu/N)^{-1}$.

3.13 A uniform scaffold-pole of length l is leant against a vertical wall, the foot of the pole being in a small pot-hole, which is distant $l/2$ from the bottom of the wall. The coefficient of friction between the pole and the wall is $\frac{1}{4}$. If the pole is to remain in position, what may be the largest angle between the vertical and the projection of the pole on the wall?

3.14 A symmetrical rotor weighing 1000 kg has a shaft which fits loosely in 0·15-m diameter bearings. If the coefficient of friction between the shaft and bearings is 0·1, find the magnitude of the torque which is just adequate to turn the rotor in its bearings.

3.15 A thrust rod carries pins of diameter $l/5$ and with centres A and B distant l apart. The pins bear loosely in holes in blocks, and these are guided by a pair of plane surfaces at right-angles, the lines of movement of A and B meeting at O. A force F in the direction AO acts on the block at end A and overcomes a force nF in the direction BO on the block at end B. At the instant when $\angle$ BAO $= 60$ degrees determine the value of n if the angle of friction between the blocks and the planes is 20 degrees,

(a) assuming frictionless pins,
(b) assuming the same angle of friction at the pins.

3.16 The figure shows a linkage which supports a vertical load P. The linkage stands on a smooth surface and is prevented from collapsing by the forces F. Each of the four hinge-pins is of radius r and the coefficient of friction at each pin is μ.

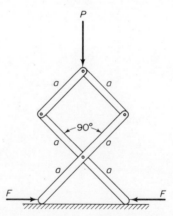

Exercise 3.16

Show that F is given approximately by the expression $P[1\cdot5 - (3 + 2\sqrt{2})\mu r/a]$. It may be assumed that $\mu r \ll a$. The load P is applied to the linkage through the upper hinge-pin.

EXERCISES

Chapter 4

4.1 Two vectors **A** and **B** are given by:
$$A = 2i + 3j - k$$
$$B = i - 3j + 4k$$
Find: (a) A, (b) B, (c) **A** + **B**, (d) **A** − **B**, (e) **A**.**B**, (f) **A** × **B**, (g) angle θ between **A** and **B**.

4.2 (a) With reference to the vectors:
$$A = A_x i + A_y j + A_z k$$
and
$$B = B_x i + B_y j + B_z k$$
in which the components are functions of t, verify that

$$\frac{d}{dt}(A \times B) = \frac{dA}{dt} \times B + A \times \frac{dB}{dt}$$

(b) A moving point has position, velocity, and acceleration vectors **r**, **v**, and **a** respectively. Show that

(i) $\quad\quad\quad\quad\quad\quad\quad\quad \dfrac{d}{dt}(\tfrac{1}{2}\mathbf{v}.\mathbf{v}) = \mathbf{v}.\mathbf{a}$

(ii) $\quad\quad\quad\quad\quad\quad\quad\quad \dfrac{d}{dt}(\mathbf{r} \times \mathbf{v}) = \mathbf{r} \times \mathbf{a}$

and (iii) $\quad\quad\quad\quad\quad\quad \mathbf{r}.\mathbf{v} = r\dfrac{dr}{dt}$

4.3 In the mechanism shown in the figure, which is not to scale, link (1) is fixed and link (2) rotates anticlockwise with uniform angular velocity ω_2. Describe the motion of link (4) during a cycle, giving the maximum and minimum values of ω_4 and the positions at which $\omega_4 = \omega_2$.

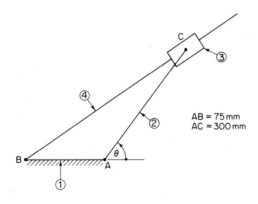

Exercise 4.3

What is the maximum velocity of sliding of the block (3) relative to the link (4)? Locate the instantaneous centre of the block (3).

4.4 The figure shows the mechanism of a vertical slotting machine. Crank AP turns at 60 rev/min.

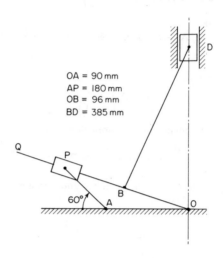

```
OA = 90 mm
AP = 180 mm
OB = 96 mm
BD = 385 mm
```

Exercise 4.4

(a) Determine the velocity of the tool box D.
(b) State velocity of sliding of P along OQ.
(c) Calculate rubbing velocity on the 36 mm diameter pin at B.
(d) Determine the torque on AP to overcome a resistance of 500 N at D.

4.5 In an offset slider-crank chain using an eccentric sheave for crank, P represents the centre of the sheave, C the centre of the shaft, and D the centre of the pin connecting the eccentric rod to a slider. Show that the velocity of sliding of the strap on the sheave is given by

$$\frac{TD}{PD}\pi n d$$

where T is the point where PD cuts the line through C perpendicular to the line or stroke of D, d is the diameter of the sheave, and n the speed of the shaft.

4.6 The mechanism shown in the figure lies in one plane and uses a crank CB driving a connecting-rod BA to which is pivoted at D a rod DE moving a slider E on a fixed slide parallel with the slide at A.

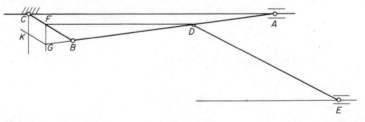

Exercise 4.6

Prove that the velocity of E is given by $\omega \cdot CK$ in the construction indicated, ω being the angular velocity of CB. DF is parallel to AC, FG and CK are perpendicular to AC, G lies on AB and GK is parallel to DE.

4.7 A four-bar space linkage ABCD has AD as the fixed link, AB and DC as cranks, and BC as the connecting-rod. AB is defined by a vector $\mathbf{r}_1$ and rotates about a fixed axis through A with an angular velocity ω_1; DC is defined by $\mathbf{r}_2$ and rotates about a fixed axis through D with an angular velocity ω_2. The connecting-rod BC, which is

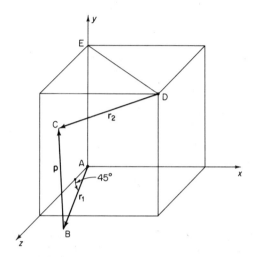

Exercise 4.7

attached to the cranks by spherical joints at B and C, is defined by a vector $\mathbf{p}$. Show that

$$\mathbf{p} \cdot (\omega_2 \times \mathbf{r}_2 - \omega_1 \times \mathbf{r}_1) = 0$$

In a particular case, A and D are located at the corners of a cube with edges of unit length as shown. AB is of unit length and is perpendicular to the axis Ax about which it rotates. DC is of unit length and is perpendicular to ED about which it rotates. In the given configuration, AB is at 45 degrees to the axis Az as shown in the figure, and DC is parallel to the Axz plane. Determine ω_2 given that ω_1 is 1 rad/s.

4.8 A space mechanism consists of a crank AB, a connecting rod BC, and a slider C, there being spherical joints at B and C. The crank rotates about A in a plane which is perpendicular to a unit vector $\mathbf{n}$. The direction of the slider is defined by a unit vector $\mathbf{i}$. If AB is represented by the vector $\mathbf{r}$ and BC by $\mathbf{b}$, and if the angular velocity of AB is $\omega\mathbf{n}$, show that the velocity v of the slider is given by

$$v = \frac{\omega(\mathbf{n} \times \mathbf{r}) \cdot \mathbf{b}}{\mathbf{i} \cdot \mathbf{b}}$$

and hence deduce that the velocity of the slider is zero when $\mathbf{n}$, $\mathbf{r}$, and $\mathbf{b}$ are coplanar.

In a particular case A lies on the axis of the slider and may conveniently be considered to be at the origin of a set of coordinate axes for which the relevant unit vectors are $\mathbf{i}$, $\mathbf{j}$, and $\mathbf{k}$, $\mathbf{i}$ being aligned with the axis of the slider. If $r = 0.1$ m, $b = 0.3$ m and $\mathbf{n} = (1/\sqrt{2})(\mathbf{i} + \mathbf{j})$, determine v at the instant when AB is

(i) in the direction $\mathbf{k}$.
(ii) at 45 degrees to the inner dead centre position.

(See also exercise 4.17.)

4.9 Determine for the hoisting mechanism shown:
(a) the crank force needed to lift 750 kg.
(b) the fixing torque applied to A.

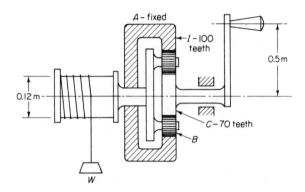

Exercise 4.9

4.10 The figure shows an epicyclic gear train for which the overall gear ratio may be varied as follows: (a) by fixing gear C whilst D is left free to revolve, (b) by fixing D and freeing C, (c) by locking C and D so that they rotate together.

Determine the speed ratio for case (a). If a 2:1 reduction with the input and output shafts rotating in the same direction is required for case (b), what must be the ratio

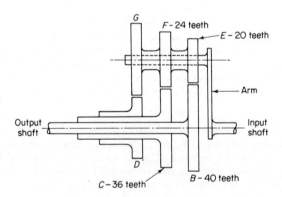

Exercise 4.10

of the pitch-circle diameters for gears G and D? What is the speed ratio for case (c)?

Calculate the torque between the locked gears C and D in case (c) for an input torque of 100 N m when G and D have the gear ratio calculated in (b). Assume that friction is negligible.

4.11 The epicyclic gear train shown in the figure is part of a computer mechanism. The angular displacement of the output shaft is $z = Ax + By$, where x and y are the displacements of the two input shafts, all rotations being positive in the same direction. Determine the constants A and B.

Note that the internal gear N_3 is free to rotate.

N_1 has 64 teeth
N_2 has 16 teeth
N_3 has 96 teeth
N_4 has 32 teeth

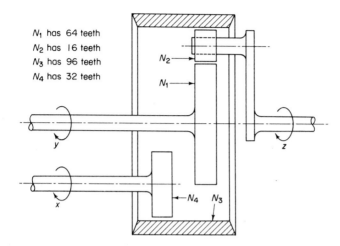

Exercise 4.11

4.12 The figure shows a limacon cam whose profile is, in polar coordinates,

$$r = 50 - 25 \cos \phi$$

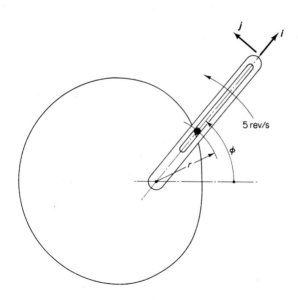

Exercise 4.12

where r is in mm. The arm rotates steadily at 5 rev/s. Find the acceleration of the follower when the arm makes an angle of 30 degrees with the $\phi = 0$ position.

4.13 The figure indicates part of an epicyclic gearbox in which the centre Q of the gear wheel of radius r, moves with angular velocity Ω about a fixed centre O. The wheel has angular velocity ω relative to the rotating 'spider' which includes the arm OQ, the length of which is R.

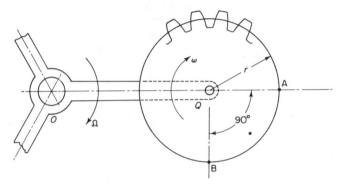

Exercise 4.13

Show that the absolute accelerations of points A and B are respectively

$$\Omega^2(R + r) + \omega^2 r + 2\omega\Omega r \quad \text{and} \quad [\Omega^4 R^2 + (\Omega + \omega)^4 r^2]^{\frac{1}{2}}$$

Do this:
(a) by the use of eqn. 4.19 taking the absolute angular velocities of the spider and gear wheel to be Ω and $(\Omega + \omega)$ respectively, and
(b) by use of eqn. 4.20 taking the spider to be the moving plane, whose angular velocity is Ω, and considering the gear to have an angular velocity relative ω to the moving plane.

4.14 The figure shows one of the vanes of a centrifugal pump impeller which turns with a constant clockwise angular velocity of 300 rev/min. R is the radius of curvature of the blade surface at the position shown. The fluid is observed to have an absolute velocity whose radial component is 3 m/s at discharge from the vane. The velocity of the elements of fluid measured relative to the vane is along the vane surface and

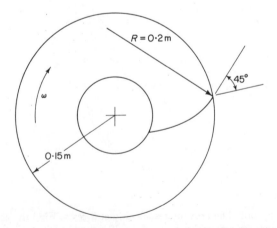

Exercise 4.14

their speed increases at the rate of 24 m/s² just before they leave the vane. Find the total acceleration of an element of fluid just before it leaves the impeller.

EXERCISES 475

4.15 An engine has a crank CB with a throw of 0·3 m, and its connecting-rod is 1·5 m long between the centres A and B. The crank rotates with a constant angular velocity of 180 rev/min. Draw velocity and acceleration diagrams for crank angles ACB of (a) 60 degrees and (b) 225 degrees, and for each case record the piston acceleration and the angular acceleration of the connecting-rod.

4.16 An aircraft is at latitude 60°N. and longitude 45°W., and is flying due west on a great circle course at constant velocity V relative to the earth's surface. The angular velocity of the earth is ω and its radius is R.

Taking the origin of coordinates at the earth's centre, and fixed axes OX through the instantaneous position of the Greenwich meridian and OY through the North Pole, find the acceleration of the aircraft relative to OXYZ.

4.17 Determine the acceleration of the slider, and the angular acceleration of the connecting-rod, of the mechanism studied in exercise 4.8 for the two given configurations.

4.18 At a particular instant the ends of a rod AB, 1·2 m in length, have the accelerations shown in magnitude and direction in the figure. Sketch an acceleration diagram for the rod, and hence find the acceleration of its mid-point and its angular velocity and angular acceleration.

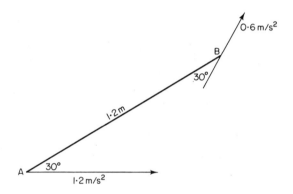

Exercise 4.18

4.19 AB, BC, and CD are three rods connected by pins at B and C, and hinged to a fixed frame at A and D, as shown; AB = r, BC = $2r$, CD = $2r$, and G is the mid-point of BC. AB is rotating with constant angular velocity ω at the instant depicted in the diagram. What is the acceleration of G and the angular acceleration of BC?

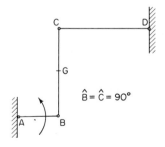

Exercise 4.19

4.20 In the mechanism shown in the figure, the pivots A and D are fixed, the crank AB rotates with uniform angular velocity ω and the link DE passes through a slider pivoted to B. C is the point in DE instantaneously coincident with B, and H is the foot of the perpendicular from A onto DE.

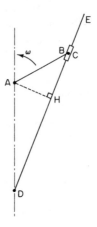

Exercise 4.20

Sketch (not to scale) the velocity and acceleration diagrams, labelling the vectors with their magnitudes in terms of ω, AB, AH, CD, and CH. Hence show that when CD = 3CH, the directions of the accelerations of B and C coincide, and find the ratio of the magnitudes of these accelerations at that instant.

4.21 The plane four-bar mechanism ABCD shown in the figure is in a vertical plane, and at the instant when the configuration is as depicted BC has an angular velocity ω

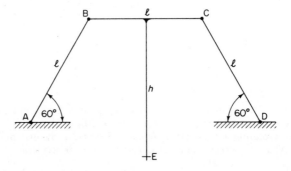

Exercise 4.21

and zero angular acceleration. From sketches of the velocity and acceleration diagrams at the given instant, determine expressions for:

(a) The velocity of point E which is at a perpendicular distance h from the centre of BC, as shown in the figure.
(b) The acceleration of E.
(c) The radius of curvature of the path of E.

Hence or otherwise deduce that if a concentrated mass is attached at E, and if the masses of the links are negligible, the system will be in stable equilibrium in the given position if

$$h > \frac{5}{2\sqrt{3}} l$$

4.22 A heavy particle A is suspended by a frictionless linkage consisting of two light rods AC, BD, as shown. The end C can slide freely on a vertical axis through the fixed

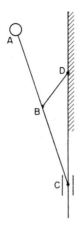

Exercise 4.22

point D. AB = BC = 150 mm, BD = 75 mm. Determine the radius of curvature of the path of A at the equilibrium position and hence deduce the frequency of small oscillations about the equilibrium position.

4.23 Link OB of the mechanism shown in the figure rotates about O at a constant speed of 100 rev/min and is pinned at B to the rigid link AC. For the position indicated in the diagram find the velocity and acceleration of the point D.

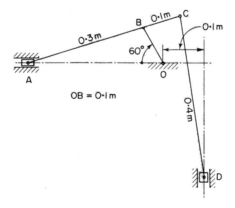

Exercise 4.23

4.24 A structural column is being erected as shown in the figure. The base A is pulled along the ground with a uniform velocity v whilst a rope attached to the column at B and passing over a small fixed pulley at C is wound in also at a uniform velocity v. C is at a height l above the ground, AB and C are in the same vertical plane, and the length AB is l.

16*

478 Dynamics of Mechanical Systems

When the column is inclined at 30 degrees to the ground and A is vertically below C, find the magnitude and sense of its angular acceleration. Also find the angular acceleration of the rope BC.

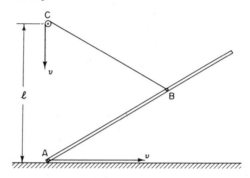

Exercise 4.24

4.25 The piston of a reciprocating engine moves along a straight line which does not pass through the axis O of rotation of the crank. The crankpin and gudgeon-pin centres are C and P respectively. Provided the crank rotates uniformly, the acceleration of the piston can be found graphically as follows: through O draw a line perpendicular to the line of stroke of the piston to meet PC in T; through T draw a line parallel to PO to meet OC in K; through K draw a line perpendicular to the line of stroke to meet PC in N; through N draw a line perpendicular to PC to meet a line through O parallel to the line of stroke in Z. All lines may be produced if this is necessary to obtain an intersection.

Show that, if the angular velocity of the crank is ω, the acceleration of the piston is $ZO \cdot \omega^2$ in the sense $\overrightarrow{ZO}$.

4.26 The figure shows a plane mechanism in which the jack AB operates between a fixed hinge A and the centre point B of the rigid link CD.

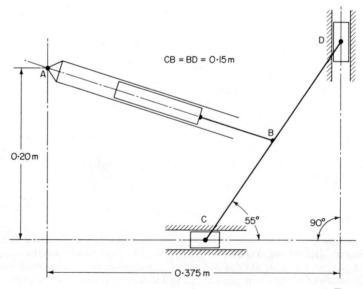

Exercise 4.26

Determine, for the given configuration, (i) the velocity of C, (ii) the acceleration of C, if oil is pumped into the jack at a constant rate to cause the jack piston to move relative to the cylinder at a speed of 0·5 m/s.

4.27 The figure shows a mechanism in which two cranks OA and OB at right-angles rotate together about the fixed centre O with a uniform angular velocity of 30 rad/s. The point E on link CD moves on a straight line through O and the dimensions of the links are OA = 0·2 m, OB = 0·15 m, AC = 0·9 m, BD = 1·05 m, CE = 0·2 m and ED = 0·3 m.

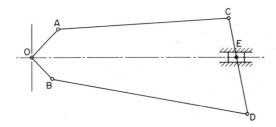

Exercise 4.27

By considering two auxiliary points rigidly attached to link CED and located where the perpendicular to the line of stroke at E meets the centre lines AC and BD, or otherwise, draw velocity and acceleration diagrams for the mechanism at the instant when angle AOE = 45 degrees.

4.28 The figure shows a cam with its profile formed of four circular arcs. Arcs AB and CD, each of radius a, have centres O and Q distant a apart. BC and DA each have

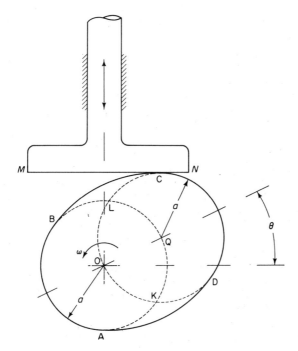

Exercise 4.28

radius $2a$ and their centres coincide with the intersections K and L shown dotted. The cam rotates with constant angular velocity ω about the centre O and drives a flat-footed follower which reciprocates vertically.

Show that the upward acceleration α of the follower is as given in the following table:

$$-\pi/6 < \theta < \pi/6, \qquad \alpha = -\omega^2 a \sin(\theta - \pi/3)$$
$$\pi/6 < \theta < 5\pi/6, \qquad \alpha = -\omega^2 a \sin\theta$$
$$5\pi/6 < \theta < 7\pi/6, \qquad \alpha = -\omega^2 a \sin(\theta + \pi/3)$$
$$7\pi/6 < \theta < 11\pi/6, \qquad \alpha = 0$$

4.29 Two parallel horizontal shafts are placed with their axes in the same vertical plane but 6 mm apart, as shown in elevation in the figure. The shafts are circular, 100 mm diameter. A circular pin 25 mm diameter is fixed to the end of shaft A, the centre of the pin being 25 mm from the axis of A, and engages in a straight slot 25 mm wide cut symmetrically in the end of shaft B. Shaft B rotates uniformly at 100 rev/min and rotation is transmitted through the slot and pin to shaft A.

By considering the equivalent crank mechanism or otherwise, determine graphically, for the orientation when shaft B has rotated 150 degrees from the position in the figure;

(a) the rubbing velocities between the sides of the slot and the surface of the pin;
(b) the angular acceleration of shaft A.

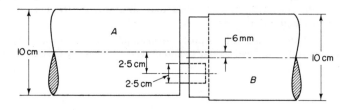

Exercise 4.29

4.30 The details of a straight-flanked cam with roller follower are shown in the figure.
   (a) Sketch the equivalent mechanisms which represent the cam and its roller when the roller is in contact with the portions BC and CD of the profile.
   (b) Determine the vertical velocity and acceleration of the follower for the position shown in the figure when $\omega = 1$ rad/s.

4.31 The figure shows a reciprocating cam at the centre point of its travel, which extends 75 mm on either side of the given position. The cam moves with simple harmonic motion at a frequency of 10 Hz. The rocker AC carries a 25-mm diameter roller at A, and turns about the fixed point C, the distance AC being 75 mm. When the cam is in its central position, AC is parallel to the line of stroke of the cam.

(a) Determine the angular acceleration of the rocker for the given position.
(b) Determine the angular velocity and angular acceleration of the rocker when the cam is 37·5 mm to the right of the given position.
(c) Determine the angular velocity and angular acceleration of the roller when the cam is 37·5 mm to the right of the given position, assuming that there is no slip between the roller and the cam.

Assume in parts (b) and (c) that the 62·5 mm radius arc of the cam profile is sufficiently long for the point of contact to be on it.

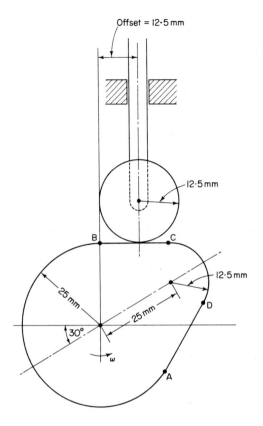

Exercise 4.30

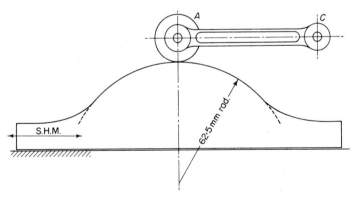

Exercise 4.31

4.32 Crank OB of length r rotates with angular velocity ω. It drives rod BC which slides through a block pivoted at the fixed point A. Derive an expression for the angular velocity of the rod $d\psi/dt$ in terms of r, θ, s, and ω.

Check your solution for the four points where the result can be deduced almost immediately, namely, for $\theta = 0$ degrees and 180 degrees, and for the two positions where $d\psi/dt = 0$.

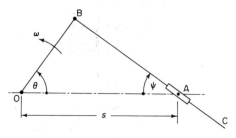

Exercise 4.32

4.33 The figure shows a quick-return mechanism. Derive expressions for the velocity and acceleration of the block B in terms of the given parameters and the crank angle θ.

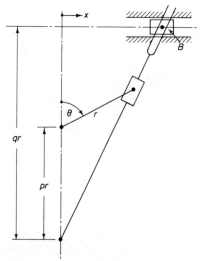

Exercise 4.33

4.34 Show that the angular velocity and acceleration of the connecting-rod of the given mechanism are given approximately by

$$\frac{d\phi}{dt} = \frac{\omega \cos \theta}{n} \quad \text{and} \quad \frac{d^2\phi}{dt^2} = -\frac{\omega^2 \sin \theta}{n}$$

$$n \gg 1$$

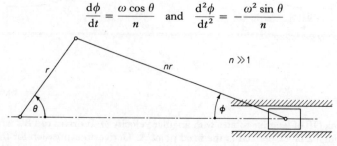

Exercise 4.34

4.35 A crank and piston mechanism has the line of stroke offset from the crankshaft centre line as shown in the figure.

Show that, if ω is constant and $n \gg 1$, the piston acceleration is approximately

$$\frac{d^2x}{dt^2} = -r\omega^2\left(\cos\theta + \frac{p\sin\theta + \cos 2\theta}{n}\right)$$

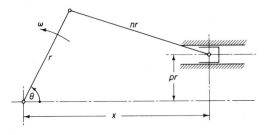

Exercise 4.35

What are the crank angles at inner and outer dead centres and what is the length of the stroke?

4.36 The figure shows a mechanism used for giving an intermittent motion to the film in a cine-projector. A pin C, fixed to a wheel which rotates about an axis A with uniform angular velocity ω, engages with a slotted piece which rotates about a parallel

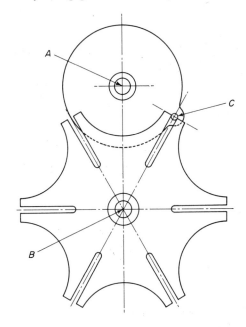

Exercise 4.36

axis B. The slots are spaced at equal angles of 60° and $AC = \frac{1}{2}AB$. Show that during the motion of the slotted piece its angular velocity is given by

$$\frac{1 - 2\cos\theta}{5 - 4\cos\theta}\omega$$

where θ is the angle CAB, and derive an expression for its angular acceleration.

4.37 Two points P and Q are fixed to a lamina which moves in its own plane. The positions of P and Q in relation to a fixed origin O are defined by the vectors $\mathbf{r}_P$ and $\mathbf{r}_Q$, and the angular velocity of the lamina is ω. Show, that in terms of unit vectors $\mathbf{i}$ and $\mathbf{j}$ as shown in fig. (a), the rate of change of the acceleration of P is

$$\dddot{\mathbf{r}}_P = \dddot{\mathbf{r}}_Q + PQ(\ddot{\omega} - \omega^3)\mathbf{j} - 3PQ\omega\dot{\omega}\mathbf{i}$$

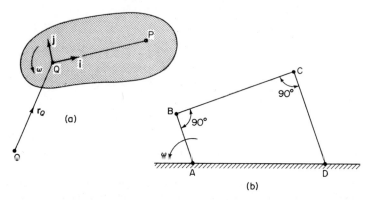

Exercise 4.37

It may be assumed that

$$\ddot{\mathbf{r}}_P = \ddot{\mathbf{r}}_Q - PQ\omega^2\mathbf{i} + PQ\dot{\omega}\mathbf{j}$$

The input crank AB of a plane four-bar mechanism ABCD, AD being the fixed link, rotates with constant angular velocity ω. The mechanism is so proportioned that when the two cranks AB and CD are parallel to each other on the same side of AD they are perpendicular to BC as shown in fig. (b).

Show, by sketching velocity, acceleration, and rate of change of acceleration diagrams, or otherwise, that in the configuration just described the magnitude of the angular acceleration of BC is a maximum.

Chapter 5

5.1 The uniform rod AB of length $2l$ shown in the figure has mass m and carries an additional concentrated mass m at the end A. It is moving freely under gravity forces and a horizontal force mg at B. When it is in the position indicated it has zero angular velocity.

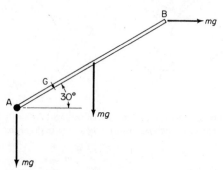

Exercise 5.1

Determine the components of the acceleration of the joint centre of gravity G, the angular acceleration of the rod and the components of the acceleration of the end B.

5.2 A thin circular wire ring of radius a is supported in a vertical plane by two pegs which are at the same level and $a\sqrt{2}$ apart, the centre of the ring being below the pegs. One peg is suddenly withdrawn. Assuming that the ring does not immediately slip on the remaining peg, derive an expression for the angular acceleration of the ring at this instant, and so determine the normal and tangential components of the reaction at the remaining peg. Hence, show that the initial motion of the ring will not involve slipping at this peg if the coefficient of friction there exceeds $\frac{1}{2}$.

5.3 A uniform disc of mass m and radius r has a concentrated mass m attached to its rim. The disc is free to roll upon a horizontal surface and is held with its plane vertical, the concentrated mass being at the same horizontal level as the centre. Show that, if the system is released from this rest position, the disc will immediately slip on the horizontal surface unless the coefficient of friction μ is greater than $\frac{1}{3}$. Determine the initial angular acceleration (a) when $\mu > \frac{1}{3}$, (b) when $\mu < \frac{1}{3}$.

5.4 A solid uniform cylinder is placed on a horizontal plane which is then given an acceleration f in a horizontal direction at right-angles to the line of contact with the cylinder. The coefficient of friction between the cylinder and the plane is μ. Show that the linear acceleration of the centre of the cylinder will be either $\frac{1}{3}f$ or μg, whichever is the less.

5.5 A smooth hoop of radius a and centre Q is rotated in its own plane which is horizontal with constant angular velocity ω about an axis O through a point on its circumference. A small ring of mass m slides on the hoop and is released from the centre of rotation O. Show that when the ring has moved to a point P on the hoop defined by $\angle POQ = 2\theta$, the torque required to rotate the hoop is

$$2m\omega^2 a^2 \sin\theta \sin 2\theta (2 \pm 3\sin\theta)$$

the sign depending on the direction of the initial movement of the ring.

5.6 A shaft rotates in two bearings, 1·5 m apart, and projects 0·3 m beyond each bearing. At each end of the shaft there is a pulley, one of mass 15 kg and the other of mass 40 kg, their centres of gravity being respectively 10 mm and 15 mm from the axis of the shaft. Midway between the bearings is a third pulley of mass 50 kg, its centre of gravity being 12·5 mm from the shaft axis. If the system is in static balance, find the forces (other than gravity) acting on the bearings when the shaft is rotating at 400 rev/min.

The shaft is to be balanced by adding masses to the outer pulleys at 15-cm radius. Find the magnitude and angular positions of the masses required. Will the shaft now run smoothly at all speeds?

5.7 A crankshaft for a six-cylinder engine, shown diagrammatically, has cranks at

Exercise 5.7

120 degrees spaced 150 mm apart between the lines of stroke. It is balanced statically but not dynamically and it is found that dynamic balance can be obtained by temporarily attaching to it masses P and Q, each of 50 g, at a radius of 125 mm, spaced 75 mm apart axially and in a place at 30 degrees to that of cranks C and D as shown.

Determine what weights of metal must be removed from cranks A, B, D, and E, at a radius of 50 mm and at the same axial spacing as the cranks, in order to balance the crankshaft completely.

5.8 Explain the terms *static* and *dynamic balance* applied to rotating machinery. The figure shows a balancing machine carrying a rotor which is to be balanced by the removal of metal in the planes marked I and II. When the rotor is spinning the scale indicates the amplitude of vibration of the carriage.

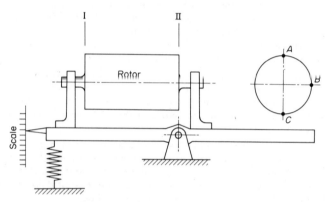

Exercise 5.8

With the rotor in the position shown the carriage vibrated with a maximum amplitude of 5·0 units. A mass of 15 g was then attached to the rotor in plane I, at 125 mm radius, in positions A, B, and C in turn, whereupon the maximum amplitudes of the carriage were observed to be 6·3 units, 11·8 units, and 10·5 units respectively.

Estimate the mass of metal which must be removed from the rotor in plane I at a radius of 75 mm, and state where the hole must be drilled. What further procedure is required to balance the rotor completely?

5.9 A wind tunnel fan with four blades is mounted on a shaft midway between two bearings which are 1·3 m apart. The resultant aerodynamic force F on each blade may be taken to act perpendicularly to the flat face of the blade at a representative section at a radius of 0·6 m. Its value is given by

$$F = 35(\alpha - 25) \text{ N}$$

where α is the angle in degrees between the plane of rotation and the flat face of the blade at the representative section. The values of α, taken in the order of rotation, are 30, 29·5, 31, and 30·5 degrees.

Calculate the magnitude of the transverse unbalanced forces acting on the bearings and their phase with respect to the first blade.

5.10 The body and blading of an electric fan, in which the rotor axis is horizontal, oscillate about a vertical axis through a total angle of 120 degrees. The motion is simple harmonic with a period of 20 s. The combined weight of the fan blades and rotor is 0·56 kg and the radius of gyration is 38 mm.

Calculate the maximum gyroscopic couple for a blade speed of 500 rev/min. Indicate, by means of a sketch, the direction in which the fan will tend to overturn if, when viewed from the front, the blades rotate clockwise whilst their centre moves from left to right. Explain briefly how you reach your conclusion.

5.11 A road-surface testing machine is shown diagrammatically in the figure. A road wheel of mass 22·5 kg, diameter 0·6 m and polar moment of inertia 1·3 kg m² is mounted on a frictionless journal bearing at the end B of a light horizontal shaft.

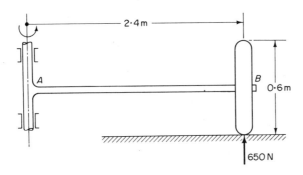

Exercise 5.11

The whole system is driven through a vertical spindle rigidly connected to the shaft at its other end A. The normal reaction between the wheel and the surface on which it rolls without slipping is adjusted by loading the vertical spindle to be constant at 650 N.

Find the bending moment at each end of the rod AB when the wheel is rolling with a uniform forward velocity of 24 m/s. Sketch the bending moment diagram for AB.

5.12 A swash-plate pump has n pistons ($n \geqslant 2$) which move parallel to the axis of the driving shaft in fixed cylinders. As seen in end view, the centres of the cylinders are distributed at equal angular intervals of $2\pi/n$ round the circumference of a circle, radius r, with the axis of the driving shaft at its centre. The swash-plate rotates with the driving shaft and its surface bears axially against the pistons without contact being lost at any time. If the bearing surface of the swash-plate is a plane inclined at an angle ($\frac{1}{2}\pi - \alpha$) to the axis of the driving shaft, show that the pistons will each be given a sinusoidal reciprocating motion of amplitude $r \tan \alpha$.

Show that the pump will be dynamically balanced if the mass-centre G of the rotating parts lies on the axis of the driving shaft and

$$(I_1 - I_2) \cos^2 \alpha = \tfrac{1}{2} nmr^2$$

where m is the mass of each piston,
I_1 is the moment of inertia of the rotating parts about an axis G1 perpendicular to the plane of the swash-plate,
I_2 is the moment of inertia of the rotating parts about an axis G2 perpendicular to G1 in the plane containing G1 and the axis of rotation.

5.13 The figure shows a light, straight, elastic shaft supported in short bearings A and B a distance l apart. A uniform, rigid rotor in the form of a perfect solid of revolution is fixed to the mid-point of the shaft. Due to errors in assembly, the mass-centre G of the rotor is displaced a small radial distance δ from the axis of the undeflected shaft. Moreover, although the axes of the rotor and of the undeflected shaft lie in the same plane, they are inclined at a small angle α. The mass of the rotor is M, its

488 Dynamics of Mechanical Systems

polar moment of inertia is I_1 and the other principal moments of inertia at G are both equal to I_2. The bending stiffness of the shaft is EI.

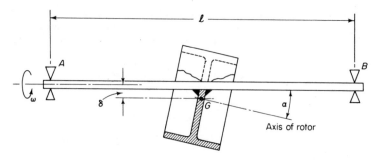

Exercise 5.13

Derive expressions for the radial displacement y of G from the centre line AB of the bearings, and for the angle θ between AB and the polar axis of the rotor, when the system rotates steadily about AB with angular velocity ω. Sketch graphs showing the variation of y and θ with ω, distinguishing carefully between the two cases (a) $I_1 > I_2$ and (b) $I_1 < I_2$.
Neglect gravity.

5.14 The figure shows the arrangement of turbine, gears, propeller and anti-roll ring fitted in a torpedo. Numbers of teeth are given on the figure; all teeth have the same diametral pitch. The turbine-shaft system has a movement of inertia of 0·1 kg m², the propeller-shaft system 0·2 kg m², and each of the three compound star wheels 0·05 kg m². Calculate the moment of inertia of the anti-roll ring necessary to neutralize inertia torques as the turbine accelerates.

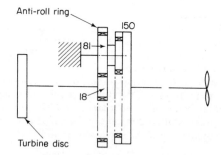

Exercise 5.14

If a constant torque of 20 Nm is applied to the turbine disc while the propeller shaft runs free, calculate the angular acceleration of the turbine in revolutions per minute per second.

5.15 The primary out-of-balance inertia force due to the mass of an engine piston can be counterbalanced by fixing a mass to the crankshaft, at crank radius, and at 180 degrees to the crank, as shown in the figure. If the balancing mass equals the mass of the piston the resultant inertia force in the direction of the cylinder axis is zero. Nothing is gained, however, because there is now an unbalanced force, equal in magnitude to the original inertia force, perpendicular to the line of stroke. The resultant force can be reduced by making the mass of the balance weight less than the mass of the piston. The mechanism is then said to be partially balanced. Show that the best that can be achieved by this method is a 50 per cent reduction of the resultant inertia force.

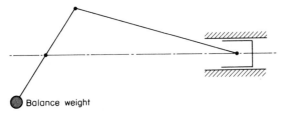

Exercise 5.15

5.16 A three-cylinder in-line engine has equal piston masses and evenly spaced cylinders. The three cranks are at 120 degrees to each other. Masses are to be attached to the two outer cranks in order to reduce by 50 per cent the primary out-of-balance couple in the plane of the cylinder axes. Determine: (a) the masses of the counter-weights relative to the mass of a piston, assuming that the former are set at crank radius; and (b) the angular positions of the counter-weights relative to the outer cranks.

5.17 Determine the out of balance forces and couples, up to and including those of the fourth order, due to the inertia of the reciprocating parts of a five-cylinder in-line engine. The engine uses a two-stroke cycle, with the firing order 1–5–2–3–4.

All the reciprocating parts are identical, the cylinders are evenly spaced, and the angular spacing of the cranks is uniform.

5.18 In a three-cylinder radial engine all the connecting-rods act on a common crank-pin. The reciprocating parts are identical for each cylinder and the centre-lines of the latter are set at 120 degrees. Determine the resultant primary and secondary forces in terms of the piston mass m, crank radius r, connecting-rod length l, and crank speed ω.

5.19 In the three-cylinder engine shown in the figure the reciprocating mass for each cylinder is m and the length of connecting-rod is l. The three connecting-rods operate on a common crank pin and the crank radius is r.

Show that the primary effects can be completely balanced by a mass $1 \cdot 5\, m$ placed at crank radius at 180 degrees to crank pin.

Show that the secondary effects are equivalent to a force $\frac{1}{2}[(mr^2\omega^2)/l] \cos 2\theta$ along O2 and a force of $\frac{3}{2}[(mr^2\omega^2)/l] \sin 2\theta$ perpendicular to it. θ is the crank angle referred to cylinder 2.

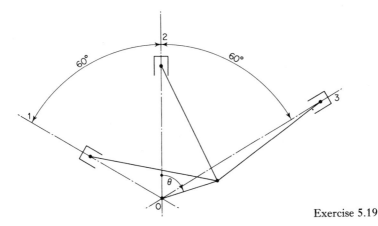

Exercise 5.19

490 Dynamics of Mechanical Systems

5.20 A V-8 engine consists of two four-cylinder blocks operating on a common crankshaft. The V-angle is 90 degrees. The four cranks are at 90 degrees in the order 1–2–4–3.

The mass of each reciprocating mass is m, the crank radius is r, and the cylinders are set at an equal pitch p.

Show that if a mass m is placed opposite each crank pin at radius r, the engine will be in complete primary and secondary balance.

What are the fourth order out-of-balance force and couple? Take moments about the central plane of the engine.

5.21 A twelve-cylinder 'Vee'-engine has two banks of cylinders at 60 degrees, and the six cranks are spaced at 120 degrees in the order 1–2–3–6–5–4. The cylinders are identical and at equal pitch.

Show that the engine is balanced for first, second, and fourth order effects.

Show also that the sixth order unbalance is equivalent to a single force, having a maximum value

$$6\sqrt{3}\ A_6\ mr\omega^2$$

acting in the plane of symmetry of the engine, where A_6 is the sixth order coefficient in the expression for piston acceleration.

$$f = -r\omega^2(\cos\theta + A_2 \cos 2\theta + \cdots)$$

and the remaining symbols have their usual significance.

5.22 The figure shows the mechanism of a shaping machine. The rods for the link motion are free to slide in the blocks at D and A, but are fixed to the upper block C, which is pinned to the ram. The lower block A is free to swivel about its centre-pin which

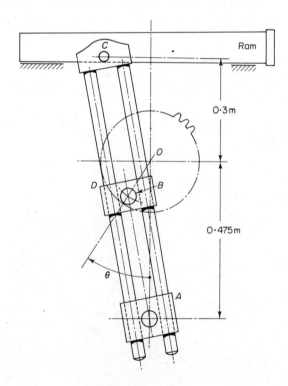

Exercise 5.22

is fixed to the main casting. The pin B is fixed to the stroke wheel OB but can be moved radially to alter the stroke of the ram. The crank speed is 180 rev/min. The mass of the ram is 160 kg.

With OB = 0·125 m, draw velocity and acceleration diagrams for $\theta = 120$ degrees and thus determine for this position the torque required at O in order to balance the inertia force of the ram.

5.23 In a light plane four-bar mechanism ABCD, the crank AB is of length a, the link BC and the crank CD are both of length $4a$, and the fixed link AD is of length $5a$, and at a particular instant angle DAB is zero. A concentrated mass m is carried by BC at E distant $4a$ from both B and C on the side of BC which is remote from D. If crank AB is driven at a constant angular velocity ω, determine:

(a) the velocity of E,
(b) the acceleration of E,
(c) the driving torque at A,
(d) the reactions at A and D.

5.24 A four-bar mechanism consists of a crank AB of length a which drives a rocking lever CD of length $3a$ through a connecting-rod BC of length $4a$. The pivots A and D are $4a$ apart. BC has mass m and moment of inertia $1\cdot6\,ma^2$ about its centre of gravity G which is at its mid-point.

Find the instantaneous value of the torque which must be applied to the crank AB to make it rotate at a constant angular velocity ω when AB and BC are in line and angle CDA is 90 degrees.

Would the torque required be independent of the inertia of (a) AB and (b) CD for other instantaneous positions of the mechanism?

5.25 The figure indicates, for the position shown, the instantaneous accelerations (calculated from an acceleration diagram) of the connecting-rod and piston of a motor car engine running at 3000 rev/min. AG = 0·1 m, and GB = 0·05 m; the piston has mass 0·6 kg and the connecting-rod has mass 1·0 kg, and moment of inertia 0·004 kg m² about the axis through G. Gas pressure in the cylinder produces a force P of 2200 N on the piston.

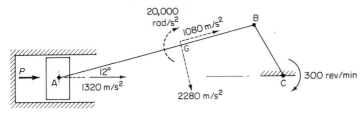

Exercise 5.25

Assuming the friction is negligible, find the magnitudes of the instantaneous loads carried by the bearings at A and B.

5.26 A cam which operates the valve of a motor car engine has a base circle radius of 15 mm and provides a lift of 6·25 mm through a plane-faced tappet. At zero speed the contact force between the tappet and the cam profile is 350 N when the valve is fully open.

Calculate the minimum nose radius for the cam if contact is just to be maintained between the cam and the tappet at a cam speed of 1800 rev/min. The effective mass of the tappet, push-rod, valve, etc., is 0·54 kg at the tappet.

5.27 A gear wheel A shown in the figure drives the larger of a pair of gear wheels B, the smaller gear of the pair driving a rack gear C. The rack can move freely in the direction of its length. Wheel A and the smaller wheel of B have effective radii r, while the larger wheel of B has an effective radius $3r$. The moments of inertia of the

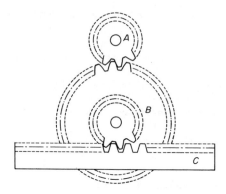

Exercise 5.27

gears A and B together with their shafts are I and $4I$ respectively. The rack has a mass M. Calculate the torque which must be applied to A to give the rack a linear acceleration f.

5.28 In the mechanism shown in plan in the figure, AB is a rigid rod hinged to a fixed point at A and to the mid-point of a second rigid rod CD at B. The rods are each of length l and uniformly distributed mass M. The motion of CD is constrained by a light tie-rod CE also of length l.

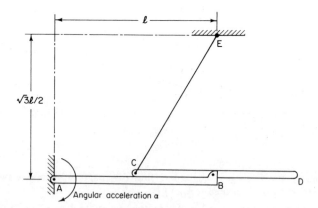

Exercise 5.28

For the mechanism initially at rest in the position indicated, with CD parallel to AB and the angle DCE equal to 60 degrees obtain an expression for the torque which must be applied to AB to give it angular acceleration α. Find also the corresponding force in CE.

Assume that AB and CD are parallel and that the distance between them is negligible.

5.29 Show that the kinetic energy of a system of particles is equal to

$$\tfrac{1}{2}MV^2 + \sum_i \tfrac{1}{2}m_i v_i^2$$

where m_i is the mass of the ith particle, $M = \sum_i m_i$ is the total mass of the system, V is the velocity of the mass-centre of the system and v_i is the velocity of the ith particle relative to the mass-centre of the system.

A caterpillar tractor of total mass 5000 kg has two tracks each of mass 250 kg. The effective moment of inertia of the moving parts of the engine is 2 kg m² about the axis of the crankshaft. At a certain instant the vehicle is moving at 3 m/s on level ground and the engine is developing 15 kW at 1000 rev/min. Calculate the instantaneous acceleration of the tractor if its motion is resisted by forces equivalent to 900 N acting at the tracks.

5.30 The figure shows a mechanism whereby simple harmonic motion is imparted to a mass m which is constrained to move along the axis OX. Derive an expression for the torque needed to overcome the inertia effects due to m when crank OA turns at a uniform speed ω.

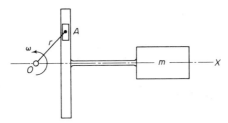

Exercise 5.30

It is proposed to dispense with the externally applied torque and to limit fluctuations in the speed of the crankshaft by attaching to it a flywheel of mass $2 \cdot 5m$. Estimate the radius of gyration needed if the fluctuation in speed about the mean is not to exceed 5 per cent of ω. Only inertia loading is to be considered. State briefly any assumptions implicit in your analysis.

5.31 A particular reciprocating air compressor consists of a low-pressure stage and a high-pressure stage driven through a shaft by a d.c. electric motor. The low-pressure stage is a twin-cylinder machine drawing air from the atmosphere and delivering it to the suction side of the single-cylinder high-pressure stage. All the connecting-rods are mounted on a single crankshaft. The cranks for the twin-cylinder low-pressure stage are at 180 degrees to each other and the crank for the high-pressure stage is at 90 degrees to the other two. The crankshaft is keyed to a flywheel and this in turn is keyed to the rigid shaft which is driven by the electric motor at 300 rev/min. The figure shows the machine in diagrammatic form.

Indicator diagrams for the high-pressure cylinder and for one of the low-pressure cylinders are also given in the figure. The diameters of the high-pressure and low-pressure cylinder bores are 141 mm and 200 mm respectively. The crank throws are all 87·5 mm and the connecting-rods are all 350 mm long. The moment of inertia of the flywheel is 40 kg m² and the stiffness of the driving shaft is 10^5 N m/rad between its ends.

Draw a turning moment diagram to scale and, assuming that the motor develops a uniform torque, estimate the amplitude of torsional vibrations due to that part of the torque represented by the fundamental term of the corresponding Fourier series.

The masses of the reciprocating parts of the compressor may be neglected in deriving the turning-moment diagram.

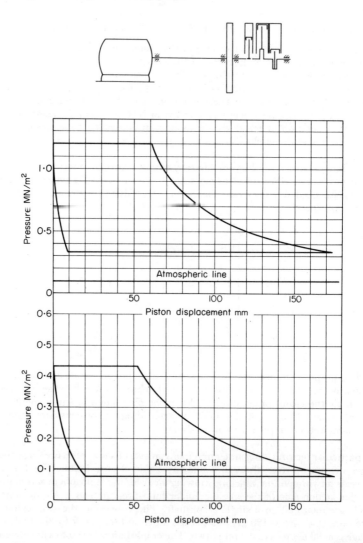

Exercise 5.31

5.32 A uniform rigid rod AB of mass m and length l lies on a smooth horizontal plane. A horizontal force P is applied perpendicular to the rod at A. Sketch the shear force and bending moment diagrams for the rod and show that the maximum bending moment in the rod is $4Pl/27$.

5.33 A diesel engine running at 2000 rev/min has a stroke of 150 mm. The connecting-rod is of effective length 300 mm and may be considered to be of uniform mass 1 kg/m. Determine the maximum bending moment in the rod due to its own inertia, when the crank and connecting-rod are perpendicular to each other.

EXERCISES 495

5.34 A rigid uniform beam of mass m and length l turns about a hinge at one end A. At B, distant αl from A, the beam is supported on and attached to a spring. The stiffness of the spring is chosen to give a specified natural circular frequency ω_0 for small vibrations about the equilibrium position.

Plot a graph to show how the amplitude of the bending moment induced at B varies with α as the beam vibrates freely at the frequency ω_0 and with a prescribed amplitude Y at the tip.

Hence or otherwise estimate the value of α for which the greatest bending moment has the least possible value, and the approximate magnitude of this bending moment. Neglect gravity.

5.35 Sketch the usual crank arrangement of a six-cylinder in-line four-stroke engine. Briefly explain the reasons for this arrangement.

Such a crankshaft has all its throws equally spaced, 150 mm apart. The out-of-balance at each throw is 3 Nm and there are no balance weights. If the shaft is rotating in two bearings at its ends at 3000 rev/min, what will be the greatest bending moment set up in the shaft by centrifugal forces?

5.36 An aircraft propeller with three blades may be assumed to have its mass distributed along the centre-lines of the blades. By considering the forces acting on an element of a blade when the aircraft makes a turn, show (a) that the root of each blade (taken to be at zero radius) is subject to an oscillating bending moment, and (b) that the aircraft itself is subject to a couple of constant magnitude.

Calculate the amplitude of the root bending moment and the magnitude of the couple on the aircraft when the aircraft turns at a rate of 10 degrees/s and the propeller turns at 1000 rev/min.

The mass per unit length of each blade varies along its length as follows:

Radius (m)	0	0.4	0.8	1.2	1.6	2.0	2.2	2.4
Mass per unit length (kg/m)	12	12	13.5	16.5	18	13.5	7.5	0

5.37 A uniform bar of length $(a + b)$ is pivoted about a horizontal axis at a distance a from one end. The bar is held in a horizontal position and is then released.

Show that, provided a is less than $b/2$, the maximum sagging bending moment in the longer part of the bar occurs at a section $(4a^2 - ab + b^2)/3(b - a)$ from the pivot.

5.38 A slender chimney stack is demolished by removing bricks from its base until it collapses by rotating as a rigid body about its base point. Assuming that the mass of the stack is uniformly distributed, show that during the subsequent motion the maximum bending moment to which it is subjected is $\frac{1}{27}Wl \sin \theta$, where W is its weight, l its length, and θ the angle which the stack makes with the vertical.

Chapter 6

6.1 The relationship between the input x_i and the response x_o of a recording instrument is governed by the equation

$$(1 + TD)x_o = x_i$$

Initially, the reading on the instrument dial is steady at zero when there is a step-wise change in the input. The instrument is required to give a reading which is within 2.5 per cent of the correct value in 0.5 s. What is the greatest permissible value of T?

6.2

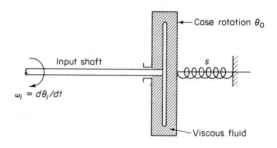

Exercise 6.2

The figure shows a simple speedometer. The torque transmitted to the case as the input shaft rotates is

$$\lambda \left\{ \frac{d\theta_i}{dt} - \frac{d\theta_o}{dt} \right\}$$

This torque is resisted by a torsion spring of stiffness s. Show that the governing equation of the system is

$$(1 + TD)\theta_o = T\omega_i$$

where $T = \lambda/s$.

The input shaft is initially stationary and the speedometer reading is zero when a constant acceleration α is suddenly applied to the input shaft and maintained. Derive an expression for the error in the subsequent recording of the speed. It is to be assumed that the case has negligible inertia.

6.3

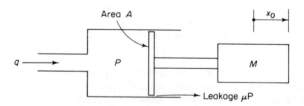

Exercise 6.3

Oil is supplied to an hydraulic ram of area A at a volumetric rate q. The ram moves a mass M at a velocity $v_o = dx_o/dt$. If oil leaks past the ram at a rate μP where P is the pressure inside the ram cylinder show that the governing equation of the system is

$$q/A = (1 + TD)v_o$$

where $T = \mu M/A^2$. The oil is assumed to be incompressible.

Sketch graphs to show the variation in v_o and P with time given that

$$q = \begin{cases} 0 \text{ for } t < 0 \\ \bar{q} \text{ for } t > 0 \end{cases}$$

and
$$v_o = 0 \text{ at } t = 0$$

Is the initial value of P what you would expect?

6.4

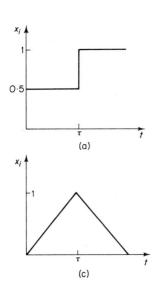
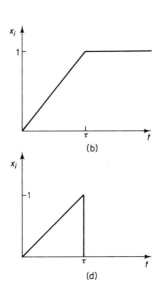

Exercise 6.4

State how step and ramp inputs to a linear system can be superimposed to give the resultant inputs shown in the figure.

6.5 A device for measuring force consists of a calibrated spring which has a stiffness of 900 N/m in parallel with a dashpot which offers a resistance of 4500 N s/m. A force of 90 N is applied when the spring is initially unstrained. After 5 s the force is reduced to 45 N. Determine the maximum force recorded. What force will be recorded after the lapse of a further 5 s? Inertia effects may be assumed to be negligible.

6.6 What is meant by the *time constant* of a simple recording device such as a thermometer?

Water flows at a steady rate through an oil-fired heater. The outlet temperature of the water is measured by a thermometer which controls the supply of oil so that the heat input is increased or decreased by an amount proportional to the difference between the measured temperature and a standard value of 35°C. Under standard conditions the outlet temperature is 35°C, the inlet temperature is 15°C and a certain quantity of water flows.

If the quantity flowing remains unchanged, but the inlet temperature of the water drops to 12°C, the steady outlet temperature drops to 34·6°C.

If conditions are standard and the inlet temperature rises suddenly from 15°C to 22·5°C, what will be the temperature recorded by the thermometer after 5 s if its time constant is 30 s? Assume that all other lags are negligible compared with that in the thermometer.

6.7

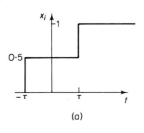
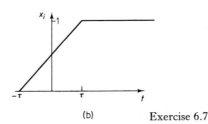

Exercise 6.7

498 Dynamics of Mechanical Systems

The relationship between the input x_i and the output x_o of a system is governed by the equation

$$(1 + TD)x_o = x_i$$

Derive expressions for the response of the system for each of the inputs shown in the figure. Show in each case that if $\tau \ll T$ the response approximates to that due to a step input at time $t = 0$. What do you conclude from this result?

6.8

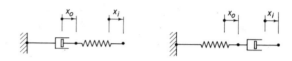

Exercise 6.8

In each of the spring and dashpot systems in the figure the spring is of stiffness s and the dashpot force is λv, where v is the velocity of the dashpot piston relative to the cylinder.

Noting that in each case the spring force must equal the dashpot force, derive for each system the differential equation which relates the input displacement x_i to the response displacement x_o.

If $x_i = X_i \cos \omega t$ sketch curves to show the way in which the amplitude and phase of x_o vary with frequency.

6.9 The rate of heat loss from a building of thermal capacity C is equal to $K_1(T_B - T_A)$, where T_B and T_A are the temperatures of the building and the atmosphere respectively and K_1 is constant. The rate at which heat is supplied to the building by the heating system is given by

$$Q + K_2(T_S - T_B)$$

where T_S is the 'set' temperature of the building and Q and K_2 are constants. The value of Q is such that the building is maintained at the 'set' temperature when the atmosphere is at constant temperature T_0.

Derive a differential equation relating $(T_B - T_S)$ to $(T_A - T_0)$.

The temperature of the building is required to remain within 5 deg F of the set value as the atmospheric temperature fluctuates sinusoidally about the mean value T_0 with an amplitude of 20 deg F and a period of 24 h. If $C/K_1 = 4$ h what is the minimum value of the ratio K_2/K_1?

Chapter 7
7.1

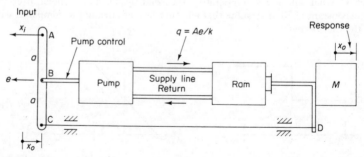

Exercise 7.1

An hydraulic pump supplies oil to an hydraulic ram at a volumetric rate $q = Ae/k$, where e is the displacement of the pump control from its neutral position, A is the area of the ram, and k is a constant. The ram controls the position of a mass M. The displacement y of the mass and the flow q to the ram are related by the equation

$$q/A = (1 + TD)Dy$$

where T is the time constant for the ram (see exercise 6.3).

The pump control and the mass M are connected by a lever ABC and a push rod CD as shown in the figure.

Show that the displacement y of the mass is related to the displacement x of point A by the differential equation

$$(1 + 2kD + 2kTD^2)x_o = x_i$$

and hence that the response to a step input will be oscillatory if $T > k/2$.

7.2

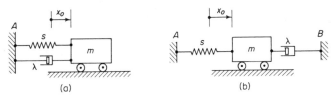

Exercise 7.2

Write down the equation of motion for each of the systems in the figure
(a) When a force $P(t)$ is applied horizontally to m;
(b) When the anchorage A is given a horizontal displacement $x_i(t)$.

Form the equation of motion for the right-hand system when B is given a horizontal displacement $x_i(t)$.

7.3 A manometer consisting of a U-tube with vertical limbs contains a column of liquid of constant cross-sectional area a, length l, and density ρ. Show that the natural circular frequency of undamped vibrations of the column of liquid is $\omega_n = \sqrt{(2g/l)}$.

A short constriction is placed in the tube as shown. The volumetric flow through the constriction is K times the difference in pressure between the two sides of the constriction.

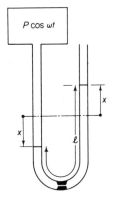

Exercise 7.3

A gauge pressure $P \cos \omega t$ is applied to the manometer. Show that the indicated pressure p is related to the applied pressure by the differential equation

$$\ddot{p} + 2c\omega_n \dot{p} + \omega_n^2 p = \omega_n^2 P \cos \omega t$$

where $2c = a/K\rho\sqrt{(2gl)}$.

If the recorded pressure variation is $\bar{p} \cos(\omega t + \psi)$, determine an expression for the value of c which makes $\bar{p} = P$ when $\omega = \omega_n$.

7.4 In a particular system of the type shown in fig. (a), exercise 7.2, $m = 0.5$ kg, $\lambda = 20$ N s/m, $s = 200$ N/m.

Show that the system is critically damped. A disturbing force of 20 N is suddenly applied to the mass. Plot the displacement x_0 of the mass for the 0.30 s period immediately after the application of the force.

Plot on the same graph the response of a system for which $m = 0$, $\lambda = 20$ N s/m, and $s = 200$ N/m when a similar disturbance is applied.

7.5 A machine weighing 50 kg is supported on springs which deflect vertically 7.5 mm under the weight. A vertical out-of-balance force of $0.1 \omega^2 \cos \omega t$ N is generated when the running speed is ω rad/s. How much viscous damping must be incorporated in the suspension if the amplitude of vibration is not to exceed 0.5 mm at a speed of 250 rev/min.

7.6 An alternating force of $50 \sin \omega t$ N applied at a particular point in an elastic structure at various values of ω gave rise to the following amplitudes of steady oscillation at that point:

Frequency (Hz)	14	16	18	20	22	24	26
Amplitude ($\pm$ mm)	0.475	0.600	0.850	1.000	0.725	0.475	0.325

Within this frequency range the structure can be regarded as a system with one degree of freedom comprising a mass attached to a spring and damped by a viscous dashpot, with the alternating force applied to the mass. Estimate the equivalent mass, the stiffness of the spring and the force exerted by the dashpot per unit velocity.

7.7 A certain system consists essentially of a mass M supported on a spring s which is in parallel with a viscous damper λ. In order to determine the vibration characteristics of the system a spring AB of known stiffness is attached to the mass at B. The variation of the amplitude of motion at B with frequency in response to simple harmonic motion of fixed amplitude of 5.0 mm at A is found by experiment to be as follows:

Frequency (Hz)	0	1	2	3	4	5	6	7	8
Amplitude at B (mm)	2.0	2.1	2.6	3.8	5.0	2.7	1.5	0.9	0.7

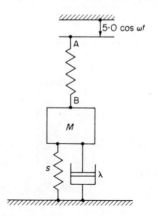

Exercise 7.7

EXERCISES 501

Estimate for the original system the undamped natural frequency, and the damping ratio, given that the stiffness of the spring AB is 225 N/m.

7.8 In a domestic heating installation a gas furnace supplies heat to the radiator water at a rate q determined by the differential equation

$$q = \frac{K_1}{1 + T_1 D}(\theta_i - \theta_o)$$

in which K_1 and T_1 are constants, θ_i is the temperature set on a regulator and θ_o is the actual temperature of the radiator water. The gas supply cuts out if θ_o exceeds θ_i.

The radiator system has a thermal capacity C and rejects heat at a rate $K_2 \theta_o$ where K_2 is a constant. Set up the differential equation expressing θ_o in terms of θ_i when the furnace is in action.

If $T_1 = 5.86$ min, $C/K_2 = 34.14$ min, and the installation is running steadily with $\theta_i = 80°$ and $\theta_o = 76°$, determine the greatest sudden increase in regulator setting that can be applied without the gas supply subsequently cutting out.

7.9 A rotor of polar moment of inertia J is suspended by means of a torsion spring of stiffness s in a tank which contains a viscous fluid. The liquid resists rotation of the rotor with a torque $\lambda \, d\theta/dt$ where θ_o is the angular displacement of the rotor.

The rotor is initially stationary and in equilibrium. At time $t = 0$ a constant torque Q_i is suddenly applied to the rotor. After a time τ the torque is suddenly removed.

Assuming the system to be underdamped, show that if $\tau \ll \omega_n^{-1}$ the motion for $t > \tau$ is the same as that due to an impulse I applied at $t = 0$, i.e.,

$$\theta_o = \frac{I}{J\beta} e^{-c\omega_n t} \sin \beta t$$

where $I = Q_i \tau$, $\omega_n = \sqrt{(s/J)}$, $c = \lambda/2\sqrt{(sJ)}$, $\beta = \omega_n\sqrt{(1 - c^2)}$.

Chapter 8

8.1 The figure shows a hydraulic relay in which the displacement of the end A of the rigid light member ABC results in movement of the actuator D. The end C of ABC is connected to a spring, which exerts a force s per unit displacement, and its movement relative to D is restrained by a viscous dashpot, which transmits a force λ per unit relative velocity. Due to the flow of hydraulic fluid the velocity of D is proportional to the displacement of B, the ratio of displacement to velocity being T.

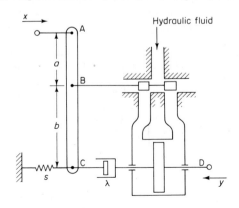

Exercise 8.1

502 Dynamics of Mechanical Systems

If the mechanism is initially stationary and A is then given a displacement x, which is small compared with a and b, derive the differential equation which describes the subsequent displacement y of the actuator.

If A is displaced at time $t = 0$ from $x = 0$ to $x = X$, find an expression for y when initial transient effects have died away.

8.2 In the control system shown diagrammatically in the figure the increase h of level in the tank, the compression x in the spring, the movement z of the control piston, and the flow rates q_1 and q_2 represent small departures from steady-state values.

The float cross-section area A_1 is small compared with the tank cross-section area A_2. The volumetric flow-rate into the control cylinder, which has cross-section area A_3, is c_1 times the port opening of the pilot valve. The volumetric flow-rate through the supply valve decreases by c_2 times the control piston movement z. The fluid density is ρ and the spring stiffness is s. Show that

$$q_1 = \frac{-k_1}{1 + T_1 D} h$$

and express k_1 and T_1 in terms of the given constants and dimensions.

Show further that

$$h = -\frac{1 + T_1 D}{1 + T_2 D + T_1 T_2 D^2} \cdot \frac{q_2}{k_1}$$

where $T_2 = A_2/k_1$.

Inertia effects and unbalanced forces on the valves may be neglected.

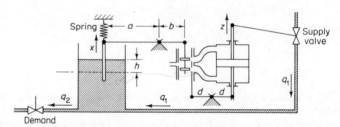

Exercise 8.2

8.3 A liquid with specific heat s flows at a steady mass rate m through a perfectly lagged tank which contains an immersion heater, the temperature of the liquid at entry being constant. The thermal capacity of the tank and contents is equivalent to that of a mass M of the same liquid. When heat is supplied at a steady rate $\bar{Q}$ the steady outlet temperature is $\bar{\theta}_o$. If the rate at which heat is supplied is Q and the outlet temperature at any instant is θ_o, derive the differential equation relating $\phi_o = (\theta_o - \bar{\theta}_o)$ and $q = (Q - \bar{Q})$. Assume that the temperature of the tank and its contents is the same as that of the outgoing liquid.

Show that the direct Nyquist diagram for $\phi_o/(q/ms)$, where $q = ms \cos \omega t$ (ω positive), is a semicircle of unit diameter.

If the difference between the actual value of ϕ_o and its desired value ϕ_i is $\varepsilon = (\phi_i - \phi_o)$, sketch the Nyquist plots for ϕ_o/ε:
(a) when $q/ms = K\varepsilon$;
(b) when $q/ms = \dfrac{m}{M} \int \varepsilon \, dt$.

8.4 The ends A and B of the system illustrated in the figure are mechanically coupled so that they have equal and opposite displacements. The mass m is connected to the spring of stiffness s_2 through a dashpot which transmits a force λ times the relative velocity between the mass and piston. All inertias other than that of m are negligible. Derive the transfer operator relating the movement θ_o of the mass to the movement θ_i of A and B.

In a particular case $s_1 = s_2 = s$, $s/m = \omega_n^2$, $\lambda/m = \omega_n$ and θ_i is a sinusoidal motion of variable frequency ω. Express θ_i/θ_o as a function of ω/ω_n of the form

$$\frac{\theta_i}{\theta_o} = f_1\left(\frac{\omega}{\omega_n}\right) + jf_2\left(\frac{\omega}{\omega_n}\right)$$

Sketch the polar locus of θ_i/θ_o as ω is varied, and describe how it could be used to determine the maximum magnification factor, the frequency at which it occurs, and the corresponding phase relationship between θ_o and θ_i.

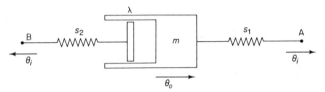

Exercise 8.4

8.5 The open loop transfer operator of a control system is $K/[D(1 + TD)]$. If $c = 1/2\sqrt{(KT)}$ is less than $1/\sqrt{2}$, show that $\log f = -\pi c/\sqrt{(1 - c^2)}$ and $M_m = 1/2c\sqrt{(1 - c^2)}$ where M_m is the greatest dynamic magnification of the closed loop, and f is the first overshoot of the output when the input is given a unit step displacement.

In a particular case $T = 0.25$ s and M is required to be 1.25. Calculate K, f and the undamped frequency of transient oscillation.

It is proposed to increase the speed of response by doubling the undamped natural frequency without altering the effective damping ratio, or the time constant T. Show that this can be done by increasing K to 20 s^{-1} and introducing derivative action so that the open loop transfer operator contains an additional factor $(1 + T_d D)$. Calculate the required value of the derivative time constant T_d and the new values of f and M_m.

8.6 The figure gives the harmonic response (Nyquist) locus relating the response y of a certain temperature control system to the error ε of unit amplitude, where ε is the difference between the actual response y and the desired response x.

(a) What is the maximum factor by which the gain may be increased if the system is to remain stable? If the gain is increased until the system just starts to oscillate what will be the frequency of oscillation?

(b) For the given system plot graphs of the amplitude of the output y and its phase relative to x against the frequency when the input is $x = 1 \cos \omega t$.

(c) A change in design introduces an additional phase lag of an amount ωT without change of amplitude. Estimate the maximum value of the constant T for the system to remain stable.

8.7 A control system using proportional control has n ($n \geq 3$) simple, non-interacting lags, each of time constant T, in the loop. Show that (a) for stability, the loop gain K must be less than $\sec^n(\pi/n)$; (b) in the limiting case $K = \sec^n(\pi/n)$ the frequency of oscillation is $\tan(\pi/n)/2\pi T$. (The open loop transfer operator is $K(1 + TD)^{-n}$.)

Derive corresponding results when an additional integrating operator is introduced into the control loop.

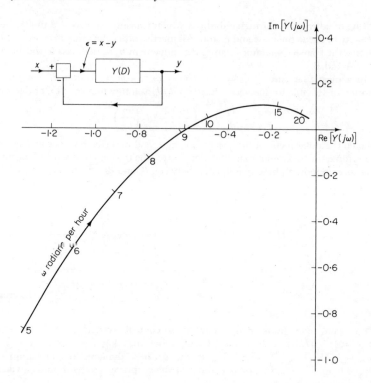

Exercise 8.6

Deduce, by a limiting process, the critical gain and frequency of self-oscillation for a system which uses integral control and has a finite time delay τ in the loop. Confirm your results by treating this case from first principles.
Note: $e^x = \lim_{n \to \infty} (1 + x/n)^n$.

8.8 Sketch the harmonic response locus and the inverse harmonic response locus for the transfer operator $y/x = F(D)$ for each of the following cases

$$F(D) \equiv \frac{1}{TD}; \quad F(D) \equiv \frac{1}{TD(1 + TD)};$$

$$F(D) \equiv \frac{1 + TD}{1 + \alpha TD} \text{ for } \alpha < 1 \text{ and for } \alpha > 1;$$

$$F(D) \equiv \frac{1}{1 + 2cTD + T^2D^2} \text{ for } 0 \leqslant c \leqslant 1;$$

$$F(D) \equiv \frac{1}{TD(1 + 2cTD + T^2D^2)} \text{ for } 0 \leqslant c \leqslant 1.$$

8.9 Compare the function of a governor with that of a flywheel in controlling the fluctuations in speed of an engine.

An engine drives a load of negligible inertia whose motion is opposed by a mean torque of 1350 Nm at the mean speed of 200 rev/min. The speed is controlled by a governor such that a permanent change in torque of 5 per cent from normal causes a 1 per cent change in speed. It may be assumed that the engine speed and torque are linearly related for small changes in speed. Determine the amplitude of speed fluctuation as a percentage of mean speed if the load torque fluctuates sinusoidally with an amplitude equal to 10 per cent of the mean torque, and at a frequency of one cycle per revolution (a) when the engine has a flywheel of moment of inertia 40 kg m², (b) when there is no flywheel.

8.10 A petrol engine is directly coupled to an electrical generator, the combined polar inertia being J and the coefficient of viscous friction B. The throttle sensitivity (increase of engine torque per unit throttle angle) is K_1 and the carburettor and inlet manifold cause a simple exponential delay, of time constant T_1. The generator voltage is K_2 times the shaft speed, there being a delay in response (due to electromagnetic effects) of time constant T_2. The voltage determines the throttle setting through a relay of negligible delay, the change of throttle angle being K_3 times the change of voltage. Show that the condition for stability of the governing system is

$$(T_1 + T_2)(J + T_1 B)(J + T_2 B) > T_1 T_2 J K_1 K_2 K_3$$

8.11 A turbo-propeller installation is to be governed by means of the fuel supply, the pitch angle of the propeller being kept constant. The variation of the net output torque Q of the gas turbine as a function of speed and of rate of fuel supply w is shown. The relation between the torque Q_L required to drive the propeller and its speed, referred to the turbine shaft, is also shown. The governor controls the fuel supply according to the relation

$$(TD + 1)w = -kn$$

where w is the increase in the rate of fuel supply due to an increase of speed n; T and k are constants, and D is the operator d/dt. The total effective moment of inertia of the system referred to the turbine shaft is J, and friction is allowed for in the torque curves.

Obtain the characteristic equation of the governed system for a small disturbance from the steady-state operating conditions in terms of the differential coefficients at the operating point, assuming linear behaviour. Hence, if $T = 0.33$ s, $J = 2$ kg m² determine k in kg/hr per rev/min if the system is adjusted to give a critically damped response to a disturbance from 12 000 rev/min.

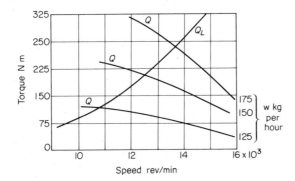

Exercise 8.11

8.12 A turbine running at 8000 rev/min drives an alternator. With the governor disconnected, a sudden increase of load torque on the alternator of Q causes the speed of the turbine to fall exponentially to 7000 rev/min, the initial rate of loss of speed being 50 rev/min/s. On introducing the governor, which may be assumed to have a proportional action, the speed drops to 7900 rev/min for the same increase of load.

To eliminate the change in steady speed resulting from an increase in load, integrating action is incorporated in the governor of such amount as to give a critically damped response with the same governor sensitivity.

By calculating first the amount of integral action, find the time lost by a synchronous clock connected to the alternator when an additional load of $2Q$ is suddenly thrown on to it. What would be the minimum speed of the turbine after this load increase?

8.13 (a) The open-loop transfer operator of a certain servo-mechanism with zero velocity error is

$$Y(D) = \frac{10}{D^2(1 + 0 \cdot 1D)}$$

If the system is stabilized by error derivative action, so that there is an additional factor $(1 + T_dD)$ in the transfer operator, what value of T will give limiting stability?

By sketching an harmonic response locus, estimate the maximum dynamic magnification M_m of the closed-loop system and the corresponding resonant frequency when T_d is five times the critical value.

(b) Show, also, that the system could be satisfactorily stabilized by using negative velocity feedback. Suggest a suitable value for the sensitivity of this feedback, i.e., the equivalent error signal feedback for an output velocity of 1 rad/s.

8.14 The figure shows a tank of horizontal area A from which the water is drawn at a constant rate irrespective of any small variation of the depth of water in the tank. The flow-rate through the inlet valve changes linearly by an amount b per unit of valve opening. The valve is operated by a cylindrical float of mass M, for which the buoyancy force changes by f per unit displacement of the float from its equilibrium position in the water. The float is attached to a light lever arm at an adjustable point distance ra from the pivot, the valve distance a from the pivot. A damping force of λ per unit velocity acts on the valve, but otherwise the valve requires negligible force for its operation and the system is frictionless.

If the instantaneous level of the water surface is a small distance h above its mean level, derive the third order differential equation which represents the motion of the surface. Hence find the maximum value of r if the system is not to hunt.

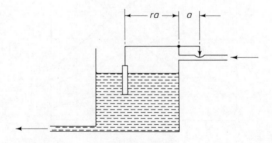

Exercise 8.14

8.15 The figure shows the block diagram for an hydraulic position control. Determine the critical value of the gain for limiting stability. What will be the frequency of oscillation if the gain is set at this value?

For the range $0 < \omega < 10$ plot a Whiteley diagram for the system, with gain adjusted to one-third of the critical value. Find the amplitude and phase of the output when the input is $x = 1 \cos 10t$.

Estimate the frequency of input for which the output lags by $\frac{1}{2}\pi$, and find the corresponding ratio of output to input amplitude.

Exercise 8.15

8.16 In a power operated aileron control, the angle θ_o of the aileron is measured electrically and compared with the desired value θ_i on the manual control lever.

The error $\varepsilon = \theta_i - \theta_o$ activates an hydraulic relay whose output angle ϕ is independent of the output torque.

The aileron is coupled to the output of the relay by a shaft whose torsional stiffness is s. The moment of inertia of the aileron about its axis of rotation is I, and its rotation is opposed by an aerodynamic couple $\lambda D\theta_o$, λ being constant.

Show that the transfer operator relating ϕ and θ_o is

$$\frac{\theta_o}{\phi} = \frac{1}{1 + \frac{\lambda}{s}D + \frac{I}{s}D^2}$$

If the transfer operator of the hydraulic relay is

$$\frac{\phi}{\varepsilon} = \frac{K}{1 + TD}$$

show, assuming Routh's criterion, that unstable oscillations will develop if

$$K > \frac{\lambda T}{I} + \frac{\lambda}{sT} + \frac{\lambda^2}{sI}$$

By sketching the steady-state vector diagram for $\theta = \Theta \cos \omega t$ show that if

$$K = \frac{\lambda T}{I} + \frac{\lambda}{sT} + \frac{\lambda^2}{sI}$$

the frequency with which the system oscillates is given by

$$\omega^2 = \frac{s + \lambda/T}{I}$$

8.17 In a thermostatic tank control, heat is supplied by passing steam through pipes immersed in the tank, which is well stirred. The thermal capacity of the pipes is C_1 and that of the fluid in the tank is C_2: the rate of heat transfer between the pipes and the fluid is K_1 times their difference in temperature and the rate of loss of heat from the tank at any time is K_2 times its temperature θ_o. If the rate at which heat is supplied by the steam is q, show that

$$\frac{\theta_o}{q} = \frac{1}{K_2\{1 + (T_1 + T_2 + T_3)D + T_1T_2D^2\}}$$

where T_1/K_1, $T_2 = C_2/K_2$ and $T_3 = C_1/K_2$.

Deduce that the response of the tank temperature to a sudden change in q will be more than critically damped.

8.18 The rate of heat supply to the tank in exercise 8.17 is controlled by means of a thermostat, the supply rate q being related to the difference $(\theta_i - \theta_o)$ between the desired value θ_i of the tank temperature and the actual value θ_o by the transfer operator

$$\frac{q}{\theta_i - \theta_o} = \frac{K}{1 + 10D}$$

The quantities referred to in exercise 8.17 have the values $C_1 = 100$, $C_2 = 500$, $K_1 = 2$, $K_2 = 5$ in consistent units.

Find the value of K such that the Nyquist locus of the system crosses the negative real axis at the point $(-\frac{1}{4}, 0)$.

8.19 The figure shows the plan view of a car-trailer, A being the point of attachment of the trailer to the car. The velocity of A has a constant component V parallel to the fixed direction XX, and a variable component $\dot{y}$ perpendicular to XX. Assuming that there is no slip between the road and the wheels of the trailer, show that the relationship between the small angle θ which the centre-line of the trailer makes with XX and y is

$$(1 + TD)\theta = \dot{y}/V$$

where $T = a/V$, a being the distance from the wheel axis to the point of attachment.

When the attitude of the trailer is as depicted in the figure, and the car is moving at constant velocity V with its centre-line along XX, the displacement y of point A on the trailer relative to the rear of the car is resisted by a spring of stiffness s.

Assuming that its wheels are light, show that the trailer will 'snake' if $k^2 > b(a - b)$, where k is the radius of gyration about a vertical axis through the centre of gravity G of the trailer and b is the distance of the centre of gravity forward of the trailer wheels. Derive an expression for the frequency of oscillation if $k^2 = b(a - b)$.

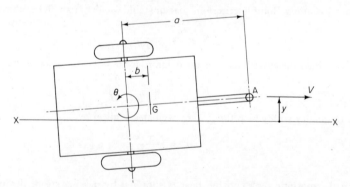

Exercise 8.19

8.20 The three-wheeled truck shown in the figure steers itself automatically to follow a straight white line painted on the ground. The front wheel can rotate about a vertical axis which intersects the axis of the wheel at O. The arm OP lies in the plane of the front wheel and a photoelectric probe at P records the horizontal deviation, e, of P from the line on the ground. A steering motor rotates the front wheel at an angular velocity $\dot{\alpha} = Ke$ relative to the body of the truck in the sense required to reduce the deviation.

If the speed, V, of the truck is constant and the wheels do not slip on the ground, show that the system is stable provided

$$(a + b)bK > V$$

where b and a are the horizontal distances from O to P and to the mid-point Q of the rear axle respectively.

It may be assumed that the deviation e is small.

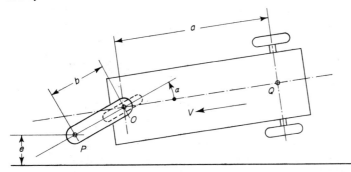

Exercise 8.20

8.21 The open-loop transfer function $Y(D)$ for a position control system is given in the following data for a steady harmonic input of angular frequency ω:

ω (rad/s)	$Y(j\omega)$	$[Y(j\omega)]^{-1}$
6	$-1\cdot75 - j1\cdot23$	$-0\cdot38 + j0\cdot27$
8	$-1\cdot17 - j0\cdot63$	$-0\cdot66 + j0\cdot36$
10	$-0\cdot82 - j0\cdot36$	$-1\cdot02 + j0\cdot45$
12	$-0\cdot60 - j0\cdot22$	$-1\cdot46 + j0\cdot54$
14	$-0\cdot46 - j0\cdot15$	$-2\cdot00 + j0\cdot63$

A modification to the system introduces a simple delay of time constant $T = 0\cdot04$ s such that the open-loop transfer function becomes

$$\frac{Y(D)}{1 + TD}$$

By constructing an appropriate locus of harmonic response, estimate the resonant frequency and the amplification at resonance for the modified closed-loop system.

Estimate, also, the value of T for which the closed loop would just become unstable.

8.22 The open-loop transfer operator of a control system is

$$\frac{K(1 + 4D)^2}{D^2(1 + 7D)(1 + D)^2}$$

Calculate the frequencies at which the open loop plane shift is 180 degrees and sketch the harmonic response locus. Hence find the range of gain for which the system is stable.

8.23 A mass m is supported at the top of a light rod of length l, which can move in a vertical plane. The rod is held in vertical balance by horizontal movements of its lowest point. Small movements of the mass are apposed by a viscous drag equal to μ times its velocity.

The horizontal movement x of the lower end of the rod is controlled by a servomechanism which operates so that

$$\frac{d^2 x}{dt^2} = k_1 \theta + k_2 \frac{d\theta}{dt}$$

where θ is the small inclination of the rod from the vertical.

Investigate how the stability of this system depends on the values of k_1, k_2 and μ.

510 Dynamics of Mechanical Systems

8.24 A block diagram for a simple position-control servomechanism with an inertia load is given in the figure. The transfer operator for the controller and load is

$$\frac{K}{D(1 + T_1 D)}$$

and there is a simple exponential delay T_2 in the feedback path.

Derive the governing differential equation for the system and show that for stable operation

$$K < K_0 = \frac{1}{T_1} + \frac{1}{T_2}$$

Determine the response to a unit step change in θ_i for the particular case $K = K_0$ in terms of suitable constants of integration, and state the initial conditions which determine these constants. Assume that the system is initially in equilibrium.

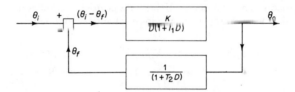

Exercise 8.24

8.25 The transfer operator of a process under automatic control can be approximated by an integration in series with a pure (finite) time delay T. Find T_p, the natural period of oscillation of the system under proportional control.

If integral action, with the time constant of the integration equal to T_p, is now incorporated in the controller, how much derivative action must be added simultaneously to the control operation for the system to remain on the border of stability with the same proportional gain?

Sketch the direct (Nyquist) and inverse (Whiteley) response loci for the original system and for the modified system.

8.26 A certain process is manually controlled. The response of the process is described by a simple exponential decay of time constant 2 s. Experiment suggests that the human transfer operator is approximately

$$k\left(1 + \frac{1}{D}\right) e^{-0.3D} \quad \text{(time in seconds)}$$

under normal working conditions.

If the peak dynamic manification is to be 1·5, estimate the required gain and the frequency of the fundamental mode of transient response.

8.27 Determine, by means of conformal mapping on the Nyquist diagram given in the figure, the frequency of the dominant mode of transient oscillations which will follow a disturbance of the system from its equilibrium state, and the rate at which the oscillations decay.

The actual open-loop transfer operator for the system is

$$\frac{0\cdot 25(2 - D - D^2)}{D(1 + D)^2}$$

Verify that the root extracted above by conformal mapping is a root of the characteristic equation for the system.

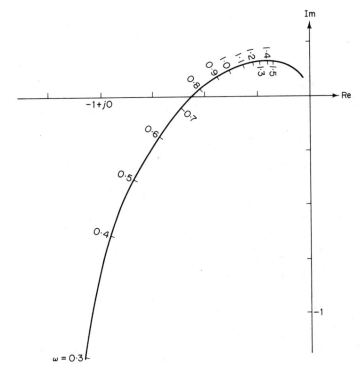

Exercise 8.27

Chapter 9

9.1 A door is closed under the control of a spring and dashpot. The spring gives a torque of 13·5 N m when the door is closed and has a stiffness of 50 N m/rad. The dashpot gives a damping torque of 100 N m/rad s^{-1}.

The moment of inertia of the door about its hinges is 90 kg m². Derive an expression for the motion of the door when it is opened 90 degrees and released, and find approximately the time it takes to close.

9.2 A moving-coil d.c. ammeter has a rotor controlled by a torsional spring; the angular displacement of the rotor and also the reading on the meter are directly proportional to the current. The undamped natural frequency of the system is 4 Hz, but the instrument is critically damped (i.e., the pointer just does not overshoot after a sudden disturbance). A current of constant amplitude and variable frequency is passed through the meter. A reading of 1 A is recorded at zero frequency; what will be the reading at 5 Hz?

9.3 A single-cylinder engine and its base of total mass 70 kg are mounted on springs in parallel with viscous dampers.

The amplitude of vertical vibration of the base has its maximum value of 1·3 mm when the engine is running at 750 rev/min. At 1500 rev/min the amplitude is 0·4 mm.

Assuming that the primary out-of-balance force is the sole excitation, and that the damping is light, determine:

(a) the stiffness of the support springs;
(b) the primary out-of-balance force at 750 rev/min;
(c) the power absorbed in the dampers at resonance.

512 Dynamics of Mechanical Systems

9.4 A mass of 5 kg is hung from a support by a light spring of stiffness 140 N/mm. The support undergoes a steady vertical oscillation of displacement

$$y_0 \sin \omega t$$

Neglecting damping, at what frequency of oscillation of the support would the mid-point of the spring be at rest?

A viscous damping force λv is applied to the mass, where v is its velocity. What must be the value of the constant λ in order to restrict the amplitude of the mass at resonance to $4y_0$?

9.5 The sprung mass of a car is 450 kg. The springs are attached to the axles which it may be assumed follow the road surface up and down exactly. Assuming that the system may be regarded as being equivalent to a single mass on a single spring of stiffness 17·5 kN/m, find the resonant speed when the car is driven on a road whose surface varies sinusoidally in level with an amplitude of 25 mm and wavelength 6 m, and with its dampers inoperative.

Determine the amplitude of vibration of the chassis when the speed is 50 km/h.

9.6 When the dampers of the car in exercise 9.5 are in operation they just prevent resonance occurring if the sprung mass is excited when the car is at rest.

Determine the amplitude at 50 km/h when the dampers are in operation and the power dissipated in the dampers at this speed.

9.7 Due to the inertia of the moving parts, a reciprocating engine is subjected to a fluctuating force at a frequency corresponding to the running speed of the engine. The engine and base have a mass of 225 kg and are supported on a vibration-isolating mounting consisting of springs of equivalent stiffness of 50 N/mm, and a dashpot which is adjusted so that the damping is 20 per cent of critical.

(a) Over what speed range is the amplitude P_t of the force transmitted to the foundation (i.e., the resultant of the spring and damper force) larger than the amplitude P_e of the exciting force? Estimate P_t/P_e at resonance.

(b) Over what speed range is $P_t < 0·2 P_e$?

9.8 An electric motor has an armature which is slightly out of balance. When a sheet of elastic packing is placed between the motor and a rigid floor vertical oscillations are set up which reach their maximum amplitude at a certain speed N. A second identical sheet of packing is then placed beneath the motor so as to double the total thickness of packing.

Show that the amplitude of the alternating force transmitted to the floor at a given speed will be less with two sheets of packing than with one, provided that the speed of the motor is greater than $N\sqrt{\frac{2}{3}}$. Damping is light.

Hint: Sketch the curves of transmitted force against frequency, and take careful account of signs.

9.9 In the engine indicator shown the pressure in a cylinder is indicated by the small vertical motion of a pointer. The piston is restrained by a spring of stiffness 20 N/mm. If the maximum error of indication for pressure variations in the frequency range 0–10 Hz is to be less than 2 per cent, show that the mass of the system must be less than about 0·10 kg.

9.10 A certain section of a building is executing steady horizontal vibration of frequency 2 Hz, as a result of exciting forces due to machinery within the building. In order to determine the amplitude of the vibration, a simple pendulum is suspended from a bracket mounted on a wall of the building. After sufficient time has elapsed for the pendulum to settle down to a state of steady forced oscillation, the amplitude of the horizontal displacement of the pendulum bob relative to its point of support is measured.

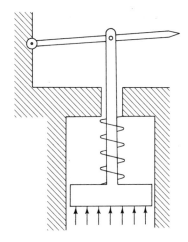

Exercise 9.9

Show that the measured amplitude of relative motion will be a good approximation to the true absolute amplitude of vibration of the building provided the pendulum is of sufficient length, and determine the minimum length of the pendulum if the approximation is to be within 10 per cent.

It should be assumed that the pendulum oscillates through small angles only.

9.11 In order to investigate the dynamics of a cam and follower it is assumed that the system is equivalent to a single concentrated mass m, representing the mass of the follower, and a single spring of stiffness s which represents the flexibility of the cam and follower. Determine the maximum amount by which the displacement $y(t)$ of the mass differs from the displacement $x(t)$ of the free end of the spring

(a) when $x(t) \equiv 0$ for $t < 0$ and $x(t) \equiv vt$ for $t > 0$, i.e., there is a sudden change of velocity at $t = 0$;

(b) when $x(t) \equiv 0$ for $t < 0$ and $x(t) \equiv a(1 - \cos \omega t)$ for $t > 0$, i.e., a sudden change in acceleration at $t = 0$;

(c) when $x(t) \equiv 0$ for $t < 0$ and $x(t) \equiv a(\omega t - \sin \omega t)$ for $t > 0$.

Notes: Cam (a) is obviously most undesirable for all but the lowest speed of operation. Cam (b) has been much used in practice because with a flat follower the cam profile is a circular arc and therefore easily constructed; it is known as an harmonic cam. Cam (c) is known as a cycloidal cam and is recommended for high-speed operation.

9.12 The disc shown in plan in the figure has a straight groove machined across it at a distance a from the centre O. A small mass slides in the frictionless groove, constrained by two equal springs attached to the disc at its periphery. The stiffness of the springs is such that, when displaced from its mid-position and released, the mass oscillates with S.H.M. of frequency $\omega_0/2\pi$ Hz when the disc is stationary.

If the disc rotates about a vertical axis through O with a constant angular velocity $\Omega (<\omega_0)$, what is now the frequency of oscillation of the mass?

If the amplitude of oscillation is $\pm b$, what is the lateral reaction between the groove and the mass as it passes through its mid-position?

9.13 For the purpose of vibration analysis, a certain shock recorder may be regarded as a simple spring-mass system; the stiffness of the light spring is 0·175 N/mm and the mass that it controls is of magnitude 9 kg. A transient force F, which is given

514 Dynamics of Mechanical Systems

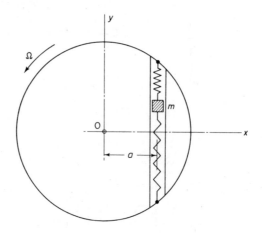

Exercise 9.12

as a function of time in the figure acts upon the mass, the system being initially at rest.

(a) Find the maximum deflection and velocity of the mass using a phase-plane construction. (Take steps in time of 0·2 s.)
(b) Find the instants at which these maxima occur.
(c) Determine the initial conditions of the era of free vibration.
(d) Repeat the solution for steps in time of 0·05 s.

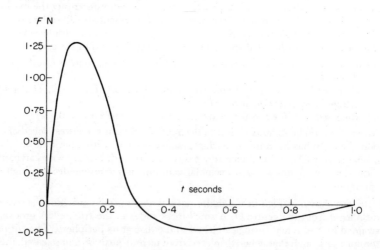

Exercise 9.13

9.14 A mass M, suspended by a light vertical spring, moves in vertical guides which exert a constant frictional force during relative motion. The spring is such that it extends δ under a pull Mg. With the mass in its position of free equilibrium, the upper end of the spring is suddenly displaced a distance δ upwards. As a consequence, the mass rises once and then falls once to come to rest in its new position of free equilibrium. Find the magnitude of the constant frictional force in terms of

Mg and show that the velocity of the mass as it first passes its new position of free equilibrium is $\sqrt{(\tfrac{1}{2}g\delta)}$.

9.15 A body of mass 1·5 kg is attached to the lower end of a light vertical spring of stiffness 900 N/m. The body is initially at rest on a horizontal surface and the spring is unstretched. If the upper end of the spring is given constant upwards velocity of 3 m/s, find the maximum tension in the spring during the subsequent motion.

9.16 The pendulum of an electric clock is given an impulse once every tenth oscillation in order to restore its amplitude from $\pm 4\tfrac{1}{2}$ to ± 5 degrees. As designed, the impulse is applied to the pendulum at the bottom of its stroke, and the clock keeps perfect time. During a repair, however, the mechanism is set so as to apply an impulse at the extreme end of the stroke; after this alteration the limits of amplitude are the same as before. Using a sketch in the phase plane, or otherwise, estimate to the nearest minute the daily error of the clock resulting from the alteration.

9.17 Provided that its angular velocity does not exceed a certain value Ω the circular cam shown in the figure imparts simple harmonic motion to its follower.

Show, that if the cam is driven at a speed $\omega > \Omega$, contact between the cam and the follower will be lost when

$$\sin\theta = \frac{\Omega^2 - \omega_0^2}{\omega^2 - \omega_0^2}$$

where θ is the angle of rotation of the cam from its position when the follower is at its mean position, and $\omega_0 = \sqrt{(s/m)}$, s being the stiffness of the spring and m being the mass of the follower.

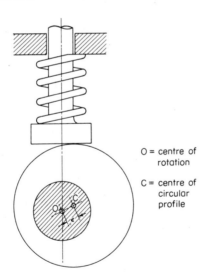

O = centre of rotation

C = centre of circular profile

Exercise 9.17

Show from a phase-plane diagram, or otherwise, that if $\omega = 2\omega_0$ and $\Omega = \omega_0\sqrt{3}$, the angle through which the cam rotates during the period when the cam and follower are out of contact is about 200 degrees.

9.18 A massive particle lies in a light pan which is attached to the lower end of a light elastic spring. The frequency of small vertical oscillation of the system is 2 Hz when the upper end of the spring is fixed.

Starting at time $t = 0$ with the system hanging in equilibrium and at rest, the upper end of the spring is given a vertical displacement $y = a \sin 2\pi f t$ where $a = 75$ mm, $f = 2$ Hz and the positive sense of y is upwards. Draw a trajectory in the phase plane to represent the motion of the particle during the next $\frac{1}{2}$ second and record the times at which the particle loses and regains contact with the pan.

9.19 The figure shows the block diagram for a simple position control mechanism with zero damping. The feedback link, which has negligible inertia, has backlash of an amount $2b$ so that when all the free-play has been taken up

$$\theta'_o = \theta_o - \text{sgn}(\dot{\theta}_o)$$

where $\text{sgn}(\dot{\theta}_o) = +1$ when $\dot{\theta}_o$ is positive, and -1 when $\dot{\theta}_o$ is negative, and within the free-play band

$$\theta'_o = \text{constant}$$

Sketch a phase-plane portrait relating the 'error' ε ($=\theta_i - \theta'_o$) to the rate of change of error $\dot{\varepsilon}$, and show that the system is unstable.

Coulomb damping of an amount $fJ \, \text{sgn}(\dot{\theta}_o)$ where J is the moment of inertia of the load, is applied to the output shaft. Show that the oscillations which follow an initial disturbance $\varepsilon = \varepsilon_1$, $\dot{\varepsilon} = 0$, are damped out if $f > b/T^2$, and that if $f < b/T^2$, the oscillations decay provided that ε_1 is less than a certain critical value ε_c, but increase indefinitely for $\varepsilon_1 > \varepsilon_c$.

Sketch phase plane portraits to illustrate the different sets of conditions.

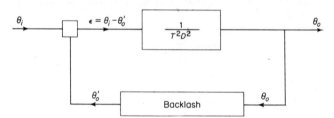

Exercise 9.19

9.20 Determine the natural frequencies and normal modes of vibration for the two systems shown in the figure.

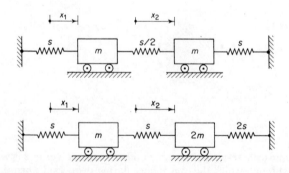

Exercise 9.20

9.21 The figure shows a vibration test platform and foundation. Derive general expressions for the amplitudes X_1 and X_2 and hence show that if ω is large compared with the natural frequencies, $X_1 \approx mr/m_1$ and $X_2 \approx 0$.

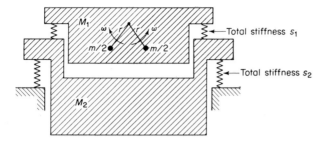

Exercise 9.21

9.22 Two circular discs each of polar moment of inertia J are mounted on a shaft as shown in the figure. The torsional stiffnesses of the sections of the shaft, which is held rigidly at both ends, are s, $3s/2$, and s as indicated.
   (a) Deduce expressions for the natural frequencies of torsional oscillations of the system.
   (b) A torque $Q \cos \omega t$ is applied to one of the discs. At what value of ω is the amplitude of vibration at that disc zero, and what is then the amplitude of vibration at the other disc?
   (c) Torques $Q \cos \omega t$ are applied at both discs, the two torques being in phase and the frequency $\omega = \sqrt{(2s/J)}$. What is the amplitude of vibration at each disc?
   (d) What is the amplitude of vibration in the shaft at the section midway between the two discs, when a torque $Q \cos \omega t$ is applied to one disc, and $2Q \cos \omega t$ is applied to the other? Again the torques are in phase and $\omega = \sqrt{(2s/J)}$.

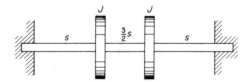

Exercise 9.22

9.23 The pivot of a simple pendulum is held by two springs as in the figure. Show that the natural frequency is the same as that for a simple pendulum with a fixed point but with length $(l + mg/2s)$.

How do you account for the fact that this system which apparently requires two coordinates x and θ to define its configuration relative to the equilibrium position has only one natural frequency?

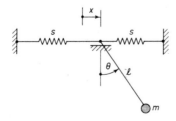

Exercise 9.23

9.24 The figure shows two discs mounted on a shaft with fixed ends. The moments of inertia of the discs, and the stiffness of the three sections of shaft are as shown. The length of the centre section of shaft is a, and a concentrated torque $Q \cos \omega t$ about the shaft axis is applied to it as indicated.

Show that, for steady vibrations,

$$\theta_2 = \theta_1 = \frac{Q \cos \omega t}{3(s - J\omega^2)}$$

The system has two degrees of freedom; how do you account for there apparently being only one resonance frequency?

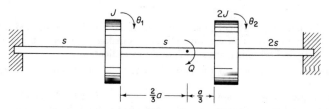

Exercise 9.24

9.25 A machine rotor of moment of inertia J rotating at a mean speed Ω is subject to a torque $Q \cos \omega t$. A mass held in a radial slot by a spring of stiffness s is in equilibrium at radius R when the machine is rotating at a constant speed Ω, and without the fluctuating torque $Q \cos \omega t$.

By forming the equation of motion for *small* radial movements x of the mass in the slot, show that the rotor will run steadily, even when $Q \cos \omega t$ is applied, if

$$\frac{s}{m} = \omega^2 + \Omega^2$$

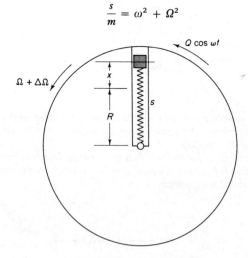

Exercise 9.25

9.26 The figure represents an engine supported on four equal rubber mountings arranged at the corners of a square. Its mass is 700 kg and its moment of inertia about the relevant axis through the centre of gravity, G, is 85 kg m². Each mounting has a vertical stiffness of 3×10^4 N/m, and the mountings are so designed that horizontal motion of the base AB is prevented.

Neglecting damping, determine the normal mode frequency (a) for vertical oscillation of the engine, (b) for a rocking motion in which the first-order movement of G is horizontal.

When running at angular velocity ω rad/s the engine is effectively subjected to a central vertical oscillating force of magnitude $1\cdot 5\,\omega^2$ N and circular frequency ω.

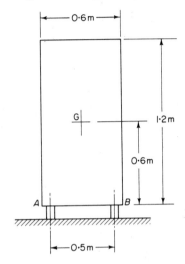

Exercise 9.26

The mountings produce damping forces which are proportional to velocity, and the damping coefficient is $0\cdot 2$ of its critical value. Estimate the amplitude of vertical oscillation of the engine when running at 150 rev/min.

9.27 A motor vehicle with wheelbase $2a$ has its centre of gravity G mid-way between the front and rear axle s. The mass of the vehicle is M and the radius of gyration is K about an axis through G parallel to the axles. The front springs have combined stiffness s and so also have the rear springs.

The vehicle is driven at constant speed V along a level road which is worn into sinusoidal corrugations of wavelength λ in the direction of motion. The height h from the hollows to the crests of the corrugations is small compared with the wheelbase. Neglecting the inertia of the wheels and elasticity of the tyres, derive equations governing rotation of the vehicle in the plane of motion and vertical movement of its centre of gravity in the absence of damping.

In a particular case $a = 2$ m, $K = 1\cdot 5$ m, $M = 1100$ kg, $s = 40$ N/mm, $\lambda = 16$ m, $h = 25$ mm.

Calculate the amplitudes of the steady translational and rotational oscillations of the vehicle when running at 80 km/h.

9.28 Two identical discs are connected by a light shaft. The natural frequency of torsional vibrations is too low and to increase it the moment of inertia of one of the discs is to be reduced. By how much should it be reduced to increase the natural undamped frequency by 10 per cent?

9.29 The rotors of a motor and generator are directly coupled by a shaft of torsional stiffness s. The two rotors have equal moment of inertia J. A torque $Q \cos \omega t$ is applied to one of the rotors. Sketch curves to show the variation in amplitude with frequency at each rotor, and indicate the phase relationships between the motion of the rotors and the excitation, (a) if Q is constant, (b) if Q is proportional to ω^2.

9.30 The figure shows, diagrammatically, a rotating mass having a moment of inertia of 50 kg m² driven through a 2·5 to 1 reduction gear by a motor whose moving parts have a moment of inertia of 8 kg m².

Determine the natural frequency of torsional oscillations of the system assuming that the inertias of the shafts and gears, and the distortion of the gears, may be neglected.

The modulus of rigidity of the shaft material is $82 \cdot 5 \times 10^6$ kN/m².

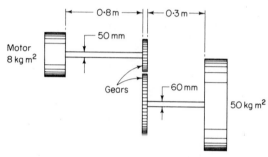

Exercise 9.30

9.31 Define the term *degree of freedom*.

Explain the terms *normal* (or *natural*) *mode* of vibration and *natural frequency* as applied to systems which have more than one degree of freedom.

A motor is connected by a short rigid shaft to a 1·5:1 reduction gear. A flexible shaft of stiffness s connects the gear output shaft to an inertia load. The effective moment of inertia of the rotating parts of the motor is J, and that of the load $4J$. All other inertia effects and all losses are negligible.

Sketch graphs to show how the amplitude of vibration due to a motor torque $Q \cos \omega t$ (a) at the motor, (b) at the load, varies with ω, and indicate the phase relationships. Derive expressions for the frequencies at which the amplitude at the motor (i) tends to infinity, (ii) tends to zero.

9.32 Three flywheels A, B, and C are attached to a shaft, with B lying between A and C. The moments of inertia of A and B are I and that of C is I_1. Each part of the shaft, between A and B and between B and C, twists one radian under a torque λ.

Find the frequency equation for small oscillations of the system and, for $I_1 = nI$, obtain an expression for the ratio of its two natural periods of oscillation. Hence show that, if n is large, this ratio is approximately

$$\frac{3 + \sqrt{5}}{3 - \sqrt{5}}$$

9.33 A uniform light shaft of length $3l$ carries four identical discs, each of polar moment of inertia 25 kg m², spaced with a length of shaft l between successive discs. The stiffness of each section of shaft is 30×10^6 N m/rad and the whole is free to rotate in bearings.

Estimate the lowest natural frequency of torsional vibration of the system. The derivation of any equations used should be explained.

9.34 Determine the natural frequencies for torsional vibrations of the given system.

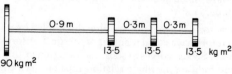

Exercise 9.34

9.35 Figure (a) shows the idealization of the crankshaft of a three-cylinder engine and flywheel. The natural frequency calculations for this system have already been made and it is proposed to add an extra length of shaft and rotor as indicated. The natural frequencies of the composite system have now to be determined.

The relationship between the receptance at A and ω^2 for the original system is shown in fig (b). Determine the lowest non-zero natural frequency for the modified system if $J_B = 0.8$ kg m^2 and $s = 2 \times 10^5$ N m/rad.

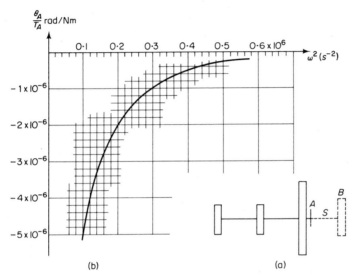

Exercise 9.35

9.36 A uniform light cantilever with bending stiffness EI and of length $2l$ carries equal concentrated masses m at distances l and $2l$ respectively from the support. Determine the natural frequencies for transverse vibrations of the masses.

9.37 Two coaxial shafts A and B, each of torsional stiffness s, are connected by a simple epicyclic gear. Three planet wheels carried in a spider keyed to shaft B mesh with a sun wheel keyed to shaft A and also with an annular wheel which is fixed. The sun wheel has 20 teeth, the annular wheel 40 teeth and the radius of the annular wheel is r. The planet wheels each have mass m_P and the moments of inertia about their centres of the sun wheel, the spider and each of the planets are I_A, I_B, and I_P respectively.

Determine the natural frequency of the system when the outer ends of the shafts are fixed.

9.38 A simple engine mechanism is arranged in a vertical plane as shown in the figure. It is frictionless, and the system may be regarded as consisting of a mass m concentrated

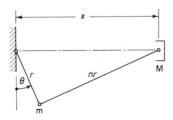

Exercise 9.38

at the big-end, and a mass M at the gudgeon-pin. Given that the approximation

$$x = r \sin \theta + nr\left(1 - \frac{1}{2n^2}\cos^2\theta\right)$$

may be used, find the natural frequency of oscillation of the system for small values of θ.

9.39 In the figure, AOB is a light rigid bell-crank lever which is free to rotate about a horizontal axis through O perpendicular to the plane of the diagram. The lever carries a mass m at A, and a spring is connected between B and a fixed point C, OC being horizontal. The distances OA, OB, and OC are each equal to l. The stiffness of the spring is equal to $2mg/l$, and its unstretched length is such that in the equilibrium position OB is vertical, as shown.

Find the natural frequency of small oscillations about this position.

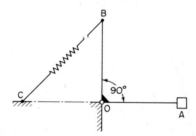

Exercise 9.39

9.40 Estimate the two natural frequencies for the beam described in exercise 9.36 by the energy method, and verify that the lower and higher frequencies are respectively over- and under-estimated.

9.41 Estimate the lowest natural circular frequency for transverse vibrations of a taut wire of tension T, length $6l$, carrying five equal masses m at points l, $2l$, $3l$, $4l$, and $5l$ from one end.

9.42 Estimate the lowest natural frequency for transverse vibrations of the given beam:

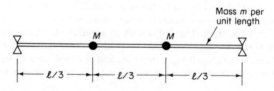

Exercise 9.42

9.43 A uniform beam of mass m per unit length carries a load of mass m per unit length uniformly distributed over the middle third of its length. The beam is simply supported at its two ends. Estimate the lowest natural frequency for transverse vibrations.

9.44 Determine the critical speeds of the uniform circular shaft shown in the figure. At A the shaft is fixed in position and direction; at B it is fixed in position only.

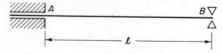

Exercise 9.44

9.45 A straight, uniform shaft of circular cross-section and flexural rigidity EI has a uniform strip of matter attached to it along a generator. This strip contributes to the mass of the system but not to its flexural rigidity. The total mass per unit length is m and the centre of mass is displaced a small radial distance e from the geometrical axis of the circular cross-section. The ends of the shaft run in short bearings a distance $2l$ apart on a vertical centre-line which coincides with the geometrical axis when the system is at rest. Show that, when the system rotates with angular velocity ω about the axis of the bearings, the bending moment at the centre of the shaft is equal to

$$\tfrac{1}{2} mel^2 \omega^2 \left(\frac{\sec\theta - \operatorname{sech}\theta}{\theta^2} \right)$$

where $\theta^4 = m\omega^2 l^4 / EI$.

At what speeds of rotation are large whirling motions to be expected?

9.46 A rocket may be idealized as a uniform rod, with mass m distributed along its length l, together with concentrated masses M at each end. Assuming that the form of the rocket at its lowest whirling speed is a half sine-wave, determine this whirling speed, taking the flexural rigidity of the rod as EI.

SOLUTIONS TO EXERCISES

Chapter 1

1.1 (a) 1; (b) 1; (c) 1; (d) 2; (e) 1

1.2

n_2	n_3	n_4
4	4	0
5	2	1
6	0	2

1.3

n_2	n_3	n_4	n_5
3	6	0	0
4	4	1	0
5	2	2	0
5	3	0	1
6	0	3	0
6	1	1	1
7	0	0	2

1.5 6

1.6 2 links, 3 hinges, 1 ball and socket joint

Chapter 2

2.2 At $\theta = \pm\pi/2$, $\rho = r$ so that (a) the profile is convex outwards at all points; (b) there is no interference if $R > 2.77a$

2.3 No restriction

2.4 $\theta_i < 82.7°$

2.5 (a) 11.4 mm; (b) 20.9 mm

2.6 16

2.7 7.4 mm, 12.3 mm

2.8 0.5 mm

2.9 $\dfrac{51 \times 50}{47 \times 20} = \dfrac{255}{94} = 2.712\,77$

2.10 $N_Q = 81$, $N_R = 18$

2.11

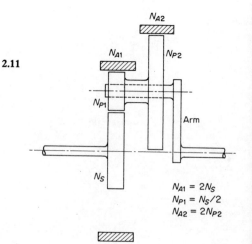

$N_{A1} = 2N_S$
$N_{P1} = N_S/2$
$N_{A2} = 2N_{P2}$

SOLUTIONS TO EXERCISES 525

2.12 $\omega_a/\omega_n = 589/189$
2.14 Crank 0·47, coupler 0·58, rocker 0·94
2.15 Crank 0·14 m, coupler 0·32 m
2.16 AE = 0·132 m, $\angle$ BAE = 20°
2.17 (a) and (b) — no unique solutions; (c) AB = 0·45, BC = 1·34, CD = 0·54
2.18 AB = 0·81, BC = 1·76, CD = 0·60, AD = 1·00
2.19 A is 0·15 m below B_2, D is 0·3 m above B_2
2.20 EP = 0·1 m, CP = 0·21 m
2.21

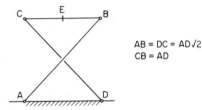

AB = DC = AD√2
CB = AD

Chapter 3

3.1 0·55 W
3.2 $R_x = \frac{1}{2}(2n + 1)P \cot \alpha$; $R_y = P/2$
3.3 0·25 m (*Note*: By virtual work the platform must remain horizontal)
3.4 6·8 N, 11·0 N, 1·05 N/mm, 29·4 N
3.5 $P = KW^r[\beta^\gamma + (1 - \beta)^\gamma]$, where $\beta = r + P_a/W_b$
3.7 0·28 W, 0·25 W
3.8 $2Q(3n + 1)/l \sin \alpha$
3.9 0·87 (*Hint*: Allow a virtual displacement at B perpendicular to the reaction at B)
3.10 47 kN
3.11 $5(0·1 - r/a)$
3.13 $\tan^{-1}(1/4\sqrt{3})$
3.14 146·5 N m
3.15 (a) 0·840; (b) 0·726

Chapter 4

4.1 (a) $\sqrt{14}$; (b) $\sqrt{26}$; (c) $3\mathbf{i} + 3\mathbf{k}$; (d) $\mathbf{i} + 6\mathbf{j} - 5\mathbf{k}$; (e) -11; (f) $9\mathbf{i} - 9\mathbf{j} - 9\mathbf{k}$; (g) 125°
4.3 $1·33\omega_2, 0·8\omega_2$; $\theta = 104·5°$ and $255·5°$; $75\omega_2$ mm/s. I lies at foot of perpendicular from B on AC produced.
4.4 (a) 0·384 m/s; (b) 0·370 m/s; (c) 0·094 m/s; (d) 30·5 N m
4.7 $1·12\omega_1$
4.8 (i) $0·0707\omega$ m/s, (ii) $0·0413\omega$ m/s
4.9 36·4 N, 25·9 N m
4.10 4:1 reduction, 1:1, 1:1, 30 N m
4.11 A = 0·2, B = 0·4
4.12 $-6·3\mathbf{i} + 24·6\mathbf{j}$ m/s²
4.14 108 m/s²
4.15 -42 m/s², 75 m/s², 61 rad/s², 49 rad/s²
4.16 $\sqrt{2(\omega V - V^2/4R - \omega^2 R/4)}(\mathbf{i} + \mathbf{k}) - (\sqrt{3V^2/2R})\mathbf{j}$
4.17 $0·018\omega^2$ m/s², $0·094\omega^2$ m/s², $0·333\omega^2$ rad/s², $0·250\omega^2$ rad/s²
4.18 0·79 m/s² at 10·9° to AB, 0·66 rad/s, 0·75 rad/s

4.19 $0.25r\omega^2$, $0.75\omega^2$
4.20 $a_b = 3a_c$
4.21 (a) $(h + l\sqrt{3}/2)\omega$; (b) $(5/2\sqrt{3}\, l - h)\omega^2$; (c) $l(h/l + \sqrt{3}/2)/(5/2\sqrt{3} - h/l)$
4.22 0·6 m, 0·64 Hz
4.23 0·58 m/s, 11·6 m/s²
4.24 $\dot{\omega}_{AB} = v^2/3\sqrt{3}\, l^2$ anticlockwise; $\dot{\omega}_{BC} = 4v^2/3\sqrt{3}\, l^2$ clockwise
4.26 1·33 m/s, 6·4 m/s²
4.27 $v_e = 1.25$ m/s, $a_e = 140$ m/s²
4.29 (a) 17 or 93 mm/s; (b) 13 rad/s²
4.30

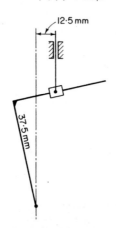
Equivalent mechanism
for contact on BC

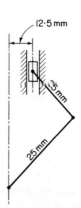
Equivalent mechanism
for contact on CD

For the given configuration velocity = 12·5 mm/s, acceleration = 37·5 mm/s²

4.31 (a) 3940 rad/s²; (b) 29·6 rad/s, 2320 rad/s²; (c) 352 rad/s, 10 000 rad/s²
4.32 $d\psi/dt = r\omega(s\cos\theta - r)/(s^2 - 2sr\cos\theta + r^2)$
4.33 $\dfrac{dx}{dt} = qr\omega\,\dfrac{p\cos\theta + 1}{(p - \cos\theta)^2}$; $\dfrac{d^2r}{dt^2} = qr\omega^2\sin\theta\,\dfrac{2 + p\cos\theta - p^2}{(p - \cos\theta)^3}$
4.35 $\sin^{-1}\{p/(n-1)\}$, $\sin^{-1}\{p/(n+1)\}$, stroke $\approx r\{2 - p^2/(n^2 - 1)\}$
4.36 $-6\omega^2\sin\theta/(5 - 4\cos\theta)^2$
4.37

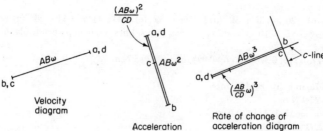

$\dot{\omega}_{BC} = 0$ ∴ ω_{BC} is a maximum

Chapter 5

5.1 $\ddot{x} = g/2$; $\ddot{y} = -g$; $\ddot{\theta} = -0.9g/l$; $\ddot{x}_B = 1.175g$; $\ddot{y}_B = -2.17g$
5.2 $\ddot{\theta} = g/2a\sqrt{2}$; $N = mg/\sqrt{2}$; $T = mg/2\sqrt{2}$

SOLUTIONS TO EXERCISES

5.3 (i) $2g/7r$; (ii) $2(1 - 2\mu)g/(3 - 2\mu)r$
5.6 750 N, 2·04 kg at each pulley
5.7 $125/3\sqrt{3}$ g at A and D, $625/18\sqrt{3}$ g at B, $25/18\sqrt{3}$ g at E
5.8 17·8 g, 29° to OB and 61° to OC
5.9 20 N at 92°, 19·4 N at 178°
5.10 0·014 N m
5.11 $M_B = 1040$ N m; $M_A = -10$ N m
5.13 $y = \delta\left(1 - \dfrac{M\omega^2 l^3}{48EI}\right)^{-1}$; $\theta = \alpha\left(1 - \dfrac{(I_2 - I_1)\omega^2 l}{12EI}\right)^{-1}$

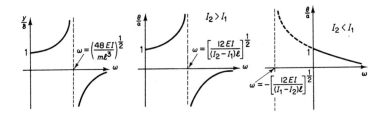

5.14 0·52 kg m², 1690 rev/min/s
5.16 (a) $\sqrt{3}$ m/4, (b) 30° to outer cranks and 180° to each other
5.17 Forces balance. $c_1 = 0.45 pmr\omega^2$, $c_2 = 5A_2 pmr\omega^2$, $c_4 = 0.45 A_4 pmr\omega^2$
5.18 Equivalent to $\tfrac{3}{2}mr$ rotating at ω, and $(3r/8l)mr$ rotating at -2ω
5.20 $4\sqrt{2} A_4 mr\omega^2 \cos 4\theta$ in plane of symmetry. $M_4 = 0$
5.22 $v_c = 2.3$ m/s, $a_c = 31.0$ m/s², 606 N m
5.23 (a) $a\omega\sqrt{3}$; (b) $2a\omega^2\sqrt{3}$; (c) 0; (d) $F_A = F_D = ma\omega^2/6$
5.24 $9ma^2\omega^2/64$; yes; (b) no, only at the other end of the stroke and for the two positions when CD has zero angular acceleration
5.25 1410 N, 2000 N
5.26 3 mm
5.27 $(13I + Mr^2)f/3r$
5.28 $\tfrac{5}{3}ml^2\alpha$; $2ml\alpha/3\sqrt{3}$
5.29 0·5 m/s²
5.30 The method of section 5.12 gives $k = r\sqrt{2}$. However, due to the simplicity of the system an exact solution can be obtained from the energy equation, $KE =$ constant. The exact solution is $k = 0.95r\sqrt{2}$.
5.31 Mean torque = 5000 N m, fundamental $\pm$ 8200 N m, 4·8°
5.32

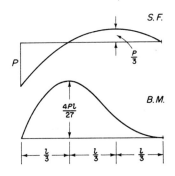

5.33 19 N m

5.34 $\alpha = 0.79$; $M = 0.04\omega_0^2 Ym/l$

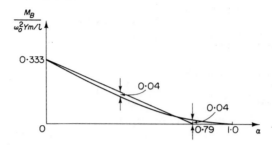

5.35 7700 N m

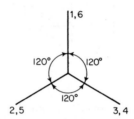

5.36 610 N m

Chapter 6

6.1 0.135 s

6.2 $-\alpha T(1 - e^{-t/T})$

6.3 $v_0 = \dfrac{\bar{q}}{A}(1 - e^{-t/T})$; $P = \dfrac{\bar{q}}{\mu} e^{-t/T}$ with $P = \bar{q}/\mu$ at $t = 0$. This is expected because initially all the flow goes in leakage.

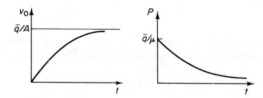

6.4 (a) (step of $\tfrac{1}{2}$ at $t = 0$) + (step of $\tfrac{1}{2}$ at $t = \tau$)
  (b) (ramp of τ^{-1} at $t = 0$) − (ramp of τ^{-1} at $t = \tau$)
  (c) (ramp of τ^{-1} at $t = 0$) − (ramp of $2\tau^{-1}$ at $t = \tau$) + (ramp of τ^{-1} at $t = 2\tau$)
  (d) (ramp of τ^{-1} at $t = 0$) − (ramp of τ^{-1} at $t = \tau$) − (step of 1 at $t = \tau$).

6.5 56.9 N, 49.4 N

6.6 35.7°

6.7 (a) for $-\tau < t < \tau$, $x_0 = \tfrac{1}{2}(1 - e^{-(t+\tau)/T})$
   for $t > \tau$, $x_0 = 1 - \tfrac{1}{2}e^{-t/T}(e^{-\tau/T} + e^{\tau/T})$
  (b) for $-\tau < t < \tau$, $x_0 = \dfrac{t+\tau}{2\tau} - \dfrac{T}{2\tau}(1 - e^{-(t+\tau)/T})$
   $t > \tau$, $x_0 = 1 - \dfrac{T}{2\tau}e^{-t/T}(e^{\tau/T} - e^{-\tau/T})$

The conclusion is, that once it is completed the exact shape of a step input does not matter provided it is over quickly enough (i.e. $\tau \ll T$).

SOLUTIONS TO EXERCISES

6.8 $\left(1 + \frac{\lambda}{s}D\right)x_0 = x_i;$ $\left(1 + \frac{\lambda}{s}D\right)x_0 = \frac{\lambda}{s}Dx_i$

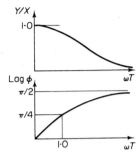
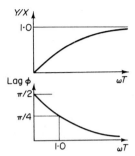

6.9 $C\dfrac{d}{dt}(T_B - T_S) + (K_1 + K_2)(T_B - T_S) = K_1(T_A - T_0); K_2/K_1 \geqslant 2\cdot86$

Chapter 7

7.2 (a) $(mD^2 + \lambda D + s)x_0 = P(t)$ in both cases
(b) $(mD^2 + \lambda D + s)x_0 = (\lambda D + s)x_i$, and $(mD^2 + \lambda D + s)x_0 = sx_i$
$(mD^2 + \lambda D + s)x_0 = \lambda Dx_i$

7.3 $c = 0\cdot5$

7.4 When $m = 0\cdot5$ kg $x_0 = 0\cdot1[1 - (1 + 20t)e^{-20t}]$ m;
when $m = 0$ $\quad x_0 = 0\cdot1(1 - e^{-10t})$ m

7.5 $c = 0\cdot63, 2270$ N s/m

7.6 $12\cdot4$ kg, 188 N/mm, 406 N s/m

7.7 $3\cdot1$ Hz, $0\cdot26$

7.8 $(D^2 + 0\cdot2D + 0\cdot1)\theta_0 = 0\cdot095\theta_i; a < 14\cdot1°$

7.9 The given solution is the limit as $\tau \to 0$ of

$$\frac{Q_i}{s\sqrt{(1-c^2)}}\{e^{-c\omega_n(t-\tau)}\cos[\beta(t-\tau) + \phi] - e^{-c\omega_n t}\cos(\beta t + \phi)\}, \text{ where } \phi = -\tan^{-1}\frac{c}{\sqrt{(1-c^2)}}$$

Chapter 8

8.1 $\left[\dfrac{\lambda T}{s}D + \left(T + \dfrac{a}{a+b}\dfrac{\lambda}{s}\right)\right]Dy = \dfrac{b}{a+b}\left(\dfrac{\lambda}{s}D + 1\right)x,$

$y = \dfrac{bs/a\lambda}{1 + \dfrac{a+b}{a}\cdot\dfrac{sT}{\lambda}}\left(t + \dfrac{\lambda/b}{1 + \dfrac{a+b}{a}\cdot\dfrac{sT}{\lambda}}\right)X$

$\approx \dfrac{b}{a}\left(1 + \dfrac{s}{\lambda}t\right)X$ if $T \ll \dfrac{a}{a+b}\cdot\dfrac{\lambda}{s}$

8.3 $q/ms = (1 + TD)\phi_0$, where $T = M/m$

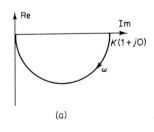
(a)

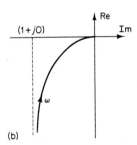
(b)

8.4 $f_1 = 1 - (\omega/\omega_0)^2$; $f_2 = 2(\omega/\omega_0) - (\omega/\omega_0)^3$

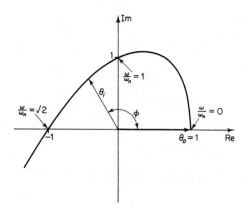

8.5 $K = 5 \text{ s}^{-1}$; $f = e^{-\pi/2}$; $\omega_n = 2\sqrt{5}$ rad/s; $T_d = 0.055$; $f = 0.234$; $M_{max} = 1.325$

8.6 (a) 1·67; 9·2 rad/hr, (c) 0·0485 hr

(b)

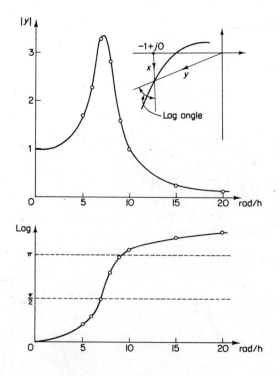

8.7 $KT < \sec^n(\pi/2n)\tan(\pi/2n)$; $\tan(\pi/2n)/2\pi T$; $K\tau < \pi/2$; $1/4\tau$

8.8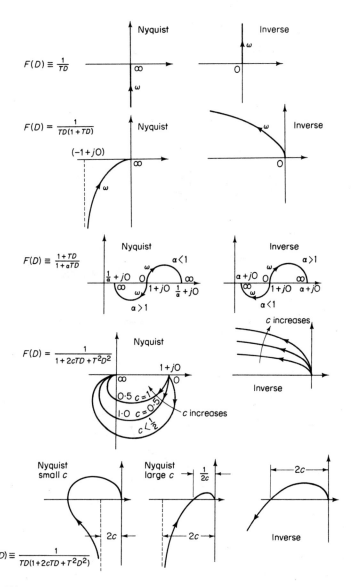

8.9 0·72%, 2%
8.11 0·027 kg per hr/rev per min
8.12 $T = 7·2$ s; 0·2 s; 7854 rev/min
8.13 (a) 0·1 s, $M_m = 1·53$ at 0·64 Hz; (b) 0·3
8.14 $(MAr^2D^3 + \lambda AD^2 + fAr^2D + fbr)x = 0, \lambda A/Mb$
8.15 $K_{crit} = 15$; 0·185, 202° lag; 4·1 rad/s, 1·85
8.18 28·9
8.19 $\omega = \sqrt{(as/mb)}$
8.21 The inverse locus is most convenient because of need to construct the harmonic response locus for the additional element. $\omega_n = 9·5$ rad/s, $M_{max} = 10$. $T^* = 0·056$ s

8.22 0·193 and 0·489 rad/s. Stable if $0·0406 < k < 0·219 \text{ s}^{-2}$
8.23 Unstable if $k_2 = 0$; if $k_1 = 0$ then for stability $k_2 > mg/\mu$; if $\mu = 0$ then for stability $k_1 > g/m$
8.24 $(T_1 T_2 D^3 + (T_1 + T_2)D^2 + D + K)\theta_0 = K(1 + T_2 D)\theta_i$;
$\theta_0 = A \cos \alpha t + B \sin \alpha t + C e^{-\beta t} + \Theta_i$ where $\alpha = (T_1 T_2)^{-\frac{1}{2}}$ and $\beta = (1/T_1 + 1/T_2)$;
at $t = 0$, $\theta_0 = 0$, $\dot\theta_0 = 0$, $\ddot\theta_0 = K\Theta_i/T_1$
8.25 $T_p = 4T$; $T_d = T/\pi^2$

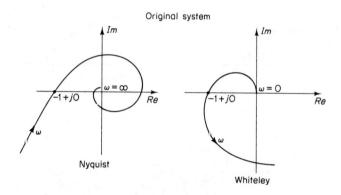

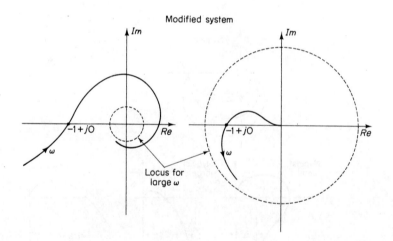

8.26 $4·31 \text{ s}^{-1}$; 0·50 Hz
8.27 $\omega_n = 0·57$; $c = 0·246$

Chapter 9

9.1 3·6 s
9.2 0·39 A
9.3 (a) 431 N/mm; (b) 131 N; (c) 6·7 W
9.4 37·5 Hz, 210 N s/m
9.5 21·5 km/h, 5·7 mm
9.6 24·6 mm, 254 W

SOLUTIONS TO EXERCISES 533

9.7 (a) 0 to 202 rev/min. At resonance $P_t/P_e = 2.7$; (b) 420 rev/min to ∞
(*Note:* These calculations can be based on direct reasoning from the vector diagram, or indirectly by showing first that
$$\frac{P_t}{P_e} = \left\{ \frac{1 + (2c\omega/\omega_n)^2}{[1 - (\omega/\omega_n)^2]^2 + (2c\omega/\omega_n)^2} \right\}^{\frac{1}{2}}$$

9.8

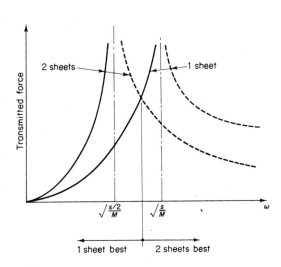

9.10 0·68 m
9.11 (a) $a\omega/\omega_n$; (b) $2a(\omega/\omega_n)^2$ for $\omega \ll \omega_n$; (c) $a(\omega/\omega_n)^2$ for $\omega \ll \omega_n$
9.12 $\omega^2 = \omega_0^2 - \Omega^2$; $-m\{\alpha\Omega^2 \pm 2b\Omega\sqrt{(\omega_0^2 - \Omega^2)}\}$
9.13 (a) -6.4 mm, 28 mm/s; (b) $t = 1.14$ s and 1.49 s
(c) -5.5 mm; -15 mm/s
(d) Very little difference should be found between the two sets of answers
9.14 $Mg/4$
9.15 129 mm
9.16 $10\frac{1}{2}$ min

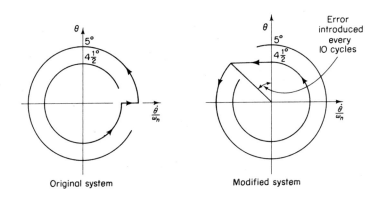

534 Dynamics of Mechanical Systems

9.17

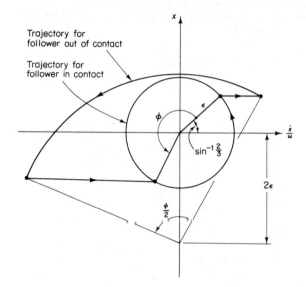

9.18 See facing page.

9.19

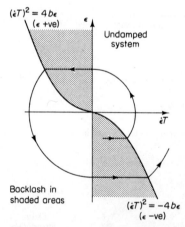

Backlash in shaded areas

$(\dot{\epsilon}T)^2 = 4b\epsilon$ (ϵ +ve)

$(\dot{\epsilon}T)^2 = -4b\epsilon$ (ϵ -ve)

Undamped system

The amplitude increases each cycle so that the system is unstable

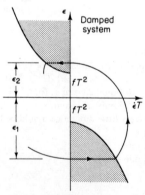

Damped system

Oscillations decay if $\epsilon_2 < \epsilon_1$. For this to happen either

$\epsilon_1 < \dfrac{bfT^2}{b - fT^2}$ or $fT^2 > b$

9.20 $\omega_1 = \sqrt{s/m},\ X_1/X_2 = 1;\ \omega_2 = \sqrt{2s/m},\ X_1/X_2 = -1$
$\omega_1 = \sqrt{s/m},\ X_1/X_2 = 1;\ \omega_2 = \sqrt{5s/2m},\ X_1/X_2 = -2$

9.21 $X_1 = \dfrac{\left(\dfrac{s_1 + s_2}{m_2} - \omega^2\right)\dfrac{U}{m}\omega^2}{\omega^4 - \left(\dfrac{s_1}{m_1} + \dfrac{s_1 + s_2}{m_2}\right)\omega^2 + \dfrac{s_1 s_2}{m_1 m_2}};\quad X_2 = \dfrac{s_1 X_1}{-m_2 \omega^2 + s_1 + s_2}$

9.22 (a) $\omega_1 = \sqrt{s/J},\ \omega_2 = 2\sqrt{s/J}$; (b) $\omega = 2.5\sqrt{s/J},\ 2Q/3s$; (c) $-Q/s,\ -Q/s$; (d) $3Q/2s$

SOLUTIONS TO EXERCISES 535

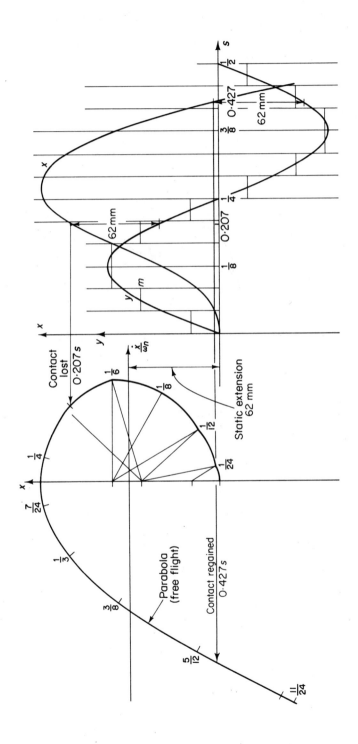

9.23 We have effectively a single mass constrained by a complicated spring; only the coordinate at the mass matters

9.24 One mode is suppressed because the excitation is applied at its node

9.26 (a) 2·1 Hz; (b) 0·5 Hz, 4·6 mm

9.27 8·2 mm, 0·7°

9.28 30 per cent

9.29

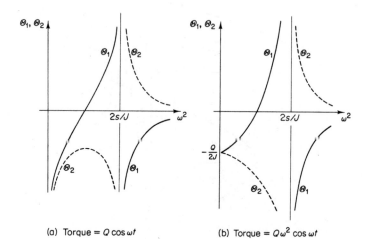

(a) Torque = $Q \cos \omega t$ (b) Torque = $Q\omega^2 \cos \omega t$

9.30 9·7 Hz

9.31 Figure as for answer to 9.29(a)

(i) $\frac{5}{6}\sqrt{s/J}$; (ii) $\frac{1}{2}\sqrt{s/J}$

9.32 $n\Omega^4 - (1 + 3n)\Omega^2 + (2 + n) = 0$ where $\Omega^2 = I\omega^2/\lambda$

9.33 135 Hz

9.34 4·1 Hz, 12·1 Hz, 19·6 Hz

9.35 90 Hz

9.36 $\omega = (0.578 \text{ or } 3.88)\sqrt{(EI/ml^3)}$

9.37 $\omega^2 = 10s/(I_B + \frac{27}{16}m_p r^2 + 27I_p + qI_A)$

9.38 $\omega^2 = mg/(M + m)r$

9.39 $\sqrt{(g/2l)}/2\pi$

9.41 $0.518\sqrt{(T/ml)}$

9.42 $\omega^2 = \pi^4 EI/(3Ml^3 + ml^4)$

9.43 $\omega^2 = EI(\pi/l)^4/1.51m$

9.44 ω given by solution of $\tan \alpha l = \tanh \alpha l$ where $\alpha^4 = m\omega^2/EI$

9.45 When $\cos \theta = 0$, i.e. $\theta = (2n + 1)\pi/2, n = 0, 1, 2 \ldots$

9.46 $\omega^2 = \pi^4 EI \bigg/ \left(\dfrac{24m^2 l^3}{\pi^2(2M + m)} + ml^3 \right)$

INDEX

Absorber, vibration, 401
Acceleration,
   relative (plane mechanisms), 142
   relative (space mechanisms), 163
Acceleration diagram, 142
Accelerometer, 380
Active device, 241
Addendum, 47
Amplifier, torque, 241
Analogue computer, 252
Angular momentum, 199
Argand diagram, 267
Automatic control, ch. 8
Automobile suspension, 79
Auxiliary point, 153

Balancing, 177
   dynamic, 187
   of long rotors, 192
   of plane rotors, 183
   reciprocating, 207
   rotary, 180
   static, 182
Base circle
   of cam, 26
   of gear, 45
Base cone, 57
Beam vibration, 426, 435, 440, 448
Bearing reaction, 181
Bevel gear, 57
Binary link, 12
Block diagram, 295, 298
Bode diagram, 356

Cams, 23
   friction effects in, 107
Centre, instantaneous, 80, 131, 151
Circular pitch, 46
Closed-loop system, 293

Complementary function, 249, 277
Composite system, 423
Constant excitation, 369
Contact ratio, 45
Continuous systems, 436
Control,
   derivative, 304
   integral, 315
Coriolis acceleration, 147
Coupler curve, 77
   symmetrical, 86
Coupling, 411
Crank and rocker, 70
Critical damping, 278
Critical speed, 451
Cross product, 133
   of inertia, 201
Crown wheel, 58

D'Alembert's principle, 178
Damping,
   critical, 278
   viscous, 275
Decibel, 356
Decrement, logarithmic, 282
Dedendum, 47
Degenerate system, 406
Derivative control, 304
Diametral pitch, 47
Displacement
   analysis, ch. 2
   diagram, 96
Double rocker mechanism, 70
Drag link mechanism, 70
Dwell mechanism, 79
Dynamic
   balance, 187
   equivalent, 215
   magnification, 264, 287
   response of mechanism, 220

538 Index

Effective inertia, 221
Efficiency,
 upper and lower bounds on, 116
Energy method, 429
Epicyclic gear, 63, 139
Equivalence, dynamic, 215
Error, 299
Error-activated system, 273
Euler-Savary equation, 79
Euler's theorem, 67
Exponential lag, 244

Feedback, 298
 non-unity, 349
Finite lag, 353
Flywheel, 227
Force polygon, 182
Four-bar mechanism,
 classification of, 88
 friction in, 109
Freedom, degrees of,
 plane mechanisms, 6, 9
 space mechanisms, 10
Frequency,
 equation, 398
 natural, 369, 398
Freudenstein, 73
Friction, ch. 3
 circle, 100
 coefficient of, 98
 in cams, 107

Gain, 335
Gain margin, 336
Gears,
 bevel, 59
 epicyclic, 63, 139
 internal, 51
 non-circular, 55
 non-involute, 54
 Novikov, 54
 skew, 59
 spiral, 59
Gear,
 tooth geometry, 42
 train, 59
Geared systems,
 vibrations in, 409
Governor, 311

Gyroscope, 194
 vibrations of, 454

Harmonic
 input, 248
 response, 260, 282
 response locus, 266, 284
Helical gear, 53
Holzer, 418
Hooke's joint, 168
Hydraulic relay, 245, 316

Image theorem
 for acceleration, 150
 for velocity, 129
Impulse, 248
Impulse response, 397
Inertia,
 cross product of, 201
 effective, 221
 moment of, 201
 principal moment of, 202
Inertia
 excitation, 374
 stress, 232
Inflection circle, 82
In-line engine, 209
Instantaneous centre, 80, 131, 151
Integral control, 315
Integrating element, 242
Interaction, 296
Interference
 with cams, 25, 28, 33
 with gear teeth, 48
Internal gear, 51
Inverse harmonic response locus, 265
Inversion, kinematic, 24, 71
Involute
 gear, 43
 rack, 48

Lag,
 exponential, 244
 finite, 253
Linear system, 226
 first order, ch. 6
 second order, ch. 7
Logarithmic decrement, 282
Logarithmic plot, 356

Longitudinal vibration, 439
Low-pass filter, 269
Lumped parameter, 436

Mechanism, 2
  classification of, 3
  equivalent, 157
  space, 10
Modal shape, 398
Mode, normal, 398
Module, 47
Moment polygon, 189, 201, 210
Momentum, angular, 199

Natural frequency, 369, 398
  equal, 414
Node, 409
Non-linearity, 392
Normal mode, 398
Novikov gear, 54
Nyquist
  criterion, 323
  diagram, 334

Octoid teeth, 59
Open-loop
  response, 327
  system, 293

Pair, 3
  class I, 5
  class II, 9
Particular integral, 249, 283
Passive device, 241
Path of contact, 45
Phase
  advance, 346
  margin, 337
  plane, 383
Pitch,
  circle, 46
  circular, 46
  diametral, 47
  of screw, 68
Polar plot, 266
Potential energy, 430
Precession, 195
Precision point, 74
Pressure angle, 46

Primary
  force, 208
  moment, 210
Principal
  axis, 202
  coordinate, 411
  mode, 398
  moment of inertia, 202
Principle of superposition, 237
Process controller, 292
Proportional element, 241

Quick-return mechanism, 164

Ramp input, 247
Rayleigh's principle, 434
Receptance, 424
Regulator, 292
Relay, hydraulic, 245
Relative
  acceleration, 142, 163
  velocity, 123, 136
Resonance, 289, 370
Response,
  steady-state, 260
  step, 249
Roberts' theorem, 84
Rolling contact, 39
Rotary balancing, 180
Routh's criterion, 318

Scalar, 119
Scalar product, 119
Secondary force, 209
Seismic excitation, 377
Servomechanism, 273
Skew gearing, 59
Slider-crank mechanism, 166, 173
  friction in, 109
Speed regulation, 309
Spiral gear, 59
Stability, 318
Steady-state response, 260, 342
Step
  input, 247
  response, 249, 279
Stiffness coefficient, 428
Strain energy, 430
Structural analysis, 1

Superposition, 237, 253
Suppression of mode, 408
Symmetrical
    coupler curve, 86
    systems, vibrations in, 417
Synthesis
    for coupler displacements, 75
    for crank displacements, 69

Tchebichev mechanism, 86
Tertiary link, 12
Thermometer, 244
Three-line construction, 151
Time constant, 244, 275
Torque amplifier, 241
Torsional vibrations, 402, 409, 418, 431, 436, 445
Trains, gear,
    compound, 60
    simple, 59
Transfer
    function, 328
    operator, 295
    relationship, 23, 240
Transient
    acceleration, 368
    vibration, 380
Transmitted force, ch. 3

Transverse vibration of beams, 426, 435, 440, 448
Turbine blade, 233

Vector,
    diagram, 124
    product, 133
    relative velocity, 123, 136
    unit, 119
Vee-engine, 214
Velocity
    diagram, ch. 4
    feedback, 308
Vibration,
    longitudinal, of bar, 439
Vibration
    absorber, 401
    isolation, 375
    measurement, 379
    of beams, 426, 435, 440, 448
    of shafts, 402, 409, 418, 431, 436, 445
    pick-up, 271, 379
    with two degrees of freedom, 394
Virtual work, 96
Viscous damping, 275

Watt mechanism, 79
Whirling of shaft, 451
Whiteley diagram, 334